Economic Futures

Economic Futures

Climate Change and Modernity

Ruth Irwin

ROWMAN & LITTLEFIELD
Lanham • Boulder • New York • London

Published by Rowman & Littlefield
An imprint of The Rowman & Littlefield Publishing Group, Inc.
4501 Forbes Boulevard, Suite 200, Lanham, Maryland 20706
www.rowman.com

86-90 Paul Street, London EC2A 4NE

British Library Cataloguing in Publication Information available

Library of Congress Cataloging-in-Publication Data

Names: Irwin, Ruth, 1968- author.
Title: Economic futures : climate change and modernity / Ruth Irwin.
Description: Lanham : Rowman & Littlefield, [2024] | Includes bibliographical references and index.
Identifiers: LCCN 2024015253 (print) | LCCN 2024015254 (ebook) | ISBN 9781786615114 (hardcover) | ISBN 9781786615138 (epub)
Subjects: LCSH: Sustainable development.
Classification: LCC HC79.E5 I79 2024 (print) | LCC HC79.E5 (ebook) | DDC 338.9/27—dc23/eng/20240514
LC record available at https://lccn.loc.gov/2024015253
LC ebook record available at https://lccn.loc.gov/2024015254

For Leo and Belen in Sindran
John and Julie in Titirangi
Zoë and Alistair on Luing
Will and Kirsty in Glasgow
James, of course,
And friends everywhere . . .

Contents

Acknowledgements

People and places make up the fabric of writing this book. The project fermented over a lifetime, but particularly in Fiji and Scotland.

At a small local community meeting in Auckland in the mid-1980s, I remember a very good-looking speaker giving a presentation on some obscure political issue. His explanation of the planetary consequences of fossil-fuel pollution on climate change was devastating. I was a teenager, and subsequently struggled alone with existential angst for the next five years (well, maybe the next twenty-five years). As a young activist, I went on the anti-nuclear peace walk in New Zealand and then a similar one in Europe the following year. From my sheltered island home isolated in the Southern Ocean, that was my first time overseas. Culture shock overloaded my brain's default network and shook it to its baseline. I think it's great for everyone to experience that kind of paring away of normative assumptions several times in their life. I was open to learning everything afresh. Part of that shake-up was staying with a Catholic priest and a nun, Leo and Belen, in a small village in Galicia in Spain for the winter of 1988. As a lonely nineteen year old, their kindness and generosity was deep-level respite from my life up until then. Leo and Belen were grassroots radicals, and they embraced community engagement in very real and direct ways by refusing to use money. This simple but remarkable experience began a train of thought for me that is still unfolding. Leo and Belen were intimately involved with village life and very aware of seasonal abundance and scarcity. They embodied deep intra-action at multiple levels: meaningful friendships, local ecology and economy, and national connection through their political and religious fraternities.

Most of us, however, are not going to dedicate so much time to localised production, so money and its edifices are incredibly useful modes of simplifying exchange and capital. Long before there were any such courses, my sojourn throughout three degrees at university were always with an eye to the anthropocentric production of climate change and the critique of its genesis in modernity. This was completely uncharted territory. Throughout my career, there has been increasing development of climate science, and since the inception of the Intergovernmental Panel on Climate Change (IPCC) in 1987, 'models' have developed which postulate types of economic approaches and their resultant emissions profiles. But these models were superficial and frankly, facile. The IPCC has consistently upheld an important critique of business as usual (BAU) and shown how it will produce up to 6 or 7°C planetary warming over preindustrial norms. At the same time, the IPCC models held significant bias against subsistence and indigenous lifestyles. They did not acknowledge degrowth literature. With an emphasis on science and policymaking, with little input from the social sciences and humanities, the IPCC did not engage with the deep-seated assumptions that inform colonialism or modernity and continued to justify the behaviours of major polluters, governments, urban communities, industrial agriculture and individualism. Without distinguishing which elements of modernity drive climate change and what is relatively benign, it is difficult to postulate the social transition to a postcarbon society. This book is a passion project, and I realise as I am writing this acknowledgement how much my personality and life history has been caught up in writing it.

Having said that, when I began this book, I did not really hope to discover 'the answer to climate change'. I've been writing transdisciplinary work on philosophy, climate science and social transition to a postcarbon economy throughout my career, and it has always felt like the problems were unanswerable. I have long recognised that economic growth is closely correlated with emissions growth. So it is somewhat to my astonishment, that while researching thermodynamics with an eye to economics, I came across Frederick Soddy's 1926 book on thermodynamics and banking. His understanding of the consequences of ex nihilo banking and exponential growth were revelatory. Soddy did not know about climate change. But he helped me realise that banking is the *driver* of economic growth. Finally, after circling around economic growth for a couple of decades, with a focus on consumer demand, I began to get a clearer grasp on how the systematic impetus for exponential growth takes place. It is not so much the pull of consumer demand as the push from the inflationary pressures of money supply. This new understanding of finance is the fundamental answer to solving the crisis of climate change. It won't be an easy transition to legislate against ex nihilo money creation, but it is utterly necessary and finally provides a clear focal point for change.

This book has taken five years to write, as I had to come to terms with complex contemporary science and dwell on the long-term consequences of finance on modern global societies and greenhouse gas emissions. The structure of the book is still quite clumsy as it shifts amongst multiple discipline areas and attempts to bring these ideas to bear on the premises of modernity and the way modern practices produce climate change. Finding the right vocabulary has been also difficult, especially thinking about the metabolism of economics and how to find accessible and descriptive terminology for it. I am not sure if 'thermodynamic economics' and 'metabolic economics' fulfil this task, but they are certainly better than 'degrowth' with its asinine connotations of contraction, anti-progress and pastoral backwardness.

My approach to economics and finance is decidedly heterodox. It is an anthropological approach to the ways philosophy of economics, finance and macroeconomics serve to drive climate change, and what we might do about it. Sir James Steuart took eighteen years to write one of the first books on modern economics, *An Enquiry into the Principles of Political Economy* (1767). Steuart wrote, 'Had I been master of my subject on setting out, the arrangement of the whole should have been rendered more concise; but had this been the case, I should never have been able to go through the painful deduction which forms the whole train of my reasoning' (Steuart 1966/1767: 7).

I am aware that in the process of writing, I had to slough off a lot of encrusted thinking and reconceptualise approaches to multiple issues. The stereotype of 'left' and 'right', for example, disappeared as I researched deeper into economic growth. Over five years, my ideas have developed and changed and something completely different emerged from what I imagined I was writing at the outset. The result of that evolution is that this book includes a lot of material that I needed to process and let go in order to get to the ideas I have now. The book is a palimpsest of skins being sloughed off. Or a braided river with multiple pathways meandering in the same general trajectory. If I started it now, I might have a more direct route. Some of the rabbit holes might have been avoided. But at the same time, the issues and paradoxes that emerged while writing are what enabled a new 'data point' to appear, which reoriented all the material in a much more interesting and important way. So, I am offering this book to the world in the state it has now reached. Possibly not a 'finished' or precise iteration of the concepts herein but a somewhat convoluted pathway to new ideas. I hope you will bear with the clumsy shifts in gear as multiple disciplinary angles are brought to bear on this very practical and theoretical piece of work. If I had eighteen years, this book would be a lot more polished, but given the size and urgency of global warming, that luxury is not possible. Please forgive its tangents, divergences and awkward wording. They are part of the diffractive multidisciplinary approach to situating modernity in its historical and ecological context. It almost goes without saying, with considerable thanks to many conversations with amazing people, any mistakes are mine.

My nomadic life and working history in multiple countries have helped shape and grounded the global orientation of the book. In Fiji, the subsistence lifestyle, the close communities and the tangible awareness and protection of the ecological environment made a big impression on me. The ocean reefs, coral, fish, octopus and sharks were incredible. The warmth of people in Fiji is incomparable. It helped reset my well-being after an earlier serious health break.

I also lived through Cyclone Winston while I was there, which was the biggest storm in the southern hemisphere for ten thousand years. In the few days leading up to the cyclone, the ocean was consistently more than 30 degrees Celsius (°C). There was a huge fish die off, littering the beaches, because oxygen had evaporated out of the water and the fish could not breathe. Nobody in Fiji produced the CO_{2e} emissions that stimulated that storm. The planetary impact of industrialisation is starkly unfair, as vulnerable populations anywhere can be exposed to the devastating impacts of climate change. My friends and colleagues at the University of Fiji need special mention, including Rigi, Atelini, Ruveni and Prathika. Runaaz and Lalesh and their lovely family were really welcoming and supportive. I loved being the kitchen help, for Runaaz's amazing cooking. Thank you all; your friendship meant a lot to me.

Later in Scotland, my friend Sean Blenkinsop invited me to join a research group, sailing the Western Isles on a square-rigged sailing ship called *Lady Averil*. We all wrote a book together and explored the islands, currents, sea birds, otters and amazing geology of the ancient Dal Riada empire. It was magic. My own family have Celtic roots, and it was wonderful to tap into the archaeology and language and unfurl the indigenous relationship with the land and sea there. That trip was about wild pedagogies, and how landscape forges culture and subjectivity. It was an antidote to the alienation, and busy overwork, of mundane life. And a stark relief from the oil wifeys' obsession with protecting the status quo in the east. Aberdeen was an eye-opener. I met one of my besties there, though. There are some people who become part of the weft and weave of your life and Will is one of those for me. A sing out to Gordie in Glasgow. My son, James, emigrated to the United Kingdom when I went to work there, and he loves London so much that he's still there! Love permeates wherever you are, Jamie.

On the sailing trip, I met the wonderful Joyce and her partner, Doug, and Zoë and Alistair for the first time. I will always be grateful for the warmth, nurturing and generosity of Zoë and Alistair as I began writing this project. The conversations about chemistry and thermodynamics with Alistair and rambling the Isle of Luing with Zoë learning about geology and layers of archaeology were wonderful. Such a blessing! Sheila is a gem. Dugie and the archaeology crowd were loads of fun. Jo's amazing yoga was life changing.

And the Isle of Luing itself which is one of the tiniest and most fascinating places I've gotten to know. Luing was a haven.

Eventually I went to Sydney to find work and turned my hand to the pragmatic tasks of mitigation, adaptation and resiliency in a Sydney council. The work was interesting, Sydney is amazing and the people there are really dynamic. I saw Aboriginal art and artifacts all through the landscape. COVID-19 was voracious as the *Ruby Princess* offloaded hundreds of actively sick people into the community. I immediately got ill and struggled with the fatigue and breathlessness of long COVID before we knew that was a thing. There were several standout people in Sydney. Ursula, who lead Below Zero and is an incredible dynamo. Lindi, who is the connective tissue in the community and is a dear friend. Susan and Cadbury, my flatmates, and Manda, who felt like a twin flame in age, stage and outlook. Sydney was a place of green shoots.

I returned home to New Zealand in the final stages of this book. Five years is a long project, and a lot of life flows past during the making of it. Thanks to Nesta, Ollie, Maddie and Purrl for extended hospitality. After ten years away, it's been surprisingly comfortable being back in my own community. Some friends visited me while I was away. Julie and John, in particular, are the poles that have held up the tent of my life for decades now. John and Julie are like magnetic south that keeps me aligned and in tune with deep friendship, good communication and loving care for each other. Another sing out to Catherine Vile, Janet Marsden, Carl Mika, Thomas Everth, Ruth, Jeremy, Volita, Moss and Oksana, Adrienne, Miguel and Leo, Karen in Rawene, Mairi and Bridie. It's been great to come back to the Titirangi bush and the Manakau Harbour. There is nothing like *tūrangawaiwai*, especially after a nomadic period of wandering the world. The sound and flight of the *kererū*, *tūī*, *pīwakawaka* and spoonbills are a welcome. It's great getting to know the birds who live here. I hope this book goes a long way to protecting them.

Acronyms

ADHD	Attention deficit hyperactivity disorder
AI	Artificial intelligence
BIS	Bank of International Settlements
BAU	Business as usual
CBAM	Carbon Border Adjustment Mechanism
CDC, CDOs	Collateralised debt obligations; financial instruments
CO_{2e}	Carbon dioxide equivalent
COP	UNFCC Conference of the Parties
ECB	European Central Bank
EEI	Earth's energy imbalance
EROI	Energy Returns On Investment
EV	Electric Vehicle
GDP	Gross Domestic Product
GFC	Global Financial Crisis
IMF	International Monetary Fund
IOU	I owe you
IPCC	Intergovernmental Panel on Climate Change
IT	Information technology
JIT	Just in time production
JWST	James Webb Space Telescope
KBBE	Knowledge-Based Bio-Economy
LIBOR	London Interbank offered rate
NSW	New South Wales, Australia

OCR	Overnight cash rate
ppmv	Parts per million volume
NFRD	Non-Binding Guidelines on Nonfinancial Reporting
NGFS	Network for Greening the Financial System
PR	Public relations
REDD+	Reducing emissions from deforestation and forest degradation
SOFR	Secured overnight financing rate
SOP	Standard operating procedure
TOE	Theory of everything
TCFD	Task Force on Climate-Related Financial Disclosures
UN	United Nations
UNFCCC	United Nations Framework Convention on Climate Change
UNICEF	United Nations International Children's Fund
USD	United States dollar
W/m^2	Watts per square metre
WWF	World Wildlife Fund

Part I

ALIENATION FROM NATURE: GROWTH AND FINANCE

Chapter 1

Scoping New Narratives

Climate change is transforming the world. It is drastically changing the planetary ecosystem; shunting aside the placid Holocene where humanity reached the apex of civilisation and technology. The rise in planetary temperature is increasing the turbulence of the hydrological system, reshaping ocean chemistry and circulation patterns, and disrupting earth-systems that predicate human existence. Coping with global warming will take more and more resources as it gets worse. Climate change brings the problems with the modern economic system into focus. Economics have externalised pollution and pretended it is not a problem. Climate change shines a light on the normative practices of modern global society and how it differs from all previous civilisations. Sophisticated technologies, fossil fuels and mass production characterise modernity. Something about these modern elements are producing climate change – unlike any previous civilisation.

We all really feel the fragility of losing the future. It is the structural machinery of capitalism that has created climate change. This machinery needs to transition if we are to evolve and survive ecological upheaval. Exponential economic growth has recently been correlated with exponential growth in CO_{2e} emissions (Le Quéré et al. 2020; Caporale et al. 2021; Mitić et al. 2022; Onofrei et al. 2022; Hansen et al. 2023; Noll 2023). Since 2010, the rate of CO_{2e} forcing is accelerating, and planetary temperature is increasing faster (Noll 2023; Hansen et al. 2023). In figure 1.1 below, the rate of average temperature increase has been steadily rising. Reflecting an exponential rather than linear curve, after 2010 the rate of global warming began to accelerate faster than ever before. Hansen et al. (2023) "project a 50–100% increase in the rate of global warming during the few decades following 2010." Under current conditions, business as usual (BAU) is incapable of reducing global emissions. Until we find healthy alternatives to economic growth, halting the

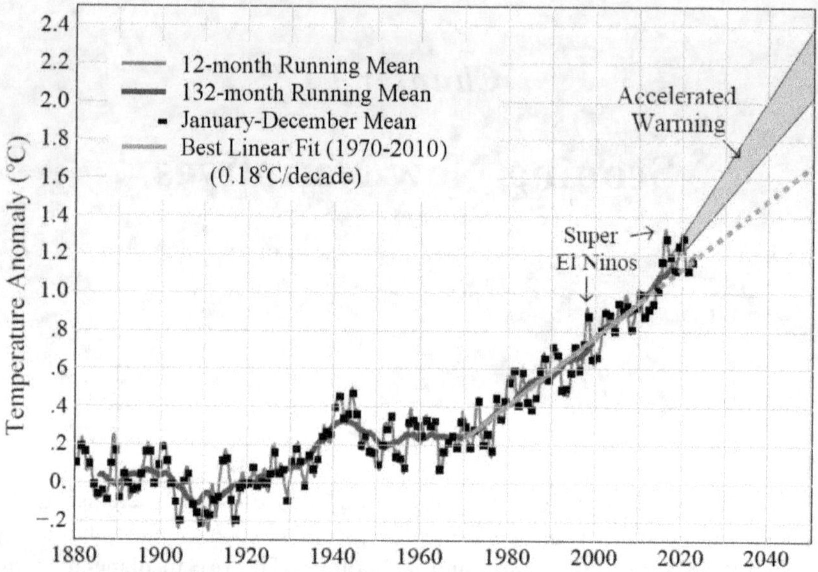

Figure 1.1 Global Surface Temperature Relative to the 1880 to 1920 Average; the Twelve-Month Running Mean Extends through May 2023 (Hansen et al. 2023)

continuous accumulation of greenhouse gases in the atmosphere is impossible (IPCC 2018; 2021).

It would be easy to direct the cause of climate change to fossil fuel emissions and consider the problem solved. Unfortunately, despite the wonderful efforts that are going into renewable energy, overall global CO_{2e} emissions continue to rise (see figure 1.1). Climate change will continue to get worse while economic growth directs modern productivity. Economic growth is directly correlated with growth in greenhouse emissions. New green technologies cannot compensate for the rapidly escalating demands of exponential economic acceleration. All the efficiency gains are absorbed, and exceeded, by the demands for more consumerism. In turn, more consumerism supports new jobs and better conditions. However, abandoning the economic growth imperative seems to condemn the poor to destitution. Governments are caught in a Catch-22 and remain committed to continuous growth. Such is the conundrum of modern economics.

Economics is an illustration of the values of a society. The narratives that organise the way economics is understood, reflects the philosophical principles that saturate public discourse. Economic growth is a product of those values, but it is not the root values themselves. Modern values emerge from the Enlightenment. Reactionary ideas against feudal despotism collected a potent mixture of ideas that put importance on freedom, equality, individual rationality, progress, and the logical order of the mechanical universe.

Enlightenment philosophy from the seventeenth and eighteenth centuries is an important part of the heady mix that underpins modern economics. Economic growth has been justified as progress. Economic growth is considered the fire that stimulates innovation, creativity, and new technologies.

Enlightenment ideology does not explain the way exponential growth structures the modern economic system. But compound growth is fundamental to the system in multiple ways. It is a fundamental criterion to the banking industry. The housing market is continuously increasing in price, as are shares, pensions funds, derivative futures trading and Bitcoin. Fulfilling consumer demand has become the supreme ethos of the modern paradigm.

Politicians, policymakers and business leaders are helplessly in the sway of economic growth. For a long time, governments ignored climate change. Leaders in government, business and finance are reluctantly succumbing to pressure from the impact of climate disasters and the subsequent swelling of public opinion, glocally (Irwin 2013). Despite new initiatives, the normative Enlightenment principles that underpin modern policy and economics continue to prevail. As a result, it is too difficult for leaders to deal with the correlation between exponential economic growth and accumulating emissions growth. The result is devastating. The cause of climate change is obviously CO_{2e} emissions, but the drivers, endemic to modern economics are more obscure. This book explores the key financial mechanisms that force emissions growth to continue accelerating. It explores a potential paradigm change that could redirect a materialist metabolic economy in the future.

Renewable energy is entering a boom period. Coal and fossil fuels are theoretically getting phased out, as wind, solar and hydro energy gradually replace the energy-dense versatility of oil. Unhappily, these initiatives are not having the desired effect of reducing CO_{2e} emissions. The expanding economy demands more energy than ever before, absorbing all the renewables without significantly retiring the use of fossil fuels.

The Enlightenment concept of 'progress' has been converted into the idea of 'growth'. Paradoxically, growth tends towards both utopia and dystopia. Under neoliberalism, incremental improvements for the poor justify growth instead of redistribution. The rich are getting richer, but in theory, the poor are better off too. That narrative is getting harder to believe, as the housing crisis bites in the West, homelessness is rife, subsistence economies are threatened by resource exhaustion, and escalating climate turbulence wreaks havoc on communities all over the planet. Traditionally, the critique of capitalism has revolved around the disparity between rich and poor, by aiming critique at the way profit syphons up (Marx 1887/1867). But in this book, I aim to look again at the structures driving economic growth which are more ubiquitous than class warfare.

Signed up to the United Nations (UN) Paris Agreement on climate change (2016), are 196 national governments. They are committing to reduce emissions to net zero by 2050. Yet so far, the commitment to macroeconomic exponential growth, with its correlated accumulating CO_{2e} emissions, remains largely unchallenged.

The housing market, banking, pensions, the stock market, derivatives futures trading, along with employment rates, all rely on economic growth for stability. Large business, institutions and governments all have enormous wage bills that depend on growth to offset inflation and generate profit. John Maynard Keynes's book *The Means to Prosperity* (1933) proposed the 'multiplier effect' which aims to level up the working classes, offering them 'progress' and the hope of gradually improving living standards. From left to right, 2 to 3 per cent exponential growth is now a fundamental aim of governments.

The key driver to economic growth is hidden in plain sight. The structure of the banking business model is to generate and proliferate debt, adding to the overall money supply. In a spiralling exponential feedback loop, greater money supply at once stimulates and requires economic growth to meet their interest and capital repayment expectations. Consumer demand is one pole of the system of economic growth, and continuous pressure to maintain value in the rapidly expanding money supply is the other. At present, economic growth is not merely a normative discourse of modernity, it is hard-baked into the structure of finance and government.

Every effort is finally going towards renewable energy generation, which is a very important step towards transforming to a postcarbon economy. Unfortunately, according to Jevon's paradox, economic growth absorbs and then exceeds all new energy inputs. More renewables do not necessarily decommission existing fossil fuel–based energy production. Lower intensity energy is simply absorbed into the growth of the economy. This not a short-term issue. It is structural to the system of global economics. Regardless of the green energy revolution, growth in consumerism leads to growth in emissions.

> Efficiency gains abound. For example, global primary energy efficiency has increased by a third since 1980. The carbon intensity of each dollar of economic output has fallen by about the same amount. However, absolute reductions in impact have been singularly elusive. Global primary energy use, carbon emissions, biodiversity loss, nutrient loadings, deforestation, and global fossil water extraction are all still increasing. Carbon dioxide emissions from fossil fuel consumption and cement production increased by 50 percent between 1990 and 2010. (Boden et al. 2013, in Victor and Jackson 2015: 241)

Decoupling economic growth from emissions growth was proposed in the 2000s, with the idea of the 'knowledge economy'. However, it has turned out that creating new markets from ideas is an energy hungry endeavour. Universities use an enormous amount of standing energy, as do libraries. Economic growth is not decoupled from emissions growth but exacerbated by energy intensive new technologies such as Bitcoin and machine learning. Information technologies (IT) use euphemisms such as 'virtual' or 'the cloud' to obscure the size of street block computer buildings required for mass information storage. Bitcoin scours the world to make use of 'excess' energy production. They need football field–size computer banks to sift for the next prime number. Enormous computing power is required to analyse vast data sets in the process of artificial intelligence (AI), or algorithmic machine learning. Decoupling knowledge or dematerialising other aspects of the economy have not resulted in more than short-term efficiency gains.

Figures 1.1 and 1.2 from James Hansen et al. (2023), ex-director of the NASA Goddard Institute for Space Studies, demonstrate the impact of the accumulation of exponential emissions on planetary warming. Hansen explains the graph in figure 1.2 shows 'a staggering increase in Earth's energy imbalance'.[1] Earth's atmosphere has long operated as a lens, where greenhouse gases, carbon dioxide, nitrous oxide, methane and hydrofluorocarbons, 'capture' a certain amount of energy from sunlight. Historically, the planet has radiated out a similar amount of energy, making a 'balance' of radiative forcing. The industrial revolution began the process of releasing buried fossil fuels into the atmosphere. The atmosphere is thickening with gasses which are energised by sunlight. The molecules capture parts of the outgoing thermal energy and do not transmit the radiation back out into space. Oceans are giant sinks for all this extra heat. They also store large amounts of dissolved carbon dioxide. But since 2005, the ocean has begun emitting carbon dioxide in the northern hemisphere summer due to rising temperatures. Warming

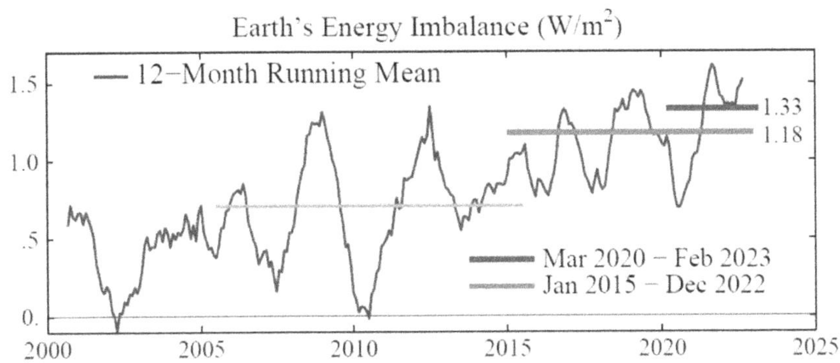

Figure 1.2 Twelve-Month Running Mean of Earth's Energy Imbalance, Based on CERES Satellite Data for EEI Change Normalised to 0.71 W/m² Mean for July 2005 to June 2015 from in Situ Data (Hansen et al. 2023)

begets more warming. Other feedbacks need to be factored in too. Hansen (2023) and others suggest the climate models of ice melt underestimate the speed and consequence of fresh water entering the ocean (Rahmstorf 2006; Hansen et al. 2016; Schloesser et al. 2019). Fresh water makes layers in the ocean and disrupts the thermohaline current. Both ocean and atmosphere have intensified hydrological turbulence as the planetary temperature rises.

ON FRAMING

The cause of climate change is straight forward; fossil fuel emissions are thickening the atmosphere. Immediate legislation against fossil fuel production could largely solve the problem. This has not occurred, because the machinery of modern economics requires immense quantities of versatile, high-energy fuel to continue economic growth. The drivers behind modern economic processes are more diffuse and difficult to grasp. Consumer demand is the 'essence' of modernity. But consumerism is not automatically essential to the human condition; it is an historical event. Nor is trade limited to modernity – trade is thousands of years old. The normative premises behind modern economics, policy and business are largely based on seventeenth- and eighteenth-century ideas in science and philosophy. Individual rationality, a sceptical alienation of humanity from natural objects and a belief in the straight forward logic of cause and effect, remain the prevailing norms across the domains of political economy.

Industrial production provides the population with the benefit of cheap consumer goods. Yet, there is almost no accountability for environmental pollution. Classical and neoliberal economics have other omissions. There is no theory of finance, no analysis of material or energy throughput, no acknowledgement of pollution or ecological and social impact. Classical and neoliberal economic theory relies on rational individual consumer choice. As the industrial revolution unfolded, intensive fossil fuel energy has replaced or leveraged labour power by multiple factors. It has completely changed the expectations of the productivity of work (Hagans 2020).

Heidegger describes storage as the 'essence' of modern technology (1977). Consumerism shifts the tempo of production from seasonal abundance to consumer demand. This profound change has reshaped the way modern people think about nature, and about ourselves, to one of resource extraction. Heidegger (1977) argues that understanding all aspects of the world as a potential 'resource' is enframing the way that we think, making it almost impossible to conceive of any other ways of knowing. Nevertheless, failure, he acknowledges, shows up the meaning of the whole in new insightful ways (1973a/1953), and it could be through the failure of climate change that we have the option to open our imagination to new modes of social, political and economic organisation.

Heidegger's (1977) important critique of the technological enframing of modern ways of knowing, determines everything in the world as a resource for potential consumption. It shows up the crucial dynamics of the wild west economics of resource extraction (Broadman 1966). Tight-knit communities that were alienated from their traditional economic territories disperse and become modern individuals, participating in the consumer-oriented global system of exchange. Globalisation, and the balancing act of modern macro-economics around the premise of growth, are completely normalised. Shifting its trajectory feels unfathomable; like King Knut trying to halt the tide.

> Modern economies are driven toward economic growth. For as long as the economy is growing, positive feedback mechanisms tend to push this system toward further growth. When consumption growth falters, the system is driven toward a potentially damaging collapse, with a knock on impact on human flourishing. People's jobs and livelihoods suffer. In a growth-based economy, growth is functional for stability. The capitalist model has no easy route to a steady-state position. Its natural dynamics push it toward one of two states: expansion or collapse. (Victor and Jackson 2015: 240)

Despite the very clear condemnation of business as usual, or BAU, by the (IPCC 2013; 2021), conventional economists are committed to the status quo. The unremitting progress of modern economic growth continues to exacerbate the pollution and emissions tragedy to the point of no return. Until now, people have found it easier to contemplate climate collapse than to contemplate a change in economic norms (Bendell 2023).

Heidegger (1977) tries hard to uncover ways to exceed and change the determinism of the technological enframing of thought. His critical analysis of Descartes and Enlightenment individualism is crucial. Heidegger argues that Enlightenment philosophy alienates individual subjective thinkers from objects in the environment (Heidegger 1973a/1953; 1973b/1954; 1993; 1996b; 1999c) and that alienation is fundamentally problematic.

Nevertheless, Heidegger remains caught up in the anthropocentrism of European thinking. Ideas circulating in feminist and indigenous philosophy, new materialism and posthumanism are proving are much more expansive. The alienation of rational individualism institutes a dualist separation between the subject versus object, technological culture versus pure nature, men versus women (Merchant 1980; Plumwood 2002). Multiple authors have been unravelling the faulty philosophy that underpins neoliberalism (e.g. Snook 1982; Peters et al. 1993; Dale 1997). The critique is well established. As yet, a cohesive new narrative to inform policy and politics in a more ecologically sustainable manner has not emerged. This book takes some early steps into this space.

BUSINESS AS USUAL

Modern technology is shaping the way that people think in profoundly different ways from earlier civilisations. Heidegger (1977) argues that modernity has a technological horizon of disclosure (Young 2002) that enframes how people think. It is based on the pace of consumer demand rather than the pace of seasons. In modernity, all aspects of the world are understood as resources waiting in 'standing reserve' to be consumed. The technological horizon of disclosure is underpinned by the modern philosophical emphasis on individual subjectivity.

The modern individual is held apart in an objective stance, from things in nature (Descartes 1980/1641). Alienation of human subjectivity from natural objects has legitimated and facilitated a shift from 'home' economics based on the local ecology, to large scale, wealth of nations (Smith 1904/1776) – mass markets which are almost entirely oblivious to the environmental consequences of production and extraction. The alienation from nature can be seen in economic theory too. Modern economics has 'externalised' pollution and monetarised scarcity without any ethos of *care* towards other species, landscapes or other peoples. Heidegger explains this phenomenon of alienation and consumerism as the modern technological enframing that produces our 'world'.

All aspects of the world are co-opted into the modern machinery of consumerism. For modernity, a forest is not a complex ecosystem with its own plethora of interactions and communication; it is lumber waiting *in situ* to be logged, a river is potential hydro energy, the mountain stores rare minerals and so on. Consumer demand disturbs the cyclic pace of seasonal production and replaces it with a twenty-four/seven global 'ordering of order' (Heidegger 1977: 17). Heidegger spent the second half of his career considering how to escape the determinism of the modern horizon of thought.

It is possible, I think, to have a sophisticated healthy society that caters for 8 billion people in more ecologically compatible ways. But it is not easy. Escape from the paradigm of resource extraction is difficult to conceive. People have very little chance of evading the neoliberal paradigm. Neoliberalism co-opts new forms of energy and accounting into the 'universal market'. More sustainable types of productivity, infrastructure, lifestyle expectations and accountability for climate change emissions are beginning to emerge. However, political commitments to reducing CO_{2e} have not translated into a challenge of systemic macroeconomic exponential economic growth.

Business as usual remains deeply entrenched. Two things are needed: identifying what systemically forces economic growth, and creating a new imaginary to shift the worlding of technology towards a more ecological outlook. Otherwise, neoliberalism redeploys sustainability towards accumulation and consumerism (N. Gough 1987; Irwin 2008b).

The first question is, How did we arrive at this point? Historically, a significant change in human tool use from the craftwork of the preindustrial era occurred with the invention of the steam engine. Fossil fuel combustion releases huge energy capacity, that changes the nature of work. The side effect, contested at the time by Ruskin and Wordsworth (Yoshikawa 2020), is dense smoke and atmospheric pollution. European cities turned black, as soot permeated stone buildings and eroded ancient statues. The problems were assumed to be site specific. Globalisation has enabled people to trade their way out of localised calamities.

There are pros and cons. The payoff from global trade is that people are not as vulnerable to food security threats from erratic local ecological disasters. The benefits of freedom from the constraints of local environment have a dark side. It also alienates modern society from awareness of the health of local ecosystems. The pollution externalities of modern industrial production were ignored for two hundred years. By now, the problems have escalated to such a scale that the impact of excessive pollution on planetary ecology is widely understood.

Once the lonely enclave of environmentalists, concern about climate change, extinction and plastic pollution is becoming mainstream. There is rising popular frustration about the lack of direction, the lack of leadership, to tackle a world marked by climate change, pollution, deforestation, soil exhaustion, extinctions and social displacement (Tong et al. 2019; Goulson 2019; Irwin 2013; 2019b). Familiar political debates fail to touch the sides of these issues. The most caring people are trapped in the worlding of modernity which entrenches oil consumption and emissions at a fundamental level. BAU is almost impossible to perceive, let alone change. The enculturation of modernity shapes our sense of status, our aspirations, expectations and notions of decent society. But the failures of modernity make its conceptual and material parameters more visible (Heidegger 1973a; 1977). Within these parameters, consumer behaviour is taking more responsibility than ever before. Uptake of electric vehicles is expanding as rapidly as manufacturers can produce them. People want change.

Classical and neoliberal economics, or BAU, produces a certain technological optimism around market choice and green growth. 'Solutions' to the issues of climate change careen across a range of isolated issues: plastic disposal, minor sops to lower nitrates, pesticides and antibiotics used in industrial agriculture, small nods to deforestation and a host of other problems.

'Carbon capture', geo-engineering, emissions trading, veganism, clean tech and the 'circular economy' are advocated. They maintain the status quo but clean up pollution. No doubt many of these things are beneficial. Reducing entropy through better design is a great idea! It reduces energy throughput, emissions and rubbish. But while consumerism keeps expanding, so do emissions. These market initiatives may contribute some minor positive impact but unfortunately, in alignment with Jevon's paradox, all efficiencies and reductions in emissions are quickly absorbed and over-run by the overall commitment to universal economic growth. More productivity means more emissions (Caporale et al. 2021; Mitić et al. 2022; Onofrei et al. 2022; Noll 2023).

Markets are incapable of solving the issue of pollution and climate change. Theoretically, the price point is set in a balance between consumer benefits and manufacturing costs. Market solutions assume that including pollution into the market equation (through pollution vouchers or an emissions trading scheme for example) will enable the universal market to accommodate the problem of climate change. The theory of market balance is far from how markets actually function. Jevon's paradox is that all innovative technologies that improve energy production are absorbed and exceeded by economic growth. The rate of emissions is climbing faster than ever (Hansen et al. 2023).

The structural imperative for growth is built into the finance industry, set up by ex nihilo mortgage lending. Ex nihilo credit increases the nominal amount of circulating national currency without altering the overall amount of exchange. The value of new credit dilutes the overall value of money and drives a demand for economic growth to offset inflationary pressure. To a certain degree, new credit increases prices, makes more profit and increases consumerism. But a population can only eat so much bread and drive so many vehicles. So, the slosh of new money creates inflationary pressure. The banks make loans ex nihilo – out of nothing – but the value of that money actually does come from somewhere. It is covered by the public's transactional fiat value in the real economy, usually calculated as gross transactions of domestic product (GDP). Erik Lindberg (2017) explains the importance of consistent money supply of inflationary and deflationary pressures.

> The amount of money needs to keep pace with the amounts of available goods and services. If not, the economy will experience either inflation or deflation. Most people think of inflation as suddenly rising prices, something that can be especially disrupting if they are living on a fixed income or have a large savings for the future: as prices rise this money, in effect, loses value. But rising prices are only a symptom of a simple issue with the money supply – namely that there is too much money in relationship to the amount

of available goods and services. Think about this in terms of supply and demand: too much money chasing too few goods and services, and the price of everything will rise.

By pumping enormous amounts of new credit into circulation, private banks are effectively stealing value from the public value of currency (Soddy 1933; Galbraith 1975). The banks are unwittingly taking value from savings, wages, and circulation. Double entry bookkeeping means both debit and credit are regarded as assets that have entered circulation – so every new ex nihilo loan counts twice.

Governments are responsive to inflation, currency values and unemployment rates, and so they attempt to stimulate economic growth. Governments are compensating for the inflationary effect of too much new money, created by private banking, in the nation's gross exchange of goods and services.

Thermodynamic economics offers a deeper set of alternatives. Beginning in 1926, Frederick Soddy (1933) took the physics of thermodynamics and applied it to social and economic processes. The thermodynamics and entropy of economic production is examined by Georgescu-Roegen in *The Entropy Law and the Economic Process* (1971), Gorz in *Ecologica* (1971), Herman Daly in *Toward a Steady-State Economy* (1973) and others. This important work advocates a fundamental shift from exponential growth and consequent increase in pollution. The focus for economics shifts to the metabolism of energy throughput. Sustainability as 'slow economics' is paced according to seasonal growth and frugal use of nonrenewable resources. The recommendation to 'reduce, reuse, recycle', emissions accounting and circular design all emerge from literature on steady state and metabolic economics.

In the 1980s, the flowering of thermodynamic economics was somehow sidelined. This was partly through the machinations of the Mount Pelérin Society (Devine 2004). Neoliberal governments, committed to the discourse of economic growth, deployed the rhetoric that there is 'no alternative' to modern capitalism. Left-wing politics has been equally committed to economic growth. Within the spectrum of conventional modern politics, there really has been no routes to think through the structural imperatives driving emissions and economic growth.

Alienation from nature got worse. Pollution and climate change got worse. Modernity forgot to ask meaningful questions and got deeper into the technological enframing of the horizon of thought. Philosophers of technology, including myself, struggled to unleash from the determinism of regarding everything as a potential resource at the demand of consumerism. Heidegger had looked to the west for insights into new pathways that might open up the future.

THE QUESTION OF TECHNOLOGY

Ancient Greek culture and philosophy has been situated as 'classical' and 'originary'. It is the *Ursprung*, or seed, of modern democratic civilisation. To better understand how technology shapes the horizon of thought, I look at the earliest examples of mechanical technology, the Antikythera mechanism from ancient Greece. This originary machine, full of complex cogs, helps to consider how their technology shaped society thousands of years before the modern period. The Antikythera mechanism is the precursor to modern clockwork but was fundamentally entrenched in much older civilisational values. Greece also struggled with many of the problems of consumer greed, overtly addressed by the ascetics of stoicism. Yet, somehow consumerism did not get out of hand and create exponential economic growth, as it does today.

Neoliberal market discourse has attempted to domesticate greed. It has the unfortunate side effect of pollution and climate change. Heidegger (1977) characterises modernity as a shift from ancient craftwork to the machinery of modern production. Craftwork constitutes an agrarian culture where workers are intimately bound with their material environment. This contrasts to the calculus, mastery, and alienation of consumer technologies today. However, I argue that the Antikythera mechanism is a sophisticated machine with cogs, levers, and careful calculus. It indicates rationality and calculus were already dominating the technological horizon of disclosure in Axial communities.

The Antikythera mechanism demonstrates a fascination with cyclic time and time-keeping (Freeth 2014). The horizon of disclosure in ancient Greece retained a cyclic, seasonal, earth-centric approach to society. The mathematics of this geocentric and ecological orientation of time, owe a debt to the record-keeping of Mesopotamia (Steele 2000a; 2000b; Irwin 2017; Freeth et al. 2021). Perhaps even more than Greece, Mesopotamia had a disposition towards large scale, hierarchical, well organised, market and consumer-oriented production. Mesopotamia invented writing, accounting, and debt. Debt included interest rates, inflation, laws, labour crises and subsequent environmental impact. Yet, despite such similar lifestyle reverberating with the worlding of our own, modernity is in crisis after merely two hundred years. Mesopotamia managed to maintain civilisation, without collapsing their delicate ecosystem, for thousands of years. Other civilisations, as Jared Diamond (2004) points out, fail to accommodate their productivity and population, to environmental limits.

Neoliberalism has an optimistic approach to technology. They believe that innovation, such as carbon capture, will enable BAU to continue. After thirty years, these innovations have failed to emerge and save BAU. Instead, the generation of emissions is accelerating (Hansen et al. 2023; Noll 2023).

Heidegger's (1977) approach to technology is very different. He argues that technology shapes the way we understand the world. Given fossil fuel technology has turbo-charged modern productivity, a critical analysis of technology is important. The essence of modern technology, as Heidegger (1977) describes it, is that everything is understood as a resource, waiting for the demands of consumerism. The question is, How do sophisticated approaches to technology in other civilisations avoid the alienation between culture and nature that pervades modernity? If technology 'worlds us', as Heidegger (1977) argues, then what shift in philosophical ethos is required for a technological worlding that is ecological in root and branch?

The alienation of modern subjectivity from the ecosystem is deeply problematic and is getting a response from a variety of discipline areas.

THE QUESTION OF ONTOLOGY

Across a range of discipline areas, there is a significant shift in understanding the underlying relationship between human subjects and the environmental context. A profound change in orientation is showing up in plant biology, animal behaviour studies, indigenous philosophy, feminist embodiment, posthumanism and new materialism, the geology of morals (Nietzsche 1989/1887; Deleuze and Guattari 1999/1980) and the agential realism of entangled subatomic particles of quantum physics (Bohr 1987/1957; Barad 2007).

Indigenous modes of technology, ownership, ecology and knowledge production emphasize the interrelationship of all things. Fundamental to the Māori worldview, is *whakapapa*, the connection of humanity with all other species and strata. *Whakapapa* is a fully explicated set of human and posthuman kinship relationships (Ngata 2011/1944; T. Smith 2000; Mika 2017). It has a similar scope to the claves of contemporary biology. Māori communities engage with landscape as a form of historical eco-technology that involves geology, biology and society. It is a palimpsest of narrative that goes back thirty generations and reaches forwards a comparable time span (Irwin and White 2019). There is no alienation of culture versus nature.

In sharp distinction from the alienation of modernity, this book considers the implications of this new technological horizon of disclosure – one shaped by the intra-relationship of all elements of the ecological field. *Whakapapa* traces kin relations that include the human, the atmosphere, ocean, soil, strata and all the ecological niches in between. New questions can arise. How would an earth, or geocentric ontology, orient technology and create a new 'world'? How might technology begin to show up in this new posthumanist enframing of thought? How might the new orientation of technology shape the assumptions, identity, and expectations of posthumanist society? How is a planetary consciousness applied in policy, infrastructure, economics, and techniques of the self? Assuming ex nihilo banking becomes illegal, how will macroeconomics reorient to a more stable system, with emphasis on ecological safety, and egalitarian redistribution instead of continuous exponential economic growth?

The current alienation of modern humanity from ecology has resulted in an economic system that is losing its relationship to reality. Nominal debt has spiralled so high, it bears no relation to the real transactional economy, nor the ability of earth systems to deliver future resources in correspondence to its staggering value. Shares, likewise, and especially derivatives future trading, is billions times higher than the resources to which they are nominally attached. Finance and trading markets have diverged into an abstract and impossible field. They have become hyperreal (Higgs 2014; Kant 1992; Baudrillard 1994; Bostrom 1998; 2003a). The finance industry is perhaps the most important driver of the expanding exponential schism between abstract nominal economic value and planetary limits (IMF 2022). Potentially, an eco-centric orientation would leave no room for the nominal schism between the broad economy and financial instruments. It could clarify that there is no legitimation for any private institution to create ex nihilo loans. The loss of new credit devaluing public exchange value of fiat currency would finish the structural impetus for continuous exponential economic growth. Finally, extinguishing ex nihilo mortgages resolves Jevon's paradox. New green technologies would successfully reduce CO_{2e} emissions. Escalating economic growth would no longer absorb the benefits of renewable technology efficiency gains.

In this book, I look at the multiple alternatives to modern economics and the alienation of culture from nature. It involves examining the very ancient, and the very contemporary, to rethink the scope and premise of economics, finance, epistemology, science and technology. A transdisciplinary approach simulates a new narrative for policy, leadership and political action, that genuinely reduces the impact of modern humanity on the environment. This is no small task – so of course this book is a sketch, an indicator of capacity, the opening of a door, a pathway – not an entire solution. Much more will need to develop. This book lays out a compelling route to finish the era of exponential growth, halt the accumulation of climate emissions and consider safer pathways to the future.

This book is divided into three parts: (I) 'Alienation from Nature: Growth and Finance'; (II) 'Time and Technology'; and (III) 'Foundational Entanglement'.

PART I: 'ALIENATION FROM NATURE: GROWTH AND FINANCE'

This part examines modernity as it presently operates – alienated from nature – with an emphasis on continuous economic growth, and struggling with the constraints of climate change, plastic pollution and resource exhaustion. The section carefully examines the meaning of 'growth' and how this correlates with emissions growth and climate change. It interrogates the Enlightenment constructs of individualism, and alienation from natural objects. It then shines a light on the obscure practices of banking and shows how the finance industry is a systemic driver of economic growth. The paradox of renewable energy is that it has not slowed down overall emissions growth – and it cannot, while economic growth absorbs and over-reaches all efficiencies in the name of progress.

CHAPTER 2: 'MODERN ECONOMIC GROWTH'

Climate change is an existential threat to global society. Leaders at all levels are finding it very difficult to imagine – let alone implement – the kinds of systemic changes that are required to mitigate the problem (IPCC 2001; 2007; 2013; 2018; 2021). Over the last hundred years or more, as the global population has grown exponentially – infrastructure in the New World has been built around cheap and abundant oil. To shift to a non-oil-based economy requires a simple, effective change in philosophical, policy and planning orientation. It requires placing ecological sustainability at the centre of planning and commerce rather than abstract hyperreality and consumerism. How do we go about this simple, yet profound, transformation in ideas?

So far, global policy initiatives such as Brundtland (1987), the United Nations Framework Convention on Climate Change (UNFCCC 1992), Kyoto (1998), the Paris Accord (2016) or the annual UN COPs on climate change have had only limited success. Changing the mindset is proving illusive. Introducing effective policies to reduce emissions and protect the environment is beginning to emerge in most nations. However, many governments remain committed to BAU.

The Intergovernmental Panel on Climate Change (IPCC) has made an incredibly important contribution to collating and popularising climate science. Under Pachauri's thirteen-year directorship, the IPCC (2007) Report made clear that climate change was 90 per cent human made and that emissions need to drop within fifteen years. But mixed messaging on policy also ensued, as the IPCC criticised BAU in one breath but assumed the continuation of neoliberal economics in the next. The recent 2021 IPCC Report mentions degrowth for the first time. In the last decade, the impact of climate change began to become obvious, as storms and flooding, drought and forest fires intensified to unheard of proportions (Van Eedon et al. 2020). Governments, cities and citizens are taking the Paris Agreement (2016) seriously and attempting to implement more renewable energy and ecological consumer choices into daily lives. The UN Sustainability Goals are confused and contradictory (Irwin 2022b; Hickel 2019; Higgs 2023). Short term relief from emissions growth has occurred but overall, unfortunately, the accumulation of CO_{2e} continues globally. A focus on energy sources alone is insufficient. Economic growth is closely tied to increased consumerism, and drives accumulating climate change emissions (Le Quéré et al. 2020; Caporale et al. 2021; Mitić et al. 2022; Onofrei et al. 2022).

People use the term 'growth' in loose and ill-defined ways. Emotional growth means gaining more emotional maturity. Physical growth has a built in optimum, and we do not keep increasing height (and hopefully girth) beyond a built-in point. Redefining 'growth' in more accurate ways may take us out of the linear or exponential growth mind set. It reintroduces a more ecologically engaged mode of understanding redistributive economics. Stern (2006; 2018) suggests that trillions of dollars need to be invested in sustainable infrastructure in the next couple of decades. There is copious opportunity for some beneficial areas of economic 'green growth'. Necessary development is not the same as structural reliance on growth as a crucial macroeconomic compensation for inflation, social stability, or a marker of success.

Commitment to economic growth comes from the 'left' and the 'right'. The Keynesian welfare state guaranteed the working classes a universal 'safety net' of health, education, and pensions for the old and infirm, paid for by progressive taxation on income and companies (Keynes 1991/1936). This meant that profits which, as Marx (1887/1867) had earlier analysed, siphon upwards to the owners of production, were redistributed to some extent. Keynes (1933) argued that if the working classes shared some of the benefits of capitalism, it makes a far more stable society.

Keynes (1933) also argued for economic growth. Governments have committed to economic growth in the name of progress or 'development'. The people can have faith that if they work hard, there will be a gradual improvement in their lot. Believing in progressive improvement has invested the working classes in the benefits of corporate consumerism. The Keynesian Settlement evolved alongside the multiplier effect, which advocates government spending to stabilise and ensure economic growth. Similarly, neoliberal theorists like Friedman (1970) argue that equity is not an issue when the overall economic 'pie' increases, and the working classes can hope to gradually gain through overall economic growth. The Keynesian discourse of working-class benefits from growth have protected the concept from rigorous critique.

Banking is most reliant on continuous exponential growth in the 'real' or 'broad' economy of products and service exchange.

CHAPTER 3: 'FINANCE AND SYSTEMIC GROWTH'

Private financial institutions create the structural imperative that forces growth into the system. It comes at the cost of devaluing the currency circulating in the nation's real economy. Economic growth is built on the foundations of alienation from ecological needs and values. Banks began double entry bookkeeping and to loan ex nihilo in the mid-1400s, just before the Medici era (De Roover 1963; Buch and Dages 2018: 4). Early exponential growth are tiny increments that are almost invisible. But late exponential growth is an escalation of large and larger amounts. By introducing large quantities of new money daily, banks today are setting up the conditions for inflation. Compensatory exponential economic growth lowers that risk.

Modern private banks, which have no overt legal permission to create sovereign currency, create online credit lines generated ex nihilo – *out of nothing*. This means (despite a lot of rhetoric to the contrary), there is no gold bullion in a back room, no deposit accounts backing loans, and no fractional reserve that justify the risk of these loans. The credit line is generated out of thin air. Banks use double entry bookkeeping – which 'counts' debt as an asset. Due to double entry bookkeeping, the new money is generated *twice* out of thin air; once as a new credit line in the seller's bank account, and once as a new loan in the lender's bank account. Doorman (2015: 2) explains how the new mortgage loan instantly enters the circulation of fiat money. 'Whenever a bank makes a loan, it simultaneously creates a matching deposit in the borrower's bank account, thereby creating new money. The money multiplier described by Keynes (1933), also explains how that first deposit then begins to circulate through the transaction system'. Theoretically, over thirty years the original loan will be repaid, shrinking the pool of money back again. But the credit line remains in permanent circulation (as do the fractional subsequent credit lines generated from further circulation of the new currency). What is more, the loan repayment, and the interest rates charged on the loan, are met by real work, which places a systemic demand for economic growth.

The new money stimulates inflation. The relatively stable use of fiat currency for exchange is vastly devalued with all the extra currency swirling through the system. At the same time, the demand to meet interest and capital repayments, forces people to contribute to economic development (over and above subsistence), through work. Growth stimulus offsets the devaluation of currency (inflation) to a certain extent.

According to the IMF (2022), the amount of global debt in 2022 is 2.5 times more than the entire amount of global GDP. The tenuous connection between new credit and real-world transactions is sheltered by the smoke and mirrors about the processes of banking. Not many people are aware that loans are created from nothing. There is no tangible reason why your aunty could not lend you capital too, with a similar double entry excel spread sheet – as long as other banks would accept her deposit.

Alienation from planetary limits benefits the finance class, because the public have no idea that the ex nihilo credit is an unauthorised devaluation from the public circulation of fiat money. Instead, inflation is blamed on rises in the consumer price index, or higher wages. The nominal value of global debt is so impossibly high, in relation to planetary limits, that it is becoming more obvious that money is a problematic chimera – a simulacra of value.

The instability of nominal credit is often discussed in relation to financial collapse, which is extremely important. However, its environmental consequences are also profoundly important. Nominal credit is a result of the alienation of modernity, and deeply implicated in the normative assumptions of progressive growth. Meanwhile, without relating it to the systemic forcing by the banking industry, macroeconomics, political economy, governments, the housing market, pension funds, and the finance industry all justify normative economic growth.

CHAPTER 4: 'CALCULATION, RISK AND FEAR'

The ubiquitous narrative of neoliberalism is familiar to billions of people across the globe. This chapter outlines the colonial philosophy underpinning modern economics. Technological optimists assume that technologies that address climate change, like carbon capture, will save the day. It shows how neoliberalism expanded to incorporate and domesticate 'risk'. But people are losing faith in technological mastery and control as sudden, catastrophic climate events occur more frequently.

The discourse of risk remains thoroughly imbued with colonial pretensions of mastery over nature. Heidegger calls this the 'ordering of order' (1977: 17). He regards it as a crucial part of the enframing of modern technological thought. The ordering of order is based on Enlightenment rationality, best outlined by Descartes's vision of mind over matter.

Descartes's (1980/1641) concept of 'subjectivity' is carefully delineated from the 'object'. The sceptical approach to rationality creates doubt about the most basic assumptions, including interpretation of the body's senses. The mind is reduced to rational deduction as the only certainty in an unknowable world. This dualism of subject from object is at the root of patriarchal and humanist assumptions of mastery (Merchant 1980).

Sceptical rationality compliments Newton's understanding of a mechanical universe. Newton's physics demonstrate forces act upon other forces. The universe obeys the laws of cause and effect. By the twentieth century, Weber (1948/1919) endorses the calculus and rationality developed by Enlightenment thinkers as the basis of science. Weber jettisons the precolonial 'unknowable' and 'mystery' of nature. The mechanical universe is completely translucent, and this makes it easy to leverage. While grieving the loss of the world's enchantment, Weber is deeply committed to human mastery over planetary resources. Rational calculus permeates science and economics, in an assumption of knowability and, thus, mastery.

However, climate change demonstrates the limited scope of mastery. Forgotten and externalised, silenced by mainstream norms, are layers of culpability and excess. The risks associated with the hubris of mastery are profound. Beck (1992/1986; 2009) was one of the first writers to directly engage with the risks of modernity. The calculation of risk is shifting the fundamentals of economic orientation, especially at the larger scale of corporates, insurance, finance, and stock market.

Risk itself has been commodified. Paradoxically, risk shows more clearly than ever before just how little is within modern control. Beck (1992) evokes 'the society of fear' in response to rising awareness of the illusion of human control. Climate denialists, doomers and preppers are saturated in this discourse of collapse. Failures of the current system are becoming visible, and fear is a neccssary first step. With that visibility comes better readiness for an onto-epistemological shift (Heidegger 2016; 1996a).

PART II: 'TIME AND TECHNOLOGY'

Modernity has a dissociative relationship to time. On the one hand, there is the seductive lure of instant gratification, couched in terms of consumer satisfaction. On the other hand, people are always waiting, and the wait time is between pay checks, or six-monthly returns, or the exponential rise of compound interest increments. The scheduling of money going in or out and the expectation of exponential economic growth creates its own pacing of consumption and regeneration. This part explores some key elements in the philosophy of time.

Kant's famous 'a priori' conditions include time as a crucial medium of experience and knowledge. This is the ultimate subjectification of time. Heidegger (1985; 1999b; 1970; 1968a/1954; 1968b; 1949) expands Kant's individualistic existential relation to time. Heidegger (2016; 1999a) considers the epoch of a civilisation as the bearer of time, in much the same way as the individual might. Heidegger (1977; 1989/1976) is also interested in how the demands of consumerism shape time for modernity, in sharp contrast to the earlier reliance on seasonal pace and the geocentric relation to time.

Kant's transcendental aesthetic is the ultimate alienation of subjectivity from the natural world. It justifies the increasingly abstract relation of humanity with the seasonal cycles of geocentric economics. His otherworldly 'noumenon' situates truth outside of the daily world of cause and effect. Nominal debt is likewise almost completely divorced from real world considerations, which is why it has escalated to 2.5 times the amount of global GDP (IMF 2020). In computer science, there is widespread belief that we are living in a simulacrum. The link between earth bound ecology and human society is completely severed.

Ancient societies have a seasonal concept of time. Time revolves in great cycles, and these cycles measure the pacing of political and economic life. Mesopotamia maintained indigenous geocentric worlding but developed the technologies of writing, debt and interest too. They had techniques to stop exponential growth spiral out of control.

Metabolic economics takes the emphasis off consumer demand, and refocuses on energy throughput, and geocentric cycles. Alienation is no longer a crucial element of the economic system, and exponential growth can be avoided.

CHAPTER 5: 'ALIENATION AND THE SIMULACRA'

The links between material ecology and wealth generation are both increasingly tenuous and irreducibly entangled. This chapter examines closely Kant's Enlightenment philosophy that promotes alienation as a fundamental goal. The transcendental aesthetic aims for the 'noumenon' outside of empirical reality. The chapter indicates the problematic results of alienation in present day leadership, and finance.

Climate change results from the alienation of modern practices from the constraints of environmental cycles. Mainstream economists and global institutions are complacent about the lack of progress towards reducing emissions. The predictable Holocene that all adults grew up with, has been ruptured. At this stage, wealth alleviates the instability and food scarcity that climate turbulence is causing. Climate stress generates extreme politics. The rise of autocrats has already begun. Wealthy individuals are 'prepping' for growing instability and civil war with elaborate bunkers in far distant lands (O'Connell 2018). Continuation of BAU, does not matter for the elite 1 per cent, who think that they can buy their way to safety.

Many of the über-rich, including Elon Musk (who owns Tesla, Twitter and SpaceX), Sam Alton (the owner of ChatGPT) and Peter Thiel (PayPal with Musk) propose that we are living in a computer simulacrum, and the world as we know it, is a complex AI-generated programme. Prepping for Armageddon combines awkwardly with the belief that we exist in a simulacrum, but both narratives are a response to the fear of climate change.

Climate denialism is also an element of the disconnect with ecology. Simulation theorist, Nick Bostrom (2013: n24) from the Oxford University Futures Institute goes as far as to say there is little chance of 'runaway climate change'. The temperate Holocene was when the planet was in the Goldilocks zone, where life is thriving. Continuously rising CO_{2e} emissions could make the planet an arid corpse, like Mars. It may be the case that life itself survives the extremely rapid global warming that industrialisation is producing. But the Goldilocks zone which includes nearly the whole planet, where mammals, such as ourselves thrive, could decrease dramatically.

The thermohaline escalator current is already stuttering, and with it the 'thermostat' of the planet (Rahmstorf 2006; 2009; Hansen et al. 2016). If cold, mineral rich water from the polar region is not carried down into the equator belt, then the equator will consistently be in the 50°C range (Hansen et al. 2016). That is much too high for humans, who have a survival limit of wet bulb point temperature of around 31.5°C at 100 per cent humidity (Vecellio et al. 2022). Furthermore, the hot equatorial water needs to flow past the northern landmasses such as the United Kingdom and northern Europe, Canada and Alaska, to keep them ice free, and available for land-based animal habitation. Without the thermohaline current, the equator will be much hotter at about 50°C, and the northern hemisphere landmass will be 5 to 10°C cooler. The southern hemisphere does not alter as much because the current is already deep and cold as it circulates the large southern oceans. Periodically over the history of the earth, changes to the thermohaline current have happened before.

> A short-lived event known as the 'Younger Dryas cooling' occurred in the Northern Hemisphere just as warming was getting underway after the last glacial, about 13,000 years ago. Something happened suddenly and dramatically to reduce or cease production of deep water in the northern seas. This catapulted the region back into near-glacial conditions, with cooling in some places estimated to be as much as 10°C in a few decades. (Manighetti 2001)

How hot the ocean needs to be for the thermohaline current to halt is unknown, but this is an important tipping point in climate warming (Hansen et al. 2016; 2023).

Furthermore, the relationship between exponential increase in economic growth is correlated with an exponential acceleration in planetary temperature, especially noticeable since 2010 (Hansen et al. 2023; Noll 2023). In these circumstances, the alienation of humanity from the environment, and the continuous drive for economic growth, which is stimulated by imaginary ex nihilo money, is deeply concerning. The justification for alienation and belief in a hyperreal simulacrum instead of the real surrounding ecological conditions emerge from Enlightenment scepticism.

Kant develops further Descartes' sceptical individual into the 'transcendental aesthetic'. Descartes (1980/1641) began the alienation of subjectivity from objects, but Kant (1992/1781) took it much further. In *The Critique of Pure Reason* (1992/1781), Kant steps out the logic of rationality through various stages of the evolution of constructivist knowledge. In a convoluted argument that is carefully detailed in this chapter, Kant alienates experiential comprehension of objects by focussing on the medium of phenomenological experience in time, space and rational thinking. Kant succeeds in severing the empirical connection of experience of real objects and turns them into an alienated and inaccessible thing-in-itself. That thing-in-itself can only be represented; it is a simulacrum (Baudrillard 1994).

AI and the information technology wizards of Silicon Valley are very enamoured with the idea that everything is a simulacrum. Since the film *The Matrix* (Wachowski and Wachowski 1999), there is widespread commitment to the idea that the earth we grew up in is a computer programme simulation (Bostrom 2003a). This extreme narrative of alienation from the forces of physics and ecological intra-action enables the discourse of alienation and mastery to ignore the warning signals of climate change (Bostrom 2013). If things go wrong, the simulacra can simply be reset, they argue. At the same time, though, the billionaires of Silicon Valley are aware that mastery is tipping over to unruly uncontrollable climate events (Beck 1992/1986). They are prepping for calamity with elite bunkers in faraway islands. They continue to calculate how much money is to be made from extractive economics, such as big data and its big energy consumption, before climate change turns from difficult to collapse (Bendell 2023; Victor and Jackson 2015).

CHAPTER 6: 'TECHNOLOGY SHAPES THE WAY PEOPLE THINK'

In his 1954 text 'On the Question Concerning Technology' (1977), Heidegger argues that modern technology shapes the way we think. Modernity signals a profound technological shift from all earlier civilisations and alienation from any other species. Modern technology is no longer simply 'handcraft'. Instead, modern technology has taken over every aspect of the globe, setting everything in 'standing reserve' for potential consumption in the overall service of global consumerism. He calls this way of thinking, the technological horizon of disclosure, or *Gestell*. Heidegger argues that in the global modern epoch, absolutely everything is co-opted into the technological horizon of disclosure. An untouched forest is waiting to be exploited as wood, a river as hydropower or a tourist destination. People are seen as 'human resource' to be at service as both labour power and consumers in the system. The smallest molecule, the largest star system, is all potentially exploitable by the technological horizon of disclosure. Concerning all things as a resource, waiting to be consumed, is all encompassing. Heidegger himself spent the rest of his career seeking subtle modes of understanding that stood in contrast or excess to the determining forces of modernity (Heidegger 1970; 1973b; 1975b/1959; 1975c; 1977; 1979; 1990/1966; 1996a; 1996c; 1996d).

The rational individual is a key motif in classical modern economics. Rather than focus on deductive logic or transcendental aesthetics, Heidegger's concept of subjectivity and the authentic mode of thinking is historical (1962; 1973a; 1996b). Heidegger's methodology looks to a person's life 'as a whole', from inception to death, in order for them to make authentic decisions in the present. He expanded this thesis to think about civilisation 'as a whole' from its beginnings to its ending (1996b). In the case of modernity, Heidegger harks back to ancient Greece for insight into the beginnings of democracy, and of philosophy, including Heraclitus 540 to 480 BC (2001), Plato 427 to 347 BC (1965), Aristotle 384 to 322 BC (1972). The 'end' of modernity was nuclear in the postwar era. Whereas now, climatic conditions are closing in on the existential capacity of humans and multiple other species.

The birth and death of modern civilisation, inevitably remain enclosed in the modern framework of thought (1996; 1977). For the birth of the modern epoch Heidegger points to ancient Greek thought, and to its end he indicates the nihilism of the technological horizon of disclosure with its looming threat of nuclear war and the death of civilisation (Heidegger 1973b; 1975c; 1977; 1996; 1999a,b,c; Nietzsche 1982; 1985; 1989; 1986; 1909; Heidegger and Jünger 2016). By examining the limits of a 'life' (albeit a life of an entire civilisation), authentic thinking can importantly reflect back the limitations onto the present moment and reveal how decisions today contribute to poor habits or normalised and invisible behaviours and (Heidegger 1996; Heidegger and Jünger 2016).

Future-oriented projections contribute consciously and unconsciously to how we act in the present. The refraction of Heidegger's (1999a) method of the beginning and end of civilisation helps people to reflect on the authentic decisions in this moment. Heidegger's methodology is unflinching in its acute observation of existential threat. Unfortunately projecting an inevitable end may also foreclose the possibility of envisaging an ongoing future. It is possible to project an Armageddon like finitude, or to project a continuous future that extends outwards as part of the life affirmative action that believes in the resilience of the environment and ongoing future generations. Questions emerge about the nature of time, the limits of technological enframing and how endemic alienation could be averted.

CHAPTER 7: 'ANCIENT TECHNOLOGY, TIME AND DEBT'

Modern consumerism is collapsing the ancient cyclic understanding of time into instant gratification on the one hand, and exponential progress on the other. But the demands on time have not always been so focused on this 'moment' or the attendant acceleration and contraction of time to meet consumer expectations. Classical scholars, including Heidegger (1973a; 1975a; 1979; 1999d; Heidegger and Fink 1979), consider ancient Greece the crucible of democratic modernity. The original invention of new ideas and techniques is when fuller consideration of their purpose and meaning occurs, according to Heidegger (1975a; 1973a). Given the importance of technology to the worlding of a civilisational epoch, it is worthwhile looking closely at early technologies that epitomise the earlier ecological worlding. Earlier worlding and the technologies that shape them are instructive because they both contrast and resonate with elements of modernity. These sophisticated civilisations last for centuries without creating climate change.

As 'the first beginning', the earliest known analogue computer dates to about 206 BC. The ancient Greek Antikythera mechanism helps to rethink the parameters of Heidegger's (1977) technological horizon. This cosmological clock dates to 206 BC, and it challenges some of the assumption inherent in the modern concept of progress, the philosophy of time and modes of governance and economics. The philosophy of time indicated by the Antikythera mechanism is quite different from the teleological or exponential model used in modern times. It is cyclic, not linear. It tracks the grand cycles of the solar system, not the hours and minutes of industrial routine. It is not engaged with the consumerist 'now' but rather of the whole cosmology of the solar system. Cosmological time contrasts sharply with the individualist Enlightenment roots of modernity (Lazar 2019).

For Heidegger (1975a), *Early Greek Thinking* is the founding discourse of modernity. But the Greeks derived much of their concepts – their mathematics, astronomy, and science – from Mesopotamia (Freeth et al. 2021). Writing, mathematics, and accounting were first invented in Mesopotamia. Their accounting technologies introduced a sophisticated system of control with normalised approximations of behaviour and productivity (Nissen et al. 1993; Foucault 1979). The city states occupied a fragile ecology of river delta, between the Euphrates and the Tigris. The first records of debt, interest, and repayment arose in the context of irrigation, managing soil salinisation, and careful stock management (Oppenheim 1967: 3–4; George 2013). Economic growth was a problem for which cosmology and governance had solutions. Mesopotamia developed early economics, and they also institutionalised the universal debt jubilee (Graeber 2011). Upon decree, all debts were void, 'wiping the slate clean' (Hudson 1993; 2002b; Charpin 2010). Debt and distribution were regarded as part of large-scale cyclic cosmology, in the same way crops, the solar year and lunar months are cyclic, along with the planets and stars. The custom of the debt jubilee persisted throughout Judaism and Christianity, but its egalitarian ethos gradually eroded into charity for the poor instead of a genuine *mišerum*, 'return to the mother', that reset ownership and debt to its equitable origin (Kaplan 2019).

CHAPTER 8: 'THERMODYNAMIC ECONOMICS'

Modern economics is alienated from the ecological constraints of the local river delta or well-being of the forest. Trade enables local populations to insulate themselves from localised pestilence, flood, or fire to a large degree. Business as usual (BAU) continues with emissions causing planetary warming and escalating the volume and frequency of hydrological turbulence. The failure of emissions reduction radically impacts communities all over the world. This chapter outlines the alternative to existing BAU. It is a subtle shift, that will not radically alter people's lifestyles or expectations. It needs three things to come about – a clear vision for the shape of a metabolic economics, tied firmly to ecology and the end of systemic economic growth.

Metabolic economics recognises that important externalities, such as pollution, are not addressed adequately in classical economics. A number of important economists' advocate 'steady state', 'degrowth' or 'thermodynamic' economics. Soddy's *Wealth, Virtual Wealth and Debt* (1933), Georgescu-Roegen's *The Entropy Law and the Economic Process*(1971), Gorz's *Ecologica* (2009), Meadows et al.'s *Limits to Growth* (1972), Schumacher's *Small Is Beautiful* (1973), Tim Jackson's *Prosperity without Growth* (2017) and Kate Raworth's *Doughnut Economics* (2018) are some of the better known. 'Reduce, reuse and recycle' emerged from this literature, along with emissions accounting and the circular economy. They are filled with good arguments – from detailed analysis of energy return on investment (EROI), to the recognition that consumerism does not make us happy. Unfortunately, the degrowth canon has not lit the popular imagination. It is equated with a regression of progress back to preindustrial, pastoral idealism. Degrowth is understood as a denial of luxury, of large homes, personal transport and imported consumer goods. It appears to constrain the smaller aspirations of working-class people too. During this time of transition, it is not yet a viable and compelling worldview that conditions the multitude of daily decisions in all their minutiae.

Nicholas Stern's *Climate Review* (2006) tries to 'decouple' growth from consumerism. It is a transitionary programme for pouring large-scale resources into generating green infrastructure along the same economic growth model that presently exists. But Stern admits that green growth is short term and unsustainable.

Finite resources are an important aspect of the steady state or degrowth literature. Metabolic or thermodynamic economics is focussed on energy throughput and the law of entropy. Policymakers are applying some ideas piecemeal but have not a sufficiently coherent narrative to replace the neoliberal story of the universal market. Yet we are not far off it. Clean energy regulations have been recently built, to enable better accounting of greenhouse gas emissions. Scopes 1, 2 and 3 emissions accounting attempt to include all energy, manufacturing and transport emissions both in-house and in import/export commodities. There is a lot more to be developed in this sphere, but it is becoming widely accepted as a component of business accountability. Notwithstanding this influence, thermodynamic economics is much more interesting than merely including CO_{2e} into accounting BAU. Caution is necessary because neoliberalism is famous for including innovations without changing the fundamental alienation of modernity from ecology.

The natural laws of thermodynamics can be applied principles of economics. Georgescu-Roegen (1971) developed the metabolics of economic throughput. Most energy available is ultimately derived from the sun. Even fossil fuels were ancient forests that condense thousands of years of photosynthesis. Fossil energy is dense and versatile but our economy needs to transfer to the lower energy intensity of renewable energy to slow down CO_{2e} emissions.

Much has been made of carbon capture as a technological solution to modern capitalist problems. Although it was preemptively incorporated into economic models (IPCC 2013), end-use carbon capture is very inefficient. Other than growing trees and improving ecosystems (Source Material 2023), large scale carbon capture has remained a technological fantasy (Hickel 2016; Thales et al. 2020). Clearly, the best way to reduce emissions is limiting them in the first place. Built-in obsolescence of consumer items is morally criminal.

The circular economy is being repositioned as another universal motif, in the manner of Adam Smith's invisible hand (1904/1776). But while circular *design* is optimal to improve recycling capacity, inevitably it is also governed by the law of entropy. No economic theory is universal. Circular design is an element of thermodynamic economics.

In my view, the unique metabolic processes of photosynthesis are core to understanding any kind of reversal of prolific CO_{2e} emissions. Marx (1887/1867) describes the metabolic rift of the land clearances, where the local villages ate and expelled food from their locale, contributing to the mineral profile of the local ecology (Foster 2000). The rift occurred after the clearances, when the food grown in the countryside was eaten and expelled in the city, where the sewerage system swept vital minerals and nutrients away. Farms now have to use greenhouse gas producing nitrogen fertilizers to compensate. Photosynthesis is the fixation of new matter from absorbing energy from the entropy of the sun. The ecosystem is built on cascades of similar relationships. It is a diverse plethora of symbiotic relationships, where one species cohabits with others and benefits from their excreta. Worms in leaf litter, fungi on rotting tree stumps or carrion animals on the dead or dying. Life creates as many deviations on the theme as possible. Biodiversity is the example par excellence of negative entropy, and along with efficient metabolics, the trophic cascades of diversity itself is a model for economics. I call this a model of 'entrophic cascades' and argue that interdependent, metabolic economics could replace economic growth.

PART III: 'FOUNDATIONAL ENTANGLEMENT'

Metabolic economics could simply be usurped into neoliberalism. It could conform to the ordering of order. The modern technological enframing of disclosure could reduce metabolic economics to standing reserve, waiting for consumer demand. Already, scopes 1, 2 and 3 emissions accounting measures are being integrated into BAU without any substantive change to the ethos of business practice. The emissions externalities are becoming acknowledged in cost-benefit exercises, but other damaging practices are still ignored and forgotten. The neoliberal claims that everything can be adequately distributed by the universal market is being applied to emissions. The instability of the carbon price in a trading market is not helping companies plan. Emissions trading relies on companies to provide offsetting carbon sinks. The inability of these companies to adequately protect or grow new forest to offset accumulating emissions is ignored (Anderson 2012; Source Material 2023).

The finance industry continues its production of new money unabated. Economic departments, finance academics and public intellectuals seem to have very little idea about how and why banking produces systemic economic growth. In this field, the decades of climate distress calls from the UNFCCC have gone unheeded.

New philosophy is required to replace the Enlightenment presumptions of the alienated, individual rational consumer. New premises of eco-social integration give an entirely new ethos to economic matters. This new ethos is important for adequately countering the dangerous alienation of continuous exponential economic growth. The following chapters work out in more detail how an ontology of intraconnection can be conceived.

CHAPTER 9: 'TECHNOLOGY, LEARNING AND DISSOLVING ALIENATION'

Technology is assumed to have developed in sophistication over time. That notion coincides with the deeper alienation of human subjectivity with environmental concerns, as though technological sophistication has improved mastery and disturbs empathy with nature. In this chapter, I unpack these assumptions and examine how knowledge and technology need not be 'enframed' by a culture of alienation.

The computer, the printing press, paper, clay tablets and forms of writing have all transformed oral knowledge transmission. Writing technologies enable memory to be externalised, so that modern people no longer recite long poems or genealogies by rote. Lotka (1925; 1945) Leroi-Gourhan (1943/1993; 1945), Derrida (1978), Heidegger (1962; 1973b; 1977), Stiegler (2018) and others explore the importance of exosomatic techniques and their impact on epistemology from the Palaeolithic to the modern.

For Heidegger (1977), exosomatic technology is another contributor to the enframing of modern thought. Technology alienates ecology from the horizon of disclosure. We are each born into an always-already existent world. The technological surroundings inform our understanding of the current norms, interpretations, and expectations of events. In modernity, everything and everyone are understood as resources waiting to be needed in the machinery of consumerism.

Stiegler (2018) argues that computers are speeding up technological processing beyond the capacity of brain synapses. Society is consequently more passive, as we allow machine learning and big data to make search decisions for us. The 'bubbles' of interest communities become isolated from one another as the internet meta data engines sift information depending on profile. Alternative viewpoints are excluded. This lack of exposure to other people's interests, combined with the physiological seclusion facilitated by our screen time, is isolating groups from one another. It diminishes the experience of compromise and communication across diverse range of interests. Like an 'opiate of the masses', TV and IT exacerbate the alienated trajectory of modernity. People are both swamped with information and deselected from alternative viewpoints. Issues such as climate change have become very divisive, and the polarised positions remain difficult to moderate. In such a void, business as usual continues to proliferate.

The theory of exosomatic memory encapsulated by progressive forms of writing technologies is a very Western lens. It fails to comprehend how the natural environment is related to indigenous modes of knowing and being. I argue that for indigenous communities, the exosomatic tool is an ecological event, because technology and ecology are the same things. This elemental pivot ingeniously dismantles the technological enframing of thought. It opens a community of engagement with eco-technical objects at the heart of human epistemology.

CHAPTER 10: 'QUANTUM PHYSICS AND ONTOLOGICAL TURN'

Locke argues that 'work' is adding value to natural raw materials. Technology exhibits the creative labour of modern effort and sets techno-culture against pure-nature. The onto-epistemology of Māori *whakapapa* includes the more-than-human (Ngata 2011/1944; T. Smith 2000; C. Smith 2000; Mika 2017). Technology is often ecological – medicine from plants, as an example – so it does not exclude nature from culture. Indigenous philosophy cuts through the alienation of nature from culture that frames modernity and opens up an ethos based on integrated, symbiotic mutualism (chapter 9).

The rationality of cause and effect dominates Newtonian science and the worldview that derives from it. The determinism of forces acting upon other forces has made the universal logic of natural laws more readily discernible. Descartes separates the knowing subject from the object under observation. Universal laws of deductive reason and the logic of cause and effect enables the subject to deduct the essential characteristics of the object. Cause and effect also eliminates the space for freedom and the unexpected, which causes Kant great concern (chapter 5).

Quantum physics challenges some of the fundamental assumptions of Newtonian science. Instead of simple cause and effect, where a particle responds to forces in a predictable and direct fashion, quantum experiments demonstrate something far more complex. Subatomic particles behave as both waves or particles, depending on the type of observation taking place. The process of observation changes the particles' behaviour (Bohr 1987/1957). Descartes's argument that the subject is strictly separate from the object is proven inaccurate. The intimate linkage between observation and object is irreducible. The alienation of objective observer from the object is dismantled and a new unavoidable entwinement between observing knower and observed object of knowledge has emerged (Barad 2007).

Newton's (1687) mechanical universe with deductive cause and effect continues the philosophical tradition of a predictable substrata with linear causal relations that are observable under objective conditions. Universal laws are proving far more complex, more symbiotic and entangled between the one and the many (Whitehead 1969).

The disparate epistemological foundations of feminist embodiment, Māori *whakapapa* and quantum physics all resonate with a similar premise of interconnection, intra-action and interdependence. These three different approaches create a diffractive interference pattern themselves, like waves, at times reinforcing each other and at other times offsetting or dampening each other (Barad 2007). Importantly though, they each offer concepts to support a fundamental notion of intraconnection across the ecological field, where human society is a subsection of broader environmental forces. The entanglement of all things shifts the moral ground of economics. Externalities like pollution are no longer enabled by a rational calculus of isolated, alienated individuals. Instead, each individual is irreducibly connected to everything. Something as noninvasive as observation impacts the fabric of reality (Bohr 1987/1957). This means that the actions of each of us are more important, not less important, than ever before.

CHAPTER 11: 'POSTCARBON FUTURES'

Change is upon us. How we manage it, is the question. Since COVID-19, people are recognising that endless busyness is not necessary – and that the spaciousness of open ecosystems do a world of good for the psyche. Because house prices are now ten times the average wage, young people cannot make enough money to buy a house. The treadmill of endless work for little gain is losing its force. On top of that, the obscured externalities of pollution that have damaged ecosystems and communities are becoming terrifyingly obvious. Plastic rubbish is everywhere. Insurance can no longer cover multiple claims generated by flooding, fire, storms, and other related 'acts of God'. Climate change is making it extremely obvious that planetary ecology is vitally important.

'Net-zero' goals aim to address these issues. A lot of wonderful initiatives are emerging, including fifteen-minute cities, better infrastructure and public transport, better planning and building regulations, renewable energy, electric cars and loads of other initiatives. People are on board and trying really hard. But addressing the systemic drive for economic growth and its consequent growth in emissions have not been addressed. This book aims to recognise what those systemic drivers are and to begin the creative process of devising a postcarbon economy without the structural imperative for continuous economic growth.

The UN Brundtland Report (1987) and Kyoto Protocol (1998) attempted to create a subtle shift in narrative from neoliberalism. But they tied 'sustainability' too closely to the ethos of profit and consumerism (Daly 1996; 2004; Irwin 2008a; 2008b). However, it is not merely a question of being more ambitious and rigorous about the ecological consequences of policy.

Until now, environmentalists have not come up with a compelling narrative that replaces the obsolete prejudices of consumerism and economic growth. However, the ideas in this text, along with similar books such as Frederick Soddy's *Wealth, Virtual Wealth and Debt* (1933); Meadows et al.'s *Limits to Growth* (1972); Jackson's *Prosperity without Growth* (2017); Kate Raworth's *Doughnut Economics* (2018); Giorgos Kallis's *Degrowth* (2018) and Jason Hickel's *Less Is More* (2020); combined with important concepts about thermodynamics and entropy touted in Georgescu-Rougen's *The Entropy Law and the Economic Process* (1971); Gorz's *Ecologica* (2009/1971); Daly's *Toward a Steady-State Economy* (1973); Escobar's *Encountering Development* (1995); Altvater's 'The Growth Obsession' (2002), Leff's 'Degrowth' (2009) and Stiegler's *Neganthropy* (2018) are introducing a new ecology of economics that make important contributions to the emerging postcarbon economy.

Metabolic economics needs to be in tandem with an eco-ontology that embraces intraconnection with the more than human world. A multiplicity of ideas from disparate discipline areas contribute to the ontology of intraconnection. Indigenous kin relations with ecological *technē*, an embodied, emotional, and thoughtful subjectivity and its biological connection to the environment, along with the physics of thermodynamics and quantum entanglement all contribute to the intricate and irreducible foundations of intraconnection. Eco-ontology can orient economics in new ways. It nudges the direction away from neoliberalism and standing reserve at the demand of consumerism. Eco-ontology aims for an egalitarian ethos of respect rather than mastery. The fundamental alienation of Enlightenment rationality that has justified the externalising of pollution from extractive production would be more difficult. The values and rights of the more-than-human need to be intimately connected to the ethos of economics.

Economic growth rarely acknowledges the role of a swelling population that inflates the consumer base. More than technological progress, stimulating consumer demand is the basis of economic growth. In turn, assumptions about the need for growth contribute to governmental monetary policy, where the primary goal of central banks is to maintain growth at approximately 2 to 3 per cent. Rising house prices and growth in pension funds ensure the middle and upper classes are committed to maintaining the current system of growth – especially the baby boom generation, who owns houses and expects benefits. Bank loans and interest rates enforce the need for economic growth, to maintain buoyancy in the global monetary system. Corporate rules legislate for profit, futures trading and stock market shares rely on upward trending in the global economic system to maintain interest in the risky trade of share ownership. Future derivative trading is a category on its own, that has nominal value thousands of times higher than planetary capability (Bank of International Settlements (BIS 2023). These are the drivers of economic growth. They are responsible for the continuous and accelerating accumulation of CO_{2e} emissions. When I started to write, the planet hovered at 416 to 419 ppmv (parts per million volume) of CO_{2e} in the atmosphere. In the last few months it suddenly leaped up to 427 ppmv. The rate of accumulation is increasing, and already, 5 years earlier than anticipated, the 1.5°C boundary has been breached (Hansen et al. 2023; Ripple et al. 2020). 'Progress' is creating dystopia. The calculation of risks has escalated to a 'society of fear' (Beck 1992).

Alternatively, we could imagine an open-ended future. The ultimate goal is to reduce the risks of climate change by aiming to return to an emissions profile of 280 ppmv. This level kept the Holocene stable for ten thousand years. The climatologist James Hansen advises that a stable atmosphere can cope with 350 ppmv of CO_{2e} at the most (Hansen et al. 2016). Atmospheric emissions levels are easily accountable (Anderson et al. 2008; NOAA Global Monitoring Laboratory, Montreal Protocol 2007). This goal requires a shift in the articulation of science, philosophy, economics and the normative drivers for action. Bearing in mind, there is no 'one' possible future, nor is any suggested future going to emerge as expected. Yet opening up our field of awareness to appreciate how supportive and abundant the ecological system already is, would initiate the changes needed. Indigenous societies make decisions for the good of the next thirty generations instead of focussing on the pace of immediate consumerism. These changes in attitude radically alter the structure and identity of society. Economics based on a fundamental eco-ontology is a necessary transition. For it to come about, we need to refamiliarise ourselves with indigenous or 'pagan' ideas that have been treated with contempt over the modern period instead of the respect and receptiveness they deserve.

Modern technology and philosophy exacerbate the alienation from material reality. Banking and the stock market have exacerbated the alienation of modernity from the reality of finite resources. There is a vast schism between the new ex nihilo currency of fractional banking, the stock market, futures trading, and nominal debt (IMF 2022) with the real economy of fiat trading. That gap between reality and fiction is getting wider and wider. There is already little relation between the two.

This book encompasses philosophy, economics and science. It is bringing about a new field, so the topics remain a little unwieldy and juxtaposed in an unexpected fashion. This book is laid out in three parts. Part I, on 'Alienation from Nature: Growth and Finance', sets out the problem. Part II, 'Time and Technology', is a philosophical analysis of the conceptual apparatus that constitutes modernity and historical alternatives. Part III, on 'Foundational Entanglement', is a diffractive combination of an eco-ontology of intraconnection that derives from a wide array of sources. Understanding the systemic imperatives for economic growth, the historical and thermodynamic alternatives and a new eco-ontology are all vital components for a new foundation for a postcarbon future.

NOTE

1. Hansen (2023) explains: The 10 years from July 2005 through June 2015 is the period used for calibration of the satellite measured Earth's energy imbalance (EEI). The calibration being provided by changes of the heat content of Earth's heat reservoirs. About 90% of the change of EEI is change of the heat content of the ocean, which is sampled by a fleet of about 4000 deep-diving Argo floats. Earth's energy imbalance was 0.71 W/m^2 during the 10 year calibration period, but EEI has subsequently increased to well over 1 W/m^2.

Chapter 2

Modern Economic Growth

Discussing economic growth is like wearing a bikini into a board room. Its met with shock and disapproval. Universities are the worst of all. Despite all the rhetoric of critical thinking and blue skies research, traditionally universities have not wanted to engage in any critique of economic growth, particularly in a business school. It is not simply that the concept is too difficult; rather, somehow, it reeks like dog poo. It is only recently becoming clear how growth is causing the world to fall apart.

Exponential growth is hard-baked into the modern economic system. In the *Daily News*, economist Alan Kohler (2023) resolves, 'It's the machinery of capitalism that must solve climate change, not the politicians and scientists who have been meeting for 26 years and fruitlessly agreeing'. The machinery of capitalism was invented by Enlightenment scientists. Newton was literally the comptroller of the British Mint. Economic growth is tied to the Enlightenment telos of cause and effect, and progress. It is entwined with optimistic approaches to technological innovation and the notion that competition encourages better productivity and profit. Growth is an expansive concept that is embedded across the spectrum, in classical, neoliberal and Keynesian approaches to economics.

This chapter looks at the ethos of growth in concepts of progress, innovation and redistributive justice. It surveys the impacts of economic growth on emissions and resource exploitation. It unpacks the discourse of 'growth' from maturity to exponential flabbiness. It checks out 'solutions' to the correlation between economic growth and emissions growth, such as decoupling. It winds up by considering Promethean technological optimism, which assumes progressive innovation will save the day.

With climate change, resource exploitation and unfair distribution have amplified consequences (Altvater 2002). The resource extraction and pollution resulting from continuous growth are impacting climate volatility. CO_{2e} emissions were ignored for most of the industrial period. With exponential economic growth, pollution is also growing exponentially (Hansen et al. 2023; Noll 2023). Unfortunately, Hansen et al. are demonstrating that, since 2010, the acceleration of *exponential* growth is now showing up in accelerating global warming. Those people who have contributed least to pollution are the most vulnerable to its impacts. In August 2022, one-third of Pakistan was underwater with flooding. Between 2016 and 2021, 135 million people were internally displaced because of weather-related disasters according to the United Nations International Children's Fund (UNICEF) (2023) report on *Children Displaced in a Changing Climate*. One-third of those people are children.

The stakes are too high to cringe away from the fundamental question of how to expunge reliance on economic growth. Climate disruption has escalated, and its turbulence is projected to get much worse. This is not merely a question of religious or political Armageddon, or the historical end of the 'modern' world (Stein et al. 2022), but rather the existential ending of entire communities, cultures, and a lot of different species. We are already in the sixth extinction event (Hansen 2016; WWF 2022; IPBES 2019; Goulson 2019; Tong et al. 2019).

The diverse range of factors which all congeal in this perfect storm requires a 'diffractive reading' that recognises resonances, patterns, and disturbances across texts and across discourse (Barad 2007). Diffraction takes cognizance of the amplification, dampening and interconnection of a range of factors. It is not merely a linear reading of cause and effect. Barad describes this methodology as 'reading texts intra-actively through one another, enacting new patterns of engagement, attending to how exclusions matter' (2010: 243). Rather than one origin, genealogical method traces the multiple historical layers of influence on contemporary meaning and discourse (Foucault 1979; 1980; Dreyfus and Rabinow 1983; Irwin 2001). The clumsy concept of 'growth' needs such a diffractive and genealogical reading. Simply adhering to economic normativity sheds no light on the very problematic concept of growth, and its implications for pollution, consumerism, population, economic flows in the banking sector and macroeconomics.

'A word must be said about the frame of mind in which one wishes the reader to approach a book such as this' (Galbraith 1975: 4). My inclination is to apologise that my own disciplinary background is socio-philosophy, with a hurried justification that I have been interested in economics since early in my career. But a transdisciplinary approach is vital, because the normative discourse within the economic discipline ignores real-world impacts. I chatted with the professor of finance at my university in 2008, and he told me proudly that there have been no new concepts in finance for more than fifty years. The global financial crisis (GFC) seemed to have no impact on the finance canon. John Galbraith opens the protective boundaries of finance and economic disciplinarity so beautifully.

> Much discussion of money involves a heavy overlay of priestly incantation. Some of this is deliberate. Those who talk of money and teach about it and make their living by it gain prestige, esteem and pecuniary return, as does a doctor or a witch doctor, from cultivating the belief that they are in privileged association with the occult – that they have insights that are nowise available to the ordinary person. Though professionally rewarding and personally profitable, this too is a well-established form of fraud. There is nothing about money that cannot be understood by the person of reasonable curiosity, diligence and intelligence. (1975: 4–5)

A diffractive and genealogical methodology engages transdisciplinary critique – bringing invisible assumptions and entitlements to light in a way that those thoroughly embedded in the discipline find difficult, or impossible. Macroeconomics is a diffuse field that is unevenly articulated. No doubt, gaping holes and errors will be evident in this text, for which I take responsibility and apologise. But emerging, as I do, from outside finance and economics is a strength, to cast fresh eyes and a broader set of concerns to the traditions of the discipline.

If economic growth and the peculiarities of money creation did not have real-world consequences, then this book might be an exercise in logic, or the journalistic exposure of a salacious scandal. But the consequences are far more serious than merely revealing truth. Climate change is directly correlated with economic growth, and it is imperative that the systemic drivers are better understood. We need fresh ideas about healthy approaches to raising capital, and the ethos of economics must have space for consideration.

THE MACHINERY OF CAPITALISM

Anthropocentric climate change is the outcome of accumulating CO_{2e} emissions that, in turn, result from more than a century of burning fossil fuels. Arrenhuis recognised the climate impacts of coal pollution in 1896 and the news was widely reported in newspapers. Finally, clean, renewable energy is starting to sweep reliance on fossil fuel out of its path. The circular economy goes some way to slowing down the runaway entropy currently coursing through the consumer system. A sea change in consciousness is taking place amongst people all over the world. At grassroots level, households, cities, companies, local, national, and global governance are taking steps to reduce their emissions profile. What is most frustrating, despite the undeniably wonderful efforts of large sectors of the global community, emissions are still steadily accumulating. When I was born, the CO_{2e} parts per million volume (ppmv) in the atmosphere, was 285, and by 2024 the cumulative pollution rose to 427 ppmv CO_{2e} – more than at any period for 4 million years (NOAA Global Monitoring Laboratory nd; Oxford Global Warming Index n.d.). Atmospheric emissions at this level were last seen during the dinosaur extinction (Hansen 2016; WWF 2022; IPBES 2019).

To draw a pathway through this multifaceted field, I want to start with a sketch of conventional political economy across the spectrum of 'right' and 'left'; the presumptions in Adam Smith, Keynes and neoliberalism. And then I want to look more closely at important elements left out of the picture. Population growth, fossil fuel energy and the changing shape of labour and finance all contribute to 'economic growth' and growth in CO_{2e} emissions. More consumerism makes emissions growth. Business as usual (BAU) has an exponential impact on CO_{2e} emissions (IPCC 2013; RCP8.5 model). Until we pick apart the basic assumptions contributing to the systemic production of exponential growth in materials and energy, we cannot begin to imagine new ways of organising economics. Nor can we reimagine new ways of constituting the monetary system (Ament 2019). These forge the beginnings of a new premise for healthy global postcarbon economics.

Crucially, there appears to be a close correlation between consumerism and emissions growth (Le Quéré et al. 2020; Caporale et al. 2021; Mitić et al. 2022; Onofrei et al. 2022; Noll 2023). This could be clearly seen in the second quarter of 2020, when COVID-19 shut down much of the world (Le Quéré et al. 2020). In March 2020, international flights were mostly grounded, shipping slowed and tourism halted, along with other non-essential services. That quarter, there was an unprecedented drop in global emissions of 17 per cent.

This one-off COVID-19 dip in emissions is the anomaly that proves the rule. Reducing energy throughput with technological efficiencies are improving per unit productivity but are not reducing overall emissions. The demand for growth keeps swelling overall productivity. There is a big disjuncture between the need to reduce overall pollution (IPCC 2021) and the aims of governments and corporations for continuous consumer growth. As technology 'progresses', energy consumption grows.

> The annual energy use of late-Palaeolithic foragers is estimated to have been around 5 GJ per person annually (Smil 2017) – the sum of food-energy metabolised plus biomass for cooking. By 1850, after nearly 10,000 years of agriculturally-supported expansion, average global primary energy consumption rose to over 20 GJ/cap (GEA, 2012). Today, after 150 years of fossil-fuelled industrial

> development, it has reached 80 GJ/cap (IEA, 2019a). In absolute terms, total global primary energy use has risen from around 1 PJ in the late-Palaeolithic to nearly 600,000 PJ today, driving changes in the composition of the atmosphere (warming) and oceans (acidification) leading to dangerous climate change (IPCC 2018). (Millward-Hopkins et al. 2020)

The growth curve in energy expenditure has escalated sharply since the postwar 1950s. This is when exponential growth and an inflation target of 2 to 3 per cent became incorporated in many national macroeconomic goals. Exponential growth is not linear. In the beginning, the doubling rate is drawing on small amounts and makes little difference to overall resource extraction, but after eighty years, the annual quantities are now vast (Soddy 1933; Higgs 2014; Hagans 2020; Noll 2023).

Classical and neoliberal economic theory has three or four large lacunae: there is no theory of finance, pollution, or the warping effects of advertising on price and fair market redistribution. The understanding of energy emissions, or throughput, has recently been borrowed from thermodynamic economics (Georgescu-Rougen 1971). At present, the impetus for efficiency gains is swallowed by the universal neoliberal functionality of profit and growth. Comprehending how and why this occurs is crucial if the transition to genuinely green economics might take place.

Energy throughput is the original component of carbon emissions. It can be seen in new national accountability measures, called scopes 1, 2 and 3. These scopes analyse where energy is used in an organisation and help focus where mitigation is needed. Energy throughput has to be understood in the context of limited extractable resources (Higgs 2014). However, efficiency gains in energy emissions will not halt the drive for exponential economic growth and commensurate emissions growth.

In what is called Jevons paradox, technological efficiencies only result in the reduction of emissions in the very short term. Efficiencies rapidly get absorbed into the unremitting need for expansion and growth on an industrial scale. Efficiencies produce greater economic returns in profit and gross domestic product (GDP).

It is becoming clear that 'sustainability' is not reducing overall emissions. Sustainability started off as an ecological framework but it got redeployed into market discourse in the Brundtland Report (1987). Carbon offsetting is a popular tool that offers organisations 'comfort' that they will succeed when committing to net-zero emissions. But carbon offsetting is also an excuse to continue to pollute (Anderson 2012; Dyke et al. 2021; Thales et al. 2020; Greenfield 2023; Source Material 2023).

Three fundamental elements are deeply problematic: a stock market that is abstracted and oblivious to real-world impacts, a financial system of private banks that have given themselves permission to create debt ex nihilo – out of nothing – and macroeconomic monetary policy that aims at continuous exponential growth. All three are very abstract and alienated from real-world impacts. Together they create structural norms that need to be brought into the light and interrogated for their contribution to year on year swelling of GDP and subsequent CO_{2e} emissions (see chapter 3). To grasp the magnitude of this problematic and generate solutions the alienation of people from ecology needs to dissolve (see section III).

EMISSIONS AND GROWTH

Times are changing, and under pressure from the 2016 UN COP26 on Climate Change, the Paris Agreement, 196 nations have made some commitment to reduce CO_{2e} emissions. The United Kingdom (UK) was the first to set legally binding net-zero targets by 2050. This was followed by the European Union, Japan, South Korea and then others. In September 2020, China committed to net zero by 2060 (but has since produced a five-year plan with immense coal use). Biden committed the United States to net zero by 2050, the first day he became president.

The net zero by 2050 messaging is tied to new markets and new opportunity for economic growth, as communities develop renewable infrastructure in energy, and transport, and retrofit buildings to reduce emissions. 'Net' zero allows trading of emissions reduction from one industry to another. A large-scale study in 2022 demonstrated that one of the largest reducing emissions from deforestation and forest degradation (REDD+) carbon offset forestry companies, Verra, have had little added value from preexisting forest (Thales et al. 2020; Greenfield 2023; Source Material 2023; Anderson 2012). Greenfield pronounced the carbon offsets as '90 per cent useless'. Kevin Anderson and the Tyndall Foundation have strongly come out against offsetting, as have other climate scientists, on the grounds that it makes the science of emissions analysis almost impossible (Anderson et al. 2008; Ripple et al. 2020; Mardirossian and Arnold 2023). The 'net-zero' discourse is technologically optimistic, using accounting measures that include the technological scrubbing of CO_2 from the atmosphere, although these technologies do not exist at scale (Dyke et al. 2021). Globally, we have left it so late to address climate change seriously that there is urgency to decrease emissions rapidly, preferably in the next decade. Scientists at the Tyndall Centre estimated at the beginning of 2020, that 500 gigatonnes (Gts) is the maximum amount of CO_{2e} that could be released to give the planet a 50 per cent chance of staying under 1.5°C (IPCC 2021). At current

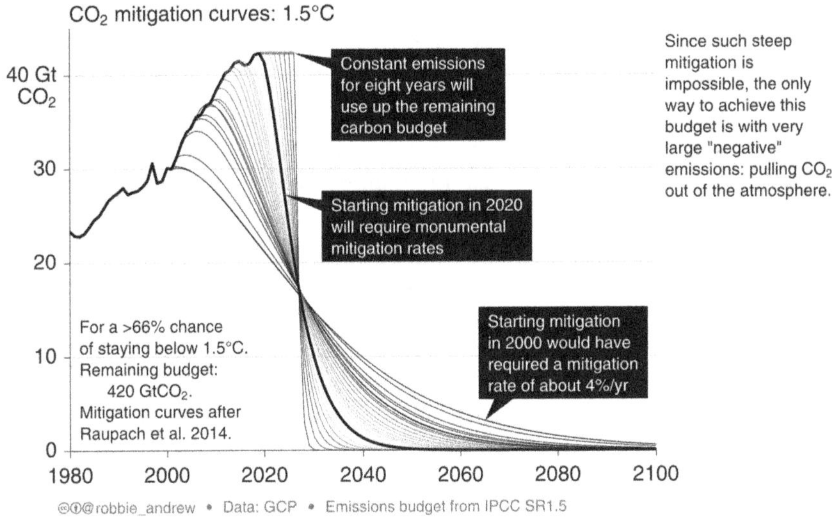

Figure 2.1 CO_2 Mitigation Curves (Robbie Andrew)

rates of emissions of 34 Gt CO_2 per year, this global, all-time allocation will be used up by 2030 (Tong et al. 2019; Rogelj et al. 2016). Kevin Anderson and colleagues at the Tyndall Centre have produced a tool that calculates the local portion of the global carbon budget. Given that there is no scrubbing technology – other than biodiversity – carbon dioxide emissions already released will continue to heat the planet for between three hundred and a thousand years (Buis 2019).

Figure 2.1 demonstrates how fast mitigation has to happen to keep to the 1.5°C threshold. Short-term investment in green infrastructure, green energy, and retrofitting insulation and passive heating will produce a large economic boom (Stern 2006). Importantly, to meet 2050 goals, capital investment in net-zero technologies needs to be frontloaded in the 2020s and early 2030s. In the UK, for example, extra investment needs to grow five-fold from around £10bn per year in 2020 to around £50bn in 2030, before peaking in 2035 (Robins et al. 2021).

The development of green infrastructure, that enables people to have a lighter footprint, is investment worth making. But if the machinery of capitalism continues to be committed to economic growth, then Jevons paradox is in play. All efficiencies will be absorbed and exceeded by the continuous expansion of consumerism. Unfortunately, global economic growth is rising exponentially (IMF 2022; BIS 2023a), and emissions continue to *escalate* (Hansen et al. 2023).

NEOLIBERAL AND SOCIALIST COMMITMENT TO GROWTH

Considering growth critically is out of bounds. Perhaps this is because economic growth has been so beneficial, especially for the postwar generation in the West. This generation remain in control, well into their seventies and eighties. Take Biden and Trump as examples. Though supposedly on opposite sides of the political spectrum, both of them are fully committed to economic growth. Nevertheless, there is no point in blaming the boomers. The system was set up *for* them, not by them.

Growth is one of the fundamental assumptions that characterises global, late-modern culture (Keyβer and Lenzen 2021; Irwin 2008a). It owes a great deal to the Enlightenment narrative of sceptical idealism and progress. The epiphany of Adam Smith to assign the 'balance' of the market to (God's) invisible hand allowed economics to resituate itself from feudalism to aggregated individualism (Adam Smith 1904/1776; Rothchild 1994; Devine 2004; Irwin 2008a; 2014). Smith's argument is that greedy selfishness in the individual is irreducible. Despite dubious morals, the overall benefit to society is administered as though by an invisible hand. Rothchild explores all three times Adam Smith mentions the invisible hand. In *The Theory of Moral Sentiments* (1759),

> Smith is describing some particularly unpleasant rich proprietors, who are quite unconcerned with 'humanity' or 'justice' but who in 'their natural selfishness and rapacity' pursue only 'their own vain and insatiable desires'. They do however employ thousands of poor workers to produce luxury commodities: 'They are led by an invisible hand to . . . without intending it, without knowing it, advance the interest of the society'. (Smith 1759/1976: 184, in Rothchild 1994: 319)

Adam Smith argues that individuals make utility maximising – or greedy decisions. In later neoliberal theory, utility maximising is designated 'rational choice' (Hayek 1960; Friedman 1970; Nozick 1974). The greedy merchants may advocate trade barriers to impede international competition, but Adam Smith (1904/1779: 453–71) argues that the market is the best arbiter of the good of society. The invisible hand aggregates all these individual consumer choices and through price makes the fairest distribution of scarce resources. 'He will thereby promote the interest "of the society"'; 'he is in this, as in many other cases, led by an invisible hand to promote an end which is no part of his intention' (Smith 1759/1976: 349, in Rothchild 1994: 319). The perfect market is a balance between demand and supply, distributing goods fairly between consumers, through the setting of price value. Progress is generated by producers aiming to create profit in the context of a static, perfectly balanced market, through innovation and novelty. Scarce resources are theoretically priced accordingly, with higher prices meaning fewer buyers. There is, however, no theory of the impact of advertising, finance, nor the impact of pollution.

The emphasis on consumerism as a signifier of status and merit makes it difficult to persuade modern societies to reduce their carbon emissions by consuming *less* when the prevailing norm is to consume *more*. The entire modern economic edifice is based on more and more consumerism, as the successful marker of 'economic growth'.

Consumerism is deeply embedded in the assumptions of all major political parties. It is not merely a neoliberal or right-wing commitment to making profit. Progress and growth are important concepts to the left too. Growth underpins the 'multiplier effect', advocated by the socialist economist Keynes in 1933. The multiplier effect overtly builds continuous exponential increase in economic growth. During the 1929 Depression, Lord Maynard Keynes was worried about how alienation, discontent and instability of working conditions created revolutionary ferment. Marx (1887/1867) had earlier argued that since being evicted from their traditional lands into dangerous industrial conditions, the resulting proletarianized workplace was a hotbed of discontent. Revolution could generate an evolution in productivity, from hierarchical capitalism to industrial socialism, and then back to small subsistence communities. Dudley says, 'Maynard Keynes promoted an active role of governments in managing aggregate supply and demand – a response to the severe shocks to the global economy during the Great Depression' (Buch and Dages 2018: 5). Keynes (1991/1936) argues that to avoid unrest, the working classes needed to be included in the consumer rewards of capitalist industrialisation, and to see year on year hope for improvement. This is the utopian discourse of 'progress' and 'growth' which results from the 'economic multiplier' (Keynes 1933). Labourers would give up the security, variety, and satisfaction of subsistence home economics for the boring, repetitive, alienating, proletarian jobs in a factory if they too could benefit from the consumer revolution.

Roosevelt is often lauded as a forward-thinking promoter of state welfare. He offered millions of unemployed farm workers a 'work for the dole' arrangement, where they planted millions of trees. He also introduced the Glass-Steagall Act of 1933 which introduced radical banking reform (more in chapter 3). But this was also a missed opportunity. Several advisors to Roosevelt, including Irving Fisher (1936), advocated the Chicago Plan, where banks are genuine intermediaries and are not allowed to create ex nihilo credit (Buch 2022). The banks can only loan the amount they have in savings deposits. Roosevelt chose to follow Keynes's (1991/1936; 1933) advice, instead, and stick with the money multiplier.

Up until 1972, following Keynesian principles, American wages tracked productivity growth. The consumer price index was less than 10 per cent of what it is today (Desilver 2018; Measuring Worth 2022; Statista 2023). Keynesian governments committed to a minimum safety net of welfare, health, education, *and* economic growth (Higgs 2014: n99 and n249). Growth is a key element for inflation targeting nations that aim for 2 to 3 per cent.

What Keynes failed to acknowledge in his promotion of an economic 'multiplier' is the impact of pollution or exponential extraction of finite 'resources' provided by the ecosystem (1933). During the 1930s a small exponential increase in overall consumerism possibly did not show up as vast quantities of extra materials as it does now. But early indicators were already there. Converting millions of hectares of wild prairie into industrial agriculture had already produced the American dustbowl that stimulated the 1929 Depression. Malthus wrote of the problems of linear growth and exponential demand in 1798. Keynes fellow British countryman, Frederick Soddy, wrote a contemporaneous critique outlining the consequences of the exponential growth curve in 1926 (Soddy 1933). By now, we know that finite resources (Higgs 2014; Stockholm Resilience Centre) and pollution writ large, results in planetary scale climate change (Hansen et al. 2016; 2023; Ripple et al. 2019).

By the 1970s, the impacts of the Depression and the world wars were beginning to fade, and Adam Smith's complacency about individual greed began to come to the forefront. Neoliberalism was on the rise. They argued that overcoming perfect market conditions – where price reflects costs and no more – requires innovation, which creates opportunities for fresh profit. This neat theoretical story ties together technological innovation, efficiency, progress and growth. By the 1950s, market saturation meant that most consumers in the West already had a fridge, vacuum cleaner and a car, and manufacturers were finding it difficult to maintain consumer interest. Innovation began to include built in obsolescence, so that gadgets have a short lifecycle and rapidly need replacing. Reducing labour costs for major corporations later saw manufacturing move offshore to newly created tax enclaves in 'third world' countries. The poster boys for neoliberal theory – Milton Friedman (1970), Hayek (1960) and Nozick (1974) – were all oblivious to the damage these new developments were doing (Higgs 2014). Friedman (1970) argued that rather than squabble over fairer resource distribution, growing the overall economic 'pie' enables more material goods and services to all participants, even if the proportions remain unequal. Profit, progress and growth are neatly tied into one theoretical package. The ethics of equality which had been a huge concern to democratic nations since the French Revolution began to disappear from political rhetoric under the guise that 'interest groups' skew market distribution and that politics should be left to the market (Dale 1989; 1997; Devine 2004; Peters 2001; Peters et al. 1993; Peters and Marshall 1996; Marshall 1994; 1996a; 1996b).

Neoliberalism took hold in the United States and the United Kingdom in the 1970s. According to the Economic Policy Institute, American wages began to diverge from tracking productivity in 1972. The Keynesian promise that wages will keep up with inflation and the growth of increasingly cheaper goods and services will improve working-class living standards only held before 1972. The neoliberal pie was supposed to grow wealth for the working classes and much more wealth for the rich. But the consumer price index in 2022 is 1.34 times higher than in 2010, whereas real wages have dropped in the same time period (Desilver 2018; Measuring Worth 2022; Statista 2023).

The consequences of modern industrial production and global consumerism are becoming obvious to see. In the Holocene, forest fires were relatively small scale and low temperature. Uncontrollable forest fires have escalated in the last few years to unprecedented levels, as the toll of planetary heating and evaporation of soil and plants has escalated. During the northern summer of 2018, eight hundred thousand hectares of forest burned in northern America. The following year, in the Amazon, more than nine hundred thousand hectares burnt in a 77 per cent proliferation in fires. Then in 2020, Australia burnt with unprecedented wildfires. 18,600,000 hectares of land was burnt through, killing an estimated 3 billion animals (Van Eedon et al. 2020) and in 2023 the devastating fire in Maui killed more than 110 people. At 1.3°C above preindustrial average temperatures, these turbulent weather events are direct results of economic growth and accumulating emissions.

In Australia and New Zealand, local councils are analysing sea level change and flooding for the extensive land loss predicted in urban areas. In the Pacific Ocean, Small Island States are already having to move villages above the encroaching sea level caused by global heating. Coral has bleached from consistent sea temperature of 30°C-plus and died, in large areas of Fiji, Vanuatu and Australia. The Arctic Circle is so warm that rain instead of snow is falling. In the Antarctic, enormous ice sheets, the Larsen B shelf, and then the Larsen C shelf broke off, taking with it 10 per cent of sea ice to break up and melt in the Southern Ocean. In 2018, an enormous hole 9.5 kilometres long was found under the land-based Antarctic Thwaites Glacier, suggesting vast amounts of land ice is melting faster than any predictions. The disappearance of sea ice takes the 'stopper' out and land-based Antarctic and Arctic glaciers are in danger of escalating their flow out into the ocean, producing noticeably higher sea levels (Schloesser et al. 2019; Lea 2023). When the Thwaites land ice sheet is melted, it is estimated to suddenly introduce two feet, or nearly one metre, of ocean level rise. Every metre depth of ocean adds 9.81 kilopascals (kPa) of pressure, creating not just more volume but significantly more turbulence and energy in wave action. Wider ocean surface increases evaporation, and warmer air carries more moisture. Evaporation is the driver for wind and storms. Sea level rise is not just land loss. It intensifies hydrological and atmospheric turbulence. Many species are not coping with the rapid change in climate, and extinctions levels are as high as each of the great extinction periods of the geological record that have drastically altered the type of life prevailing on our planet (Hansen 2016; WWF 2022; IPBES 2019; Goulson 2019; Tong et al. 2019).

Nevertheless, the key concept of economic growth, that has held together both right- and left-wing politics in the global capitalist economic system, remains largely taboo. Until 2021, there was no funding or institutional support for reconsideration and critique of the concept of economic growth. It is almost tantamount to treason.

WHAT IS GROWTH?

To address the dissonance between exponential material growth and finite planetary conditions, the core concept of growth needs to be unpacked. The notion of growth means different things in different contexts: maturity, cycles, and exponential progress. Physical maturity and seasonal cycles have inbuilt limits. But exponential growth escalates into infinity. Different macro trends impact on the underbelly of economic growth, including peak population, an aging demographic, peak oil and the end of cheap fossil fuels, resource exhaustion, new technologies and changing expectations about work, and, of course, climate change.

Growth of a child's physical body usually reaches an inbuilt maximum at maturity that is defined by our inherited genetic code. If we keep growing, its usually excessive, and we are getting flabby!

Seasonal growth also has its own inbuilt rhythms of abundance and fallow periods. Seasons turn in a repetitive cycle that tends to oscillate within a broad normative spectrum. Ecologies are intersectional, growing takes place according to environmental availability of nutrients, water, sunlight, and energy. These wax and wane with complex and repetitive patterns that interfere with one another. As examples, weather cycles like the *La Niña* event will enhance forest growth, or every fourteen years there is a cicada bonanza. Smaller and larger cycles have a set of intersecting, diffractive patterns that amplify or minimise each other. The variability of seasonal growth oscillates through smaller and larger cycles of birth and decay, rather than creating a trajectory of continuous increase.

Like a child reaching adult maturity, most businesses have a 'natural' or optimal limit at maximum returns for capital outlay. This is the marginal return on investment. Much economic activity is also seasonal in productivity and sales. Most businesses – some 97 per cent in New Zealand – are small family firms with less than twenty staff (NZ MBIE 2022). Many of those family firms are not seeking large profit margins. They are sensitive to the parameters of marginal returns. They operate under a presumption of 'enoughness' (Jackson 2017). If they have enough to maintain plant and expenses, and pay decent wages, so everyone is housed, fed and happy, they have a decent house, a car that works and the occasional local holiday, that is 'enough'. The marginal returns get much slimmer after a certain inbuilt point (expanding would create large capital outlay; a new factory, new plant, more staff with less profit per dollar of capital expense), and so once that maturity is gained, it is not cost effective to continue to grow a business. The rhetoric of progress and growth has been 'sold' to the working classes, but it actually serves a different sector of the population, notably large-scale global corporations, major institutions, banks and the stock market.

Keynes 'multiplier', which we now call 'exponential growth', is not just linear. It looks like the famous hockey-stick curve, where numbers double, quadruple, octuple and so on with alarming rates of increase. Fred Soddy illuminates the way small increments of exponential doubling can go from tiny amounts to ungainly proportions very rapidly.

> It is characteristic of the dizzy virtues of compound interest that they are not at all dizzy at the start. It is only after they have been in operation a certain time that they show any disposition to become marvellous and to transcend the bounds of the physically possible...A single grain of corn doubled as many times as there are squares on a chessboard represents more corn that the present population of the world could consume in a period longer than that covered by the records of history, whereas doubled only half that number of times it would scarcely suffice to give London a square meal. This means that a system might show no signs of breakdown for a century and yet become absolutely impossible during the course of the next. (Soddy 1933: 140)

Clearly, modernity has entered that later century.

Boom and collapse are a biological occurrence. Exponential growth can occur naturally for limited periods. Bacteria, for example, can exponentially increase when the nutrients, temperature and moisture is optimal, until they exhaust one of those contributing environmental factors, and just as suddenly, the populations collapse. Wild species have been known to do something similar; mice are a 'plague' sometimes and insects, like locusts, can exponentially increase under the right conditions for a short period. But exponential growth in population dynamics never continuously expand.

They exhaust the boundaries for resources such as nutrients, energy, or water, and then collapse to baseline again.

POPULATION GROWTH

Modernity assumes the trajectory into the future is 'progressive'. In the last few decades, this settled into technological 'optimism' which predicts technological solutions for any environmental or social problems (Feenberg 1996; Nordhaus 2018). As early as 1798, Malthus noticed the inversion of teleological progress. He made a statistical analysis of population growth, which he described as 'geometric'. Geometric – or exponential – population growth was contrasted with finite land productivity. Malthus warned that continually doubling a huge population would reach a stage when the earth's land mass could not support everyone. He was writing in the context of colonial expansion, but his example was the familiar finitude of the British Isles (Bashford and Chaplin 2016). Malthus was later refuted by other economic historians in squabbles over statistical veracity. Neoliberals reject his warning by arguing that new technologies, such as the chemical 'green revolution' will wrest more food productivity from limited land mass. Thus, enabling unabated exponential population growth to continue. Anti-Malthusian ideas continue to play an important part in obscuring the gap between exponential growth and finite resources – most recently through 'discounting' the needs of future generations. This is the astonishing claim that progressive inflationary devaluation of money is assumed to make it cheaper for future generations to address climate disasters than now (Nordhaus 2018; Masini 2021).

Human population grew very slowly between 10,000 BC and AD 1700, at an average of 0.4 per cent, from an estimated less than 4 million to one billion people by 1804. The land clearances in Britain began in the 1500s, followed by the beginning of the industrial revolution in the 1700s. By 1760 population growth started to increase swiftly. In 1925 it took a further sharp upward turn from 2 billion, and the *rate* of growth (not the overall number) peaked in 1962–63 at 2.2 per cent. Global population reached 8 billion in 2022. Fertility decline is underway in every country on the planet (Irwin 2015a; Vollset et al. 2020). However, increase in gross population is still built into the system. There is a lag, as the huge population keeps producing a gross additional 80 million people per year, but proportionally, the population is aging, infertility is getting worse, and the rate of growth is slowing.

The allergy to critique of growth permeates population discourse as well. Global bodies such as World in Data and the United Nations (UN) Population Bureau keep fudging the statistics to insist that population growth continues to rise long term. There is little public acknowledgement that the last two hundred years of rapid population increase is grinding to a halt and may decline with as much speed as it increased.

The United Nations World Population Clock has speculated that fertility rates will abruptly reverse fifty years of a downward trend and inexplicably start to rise. Their prediction refutes worsening infertility, and the average age getting older in all countries. Average age globally is now thirty-one, becoming thirty-five in the next decade. An aging population means relatively fewer people in the childbearing age bracket. With a dubious absence of analysis, most likely aimed at soothing the global demands for economic growth, the UN Population Bureau speculates that the population will continue to increase, albeit slowly.

Demographic transition from high numbers of children to a rate of replacement or less than replacement tends to occur, I believe, when people can trust in the consistency and stability of life. Stability can occur within stable subsistence economies or with state welfare provision for health, education and welfare (Irwin 2015a). High birth rates exist where the infant mortality rate is high. In precarious situations, children are a family safety net. Survival rates may be difficult, but the family is the source of security for invalid or aged family members rather than the state. Where contraception is hard to access, fertility rates are also high. Demographic transition to two or fewer children per couple can take merely a couple of generations, and this has already taken place in most countries around the world with good health care provision. Replacement rate is 2.1 children per couple and most 'developed' nations are already below that figure.

Myrskylä et al. (2009) argue that the correlation between declining fertility rates and higher scores on the Human Development Index (HDI) would reverse in the twenty-first century. In fact, as they acknowledge, most studies demonstrate that higher indexes in health, wealth and education are strongly correlated with below-replacement birth rates:

> Numerous studies point to the fact that the causes of low fertility are rooted in economic and social development: more income, lower mortality and higher life expectancy, higher education, higher levels of female labour force participation, greater gender equality, and access to advanced birth control methods have all been identified as driving forces of reduced fertility. (Myrskylä et al. 2009: 2)

Despite these well-established relationships, ongoing fertility decline and aging of wealthier populations, Myrskylä claim a sudden reversal in trends. They use cross-sectional and longitudinal analyses between total fertility rate and HDI. Supposedly these methods demonstrate that 'the well-established negative relationship between fertility and development is reversing as the global population enters the twenty-first century' (2).

> Our hypothesis that the fertility–HDI relationship reverses within the HDI window of 0.85–0.9 implies that the longitudinal country-trajectories in Figure 2 should be J-shaped. . . . While there are clear exceptions, as for instance Japan, the trajectories for the large majority of countries (18 out of 24, representing 74% of the population in the 24 countries included in Figure 2) are in accordance with our hypothesis: as development has continued and countries have attained an advanced HDI level of 0.9 or higher, the trend in the total fertility rate has reversed. (Myrskylä et al. 2009: 4)

This analysis reminds me of the climate sceptics who tried to claim that global warming was reversing between 1998 and 2004. Statistical anomalies, plateaus and irregularities are a normal feature of evidence gathering – it is the overall trends that are statistically interesting, not small fluctuations. This oft-quoted article published in the highly reputable journal *Nature* justifies the UN using a mathematical derivative J curve with a built in plateau, so the data simply gets shaped to the model. This suits the growth model of economics, which quietly relies on rising global population to buoy up continuous acceleration in demand. The evidence, I suggest, is that the hockey stick graph of exponential population growth has started to turn into an S curve. Globally, the rate of growth has now dropped to 1.05 per cent and continues to shrink (Irwin 2015a). Recent studies are confirming my view of contracting global population (Vollset et al. 2020). Any growth in fertility rates is just short-term minor aberrations, but the overall trend is steadily declining (World Bank Birth n.d.). My data crunching predicts peak population in 2068. Vollset et al. (2020) estimate 2064. The latest UN report, *World Population Prospects 2022*, has three future trajectories: low, medium, and high. The data in the low trajectory most accurately reflects actual birth rates and it is an important part of the analysis, but the headlines are consistently based on the medium trajectory, which is used by most, if not all, countries as their baseline for policy and budgeting. The low trajectory model peaks in the late 2060s and then the curve declines, indicating future contraction in global population trends. The midline plateaus around 2100, and the high trajectory keeps escalating up (UN DESA 2022: 39).

Population growth is rarely discussed in monetary theory. But politicians often recognise the role of population growth in keeping economic growth buoyant. Usually, in richer countries where the population is shrinking, migration is used to prop up youth numbers in an otherwise aging population (Irwin 2015a). Migration enlarges the numbers of capable working adults that contribute to the economy and children who contribute to economic demand. Despite the looming shortages in planetary resources, politicians in wealthy countries are starting public relation (PR) campaigns to encourage domestic childbearing. They hope to further prolong the growth bubble. Rising population pressure obviously impacts housing demand, and this has been a keystone in monetary policy and finance. Neither banks nor governments have begun to include the future impact of a steady state or contracting global population on housing. But as global population growth slows and ages, it will have repercussions on the demand for debt. And more importantly, repercussions on the feedback loop of spiralling exponential growth.

RESOURCES

There are multiple ways in which the earth's resources are taken for granted. Up until recently, the atmosphere and climate were difficult to discuss, because people rarely considered it. Clean fresh water was taken for granted in many areas but is under threat as aquifers and springs are plundered beyond capacity, and mountain ice reservoirs recede due to global warming. Rare earths and other minerals are not an endless commodity but in limited supply (Higgs 2014, Stockholm Resilience Centre nd). Yet consumerism sees fit to pump the finite supply of precious resources, such as helium into party balloons, with no thought for future generations. Soil runoff from deforestation and monoculture is a huge issue. The oceans have been plundered for fish, iron and as a CO_{2e} and rubbish sink. Bottom trawling and other problematic industrial fishing techniques continue to decimate fish populations. Since 2005, ocean temperatures are so high that the CO_2 is outgassing, or dissolving back out again, from ocean water in the northern hemisphere summer.

The advent of oil as a high-yield, transportable and stable energy source completely transformed the role of work at the turn of the nineteenth century. Never before could so much be achieved by one labourer, with the help of fossil fuel–driven machinery (Hagans 2020: 4).

Fossil fuels originate as underground strata that took millions of years of geological pressure on ancient, buried forests to form carbon deposits. A newly discovered oil field erupts once it is uncapped, because of the immense pressure it has been under. Peak oil is usually estimated once the geyser effect has dissipated, and the oil needs to be pumped out. The field is considered half exhausted when this occurs. Most fields reached peak oil in 2005. Since then, there is significantly higher expense for oil extraction. Energy returns on investment (EROI) is the amount of energy it takes to extract further energy. At the turn of the nineteenth century, EROI was approximately one barrel needed to extract one hundred barrels of new energy. Now, almost twenty years after peak oil, where low-return shale oil is making up a substantial proportion of oil reserves, the EROI is between 1/1.1 and 1/6 (Hall et al. 1986; 2014; Brockway et al. 2019). Huge government subsidies are keeping oil extraction viable. The decision by the people of Ecuador in August 2023 to halt future oil wells in the Amazon basin is a gamechanger.

WORK

The nature of work is changing. There are several counterintuitive forces making predictions of future work difficult. On the one hand, an aging global population, especially in 'developed' nations, will multiply the service and health sectors. Migration will continue to make housing demand until the 2060s, when a shrinking global population will mean demand begins to slow and reverse the trend (Irwin 2015a). Robotics and artificial intelligence (AI) are impinging on a wider variety of middle-class professions, making large sectors of traditional occupations redundant (Stiegler 2018).

Marx (1887/1867) famously critiqued the proletarianization of labour with the advent of the industrial revolution. Fossil fuels enabled the kilojoules of work per labourer to increase to unprecedented levels (Hagans 2020: 4). By the early twentieth century, Taylorism had removed the complexity of making things from start to finish and replaced it with the production line, where work is reduced to small, repetitive tasks that require little or no competence for the overall assembly of manufactured goods. The skills and knowledge of craft labour was unnecessary and became largely lost. Stiegler argues that the proletarianization of work is extending further and further as AI and robotics encroach into wider fields such as quantitative research, accounting, law, medicine, driving, and other middle-class occupations. While technological optimists will argue that new technologies change the nature of work rather than making work itself redundant, it is clear that there is less and less need for workers, and the association of merit to employment is going to become increasingly more tenuous. This is particularly true, as it is now more widely known that the way banks generate money out of nothing seems to shake the fragile association of hard work with wealth and success (see chapter 3).

DECOUPLING ECONOMIC GROWTH

Most governments have legislation that insist on growing the economy by 2 to 3 per cent per year (see Reserve Bank of Australia 1996). An annual growth of 3 per cent results in a doubling of the global economy every twenty-four years, and at that rate, by 2044, modernity will have approximately doubled what was produced in 2020. By 2100 we could be producing roughly sixteen times global production and global waste (Hickel 2016). With this level of unremitting growth, there is no possibility of achieving zero carbon emissions, nor of decreasing net emissions by absorbing carbon back into soil, forest, and plankton. Exponential material growth follows the famous hockey stick curve. Kerryn Higgs outlines the level of overshoot – which could be sustainable at 50 billion tonnes a year (Higgs 2014: 8; UNEP 2017; Hoekstra and Wiedmann 2014; Bringezu 2015; Hickel 2019). The United Nations Environment Programme (UNEP) has claimed that accelerating consumerism would affect the production curve by more than a billion tonnes a year by 2030, but Higgs (2023) explains that accelerating consumption is happening exponentially faster.

> Sources agree fairly closely on the current extent of material extraction, estimated at 87–88 billion tonnes a year in 2015, having tripled over the previous 50 years, and still growing (UNEP 2019). Circle Economy (2020) found that, from 2015 to 2017, material extraction rose by 8% to 100.6 billion tonnes a year. In line with others who quantify material extraction, UNEP (2019) and Circle Economy (2020) include in their calculations metals, non-metallic minerals (such as sand), fossil fuels, biomass, water and land conversion. Cooper at al. (2018), in addressing the extent of humans' sediment production in the mineral and construction industries, include all forms of material shifted from its original location (such as overburden, waste rock etc.) and arrive at a figure of around 316 billion tonnes in 2015, much greater than estimates that ignore waste. (Higgs 2023: 77)

Recognition of the physical constraints of resource extraction and the impact of pollution has generated some creative ideas to continue ever-expanding exponential economic growth. The aim is to 'decouple' the growing economy from material goods. In the 1990s, 'decoupling' growth and productivity from physical resources and expanding the 'knowledge economy' was thought to enable growth without emissions (Brundtland 1987). Innovation in ideas and the service sector was thought to produce value without emissions. There are three key drivers of the knowledge economy (Peters 2007) that are expected to provide growth without real resource implications: education, technology, and finance. None of the three have been able to subsidise or replace the continuing expansion of real emissions from real production. Education and technology, banking and the stock market are all associated with AI machine learning which has become one of the most intensive consumers of energy across all sectors. It is exponentially snowballing from 4 to 6 per cent of global electricity supply in 2020 to an estimated 8 to 21 per cent of global electricity by 2030 (Magubane 2023). Nevertheless, as we will see below, decoupled or not, these three key drivers are fundamental to the shift in worldview that needs to take place to bring climate change emissions to a close.

Education showed promise at decoupling emissions while contributing to economic growth. In the 1990s, neoliberal policy dominated education and fees were applied to students because credentials were understood as a private benefit to an individual's human capital rather than a social good (Marshall 1996a; 1996b). The internationalisation of tertiary education took off. But the digitalisation of learning and libraries have accentuated the carbon footprint of education. Overseas student flights and peer pressure to attend international conferences add to a raft of high-emissions educational activity. Complex financial mechanisms were invented to create money out of 'futures trading' rather than material products. Initial expectations were that profit could be made from abstract ideas, with no pollution. However, energy consumption and the impact of abstract finance on the real economy continues to grow.

Perhaps most significantly, the role of financial instruments which were promoted as abstract enhancement of economic growth. There was no analysis of how they impact the real economy or the pollution associated with it. By 2022, 'nominal' credit is 238 per cent more than the entire global GDP in 2022 (Gaspar et al. IMF 2023), down from more than 250 per cent of GDP in pre-pandemic 2020. This forces economies to keep growing to 'catch up' with the abstract conceptions of debt.

Likewise, the function of ex nihilo credit has not been examined in relation to emissions growth. I argue that ex nihilo loans and the concept of 'nominal' value are key drivers to economic growth. Finance and derivative future trading are core components contributing to Jevons paradox. All the transition to renewable energy is swallowed up and outstripped by new energy demands and, thus, more emissions.

Finance is not included in economic theory because it is assumed that money is simply another commodity that can be bought and sold on its own market. The *purpose* of money is to approximate exchange value, making barter easier across space, time, and different classes of goods. The *value* of money has been variously tied to a gold standard, the daily London Interbank offered rate (LIBOR), the overnight cash rate (OCR), the secured overnight finance rate (SOFR) or its international exchange value. Money is presumed to be a commodity. But this completely fails to accommodate the machinery of ex nihilo mortgages and fractional debt. As Kenneth Galbraith (1975: 18) notes with incredulity, 'The process by which banks create money is so simple that the mind is repelled'.

Since the US National Banking Acts of 1863 and 1864 (Dunn and Wheelock 2022), fractional reserve banking has gained some legitimacy. Banking has high expectations of the real economy to meet the repayment of new capital debt and interest payments on debt. What is rarely explored is how more new capital loans and associated new interest stimulate inflation, and to compensate, the metabolism of the whole economy needs to increase: forcing more growth into the system. Extracting finite physical resources at the rate of exponential consumerism is producing risks that far outreach the limits of human rationality and control.

Even without a direct theory of finance, neoliberal assumptions of rational utility maximising and the invisible hand permeate and justify the rhetoric of banking. The philosophy of modern technological and intellectual superiority is present too. Superiority has alienated modern society from the ecological milieu, and ultimately, this is the deep-seated issue at stake. Until we revoke this outdated caricature and challenge the alienated dualisms (Merchant 1980) at the core of Enlightenment thinking (see Part III), the illusion of mastery over nature will continue. There are other ways of managing finance capital and economic systems. Thermodynamic economics is a materialist approach that prioritises reducing the effects of entropy and waste. However, as we shall see later in this book, an onto-epistemology of intraconnection and respect for other species and ecosystems is essential. Without an ethos of intraconnection, the inclusion of metabolic energy accounting will fail to address climate change and species extinction. The issues of finance are considered more deeply in the next chapter.

Chapter 3

Finance and Systemic Growth

Days after the severe storms and landslides lashed the North Island of New Zealand, I stood on my house veranda looking at the shards of light dancing on the ocean, a woodpigeon contentedly eating berries not far from my head. Tui's Cavoodle, the sun shines. My chest palpitates with worry as I wonder how long it will take for another storm to turn the thirty metres of the fine clay my house perches upon into fluid slurry – turning the cliff into a giant landslide into the bay. Drenching rain is three millimetres an hour, but New Zealand experienced eighty millimetres in three hours in February and eighty-five millimetres in March 2023. Both storms caused huge landslides and loss of housing, roads, and infrastructure. I look deeper, past the playful surface waves in the bay, as a shag pops up from working the yellow muddy water, laden with thousands of tonnes of soil from this storm, in free form throughout the ocean harbour. Along with uprooted trees and broken branches, there are dead sea birds on the beach that didn't survive the wind. I'm surprised there are any fish with so much silt challenging their clean oxygenated water. I feel scared. And sad.

Overnight, I looked at Auckland Council's excellent new mapping of climate filters which shows they expect my home is likely to slide into the sea sometime over the next twenty-five years. I am a climate scholar; I kind of knew that. But I had not realised it would happen quite so fast. And oddly enough, for addressing modern systemic failure, this is the crux of it – housing and climate change.

This chapter looks more closely at the modern philosophy of debt. It is important to understand how modern debt dilutes the overall pool of money in the real economy (Soddy 1933; Galbraith 1975). Vast quantities of new money triggers inflation and economic growth – which are in turn, responsible for accumulating CO_{2e} emissions. Endogenous banking creates new money with every house mortgage, and then multiplies that new money many times over within the finance system by leveraging cascading interest-bearing loans. The fiat or real economy has to expand, to cover the new money which is not yet associated with like for like transactions. Likewise, economic growth is needed to meet the interest on those loans, which similarly is a new demand on the existing transactional distribution of labour, goods, and services (Tobin 1963; Galbraith 1975; Minsky 1992; Jackson and Dyson 2013; Werner 2014).

Few people are looking squarely at how ex nihilo debt works – with important exceptions: Soddy (1933/1926), Tobin (1963), Galbraith (1975), Minsky (1992), Reserve Bank of Australia (1996), Stevens (2003), Mengle (2007), Werner (2014), McLeay et al. (2014), Hickel (2016), Bundesbank (2017), Bendell (2023), Victor and Jackson (2015), Dyson et al. (2015), Doorman (2015), Jackson and Dyson (2013), Dyson et al. (2016), Lindberg (2017), Ament (2019), Graeber (2019) and Williams (2019). My critique is that by stimulating inflation, *ex nihilo* debt is largely responsible for the compensatory demand for economic growth and the consequent impact on pollution and climate change.

Classical and neoliberal economics have little theory of finance (with the notable exception of Keynes 1991/1936). This convenient silence has obscured the way in which banks generate loans out of nothing and thereby usurp the value of publicly circulating fiat money (Soddy 1933; Tobin 1963; Galbraith 1975). A theory that acknowledges how simple a process bank mortgage loans really are, would create widespread incredulity and scandal. Really, if you want to raise capital to buy a new house, why not ask your auntie to write up an excel spreadsheet with double entries on the loan and the credit – which she could then electronically transfer to the seller's bank? Essentially, that is all that is taking place. Instead, at the level of finance, several economic myths are propagated: firstly, that banks are merely an intermediary institution between customers with savings deposits and customers requiring loans; secondly, that banks deal with money which is a commodity like any other; thirdly, that credit generation is 'endogenous' and only operates within the banking fraternity (the only reason that your aunt's spreadsheet would not be accepted); fourthly, that the central bank fractional reserve guarantees loans and places limits on money creation. The lack of analysis of finance means the way banks usurp the value of publicly circulating money is ignored, and the role of loans as a driver of both inflation and economic growth is obscured. The failure to have a theory of finance is linked to the failure to have a theory of pollution.

Banking manages to slide between multiple, sometimes contradictory, concepts concerning money, making use of them as suits. Adam Smith (1904/1776: 453–71) proposed that the invisible hand aggregates individual rational utility maximising choices in the market. It requires minimal 'interference' or regulation from government. In the *Theory of Moral Sentiment* (1759), Smith writes that inadvertently, individual selfish motives meet society's overall needs (in Rothchild 1994: 319). While these assumptions about the ways markets' function can be seen within banking rhetoric, the role that finance plays in the theory of the market is ignored, or else money itself is simply seen as another market commodity (see Keynes 1991/1936). But I consider there to be a major gap between the abstract market of finance, and the 'real economy' where 'fiat' money is a convenient standard for like-to-like transaction.

As a tool of transaction, money attributes a semi-standardised value to various kinds of goods and services, so that people can trade different things on a like-for-like basis. Money in the form of salt, cowrie shells, silver, gold, and coins have been used for this purpose for thousands of years. The Viking horde of is the classic example of wealth buried, 'banked', as stable value in uniform weights of silver.

Sophisticated approaches to debt are thousands of years old, with complex accounting records found on five-thousand-year-old clay tablets from Mesopotamia, and in the quipu, which is a knotted string system that suffused the Andes. Records were kept about population, resources, taxes, labour obligations and debt. The abacus was in proficient use by earlier moneylenders in Mesopotamia, Egypt, China, Japan, the Arab world, and Europe. These long histories are interesting because they demonstrate that money itself, and indeed debt, is not producing systemic climate change. Ancient modes of trade and debt had important techniques for avoiding exponential growth and exploitation of the ecosystems beyond carrying capacity. This is something to come back to in chapter 7.

In normative modern economics, the limited theory of money tends to characterise it as *neutral*. Money is often seen as convenient standardised token that enables barter between different asset classes. There is tension between a charterist approach to money – where the state legislates value (through things like tax collection, bonds and the OCR) – and a market approach. The market approach sets value via relative currency exchange and normative pricing on large stable markets. Minsky (1992: 77) questions this notion of market neutrality, or that money is a 'veil', as phrased by Milton Friedman. Friedman (1970) here, is riffing off Adam Smith's invisible hand. This feudal idea of the hand of God behind the veil of ignorance had directed the decisions of the king but, more democratically, shifts under Adam Smith to direct the market (Devine 2004). The invisible hand of the market sets the fairest price through aggregated consumer decisions that ensure equitable redistribution. According to Friedman, price is a neutral, a-political, redistribution measure adjudicated by the invisible hand. In other words, the value of money is related to 'price' as set by the market. Minsky makes note of uneven power relations in market transactions per se. 'Non-neutrality of market price is a problematic outcome of some individuals having better access to information than others, making market equilibrium an Ideal rather than reality' (Minsky 1992: 77).

However, the real issue here is ex nihilo credit flooding monetary supply side and a relatively stable aggregated demand. That means that too much currency soaks through all transactional markets and prices go up while exchange remains roughly the same. Innovation is supposedly the creator of profit. But oversupply of money raises price and profit for traders in the short term but, ultimately, also raises inflation. The value circulating throughout the fiat economy is usurped by the banks' creation of money.

CREATING CREDIT

Postponing payment for commodities or capital outlay is built into the modern banknote, which is essentially an I-owe-you (IOU) (Ament 2019). Although Keynes (1991/1936: 4) sees a money note in opposite terms; as the discharge of debt, or the discharge of the contract. Keynes is careful to highlight the role of the state in ensuring the clarity of 'money of account' contracts in terminology and law. 'The State, therefore, comes in first of all as the authority of law which enforces the payment of the thing which corresponds to the name or description in the contract'. All the money in circulation is transactional debt, as much as it is the availability to store wealth.

I believe that in modern monetary practice, the shift from hording to modern debt began with double-entry bookkeeping, that emerged in Italy in the fifteenth century (De Roover 1963; Buch and Dages 2018: 4). The banknote is an asset *because* it is an IOU; both deposits and debts are treated as assets. Modern banks have taken the IOU to a whole new level.

> Contrary to popular belief, which assumes that only governments create money, in the existing monetary system virtually all money spent in the economy is created by commercial banks when they extend credit to borrowers. This circumstance has evolved out of the long history of modern money. Its implications for investment, and more broadly for how economies function in general, is still coming to light (e.g. Godley and Lavoie 2006; Minsky 1986; Ryan-Collins et al. 2012; Wray 2012). (Victor and Jackson 2015: 247)

Monetary policy is a dynamic system with multiple moving parts. But the simplest way of understanding the way banks create money is to understand 'fractional' banking.

A myth is promoted that there is a back room with gold bullion in it, or banking deposits, which are drawn from to make a loan. Those days are long gone (Soddy 1933; Galbraith 1975; Buch and Dages 2018). A simple example of a customer who is buying a million-dollar house helps to explain the actual process. The bank generates the money on the basis of the ability of the customer to pay it back. Rendahl and Freund (2020) make a big deal about this, as the banks vet customer suitability. But really, if the customer does not pay back the loan, the cost is born by the original dilution of value in pricing aggregated demand, not by the banks. The banks never possessed the money and have lost nothing when a customer defaults, other than the promise of interest payments coming in over the next decades. What's more, the mortgage is a registered 'interest' over the property and mortgagee sales generally cover the debt.

Tobin (1963: 1) explains that customer service representatives in banking rarely understand their own profession, 'the practical banker who is so sure that he lends only the money depositors entrust to him'. But the structure of banking is quite different. There is no relation between deposits and loans. They are two entirely different systems, both managed by the retail banks.

> The banker is shown to have a worm's eye's view, and his error stands as an introductory object lesson in the fallacy of composition. From the Olympian vantage . . . the banker's dictum must be reversed: depositors entrust to bankers whatever amounts the bankers lend. To be sure, this is not true of a single bank; one bank's loan may wind up as another bank's deposit. But it is, as the arithmetic of successive rounds of deposit creation makes clear, true of the banking system as a whole. Whatever their other errors, a long line of financial heretics have been right in speaking of 'fountain pen money', money created by the stroke of the bank president's pen when he approves a loan and credits the proceeds to the borrower's checking account. (Tobin 1963: 1)

When someone wants to buy a house, they walk into a bank with the ability to promise to repay the capital loan through their wages, with interest, over a long period of time. The bank has faith in the promise of wages but also in the value of the asset to repay any lending against it, because overall they have faith in the housing market continuing to rise. Rendahl and Freund (2020) argue that banks do not create money out of thin air – they create money on the basis of the asset and the scrutiny of the customer. The empirical facts are that the banks do not have any money at all prior to the customer walking through the door. With a sleight of hand, banks then generate the money the customer requests (Werner 2014). For the banks, profit is stimulated on three counts; double-entry bookkeeping means the credit delivered to the seller's bank is new money, the new loan counts as an asset, plus interest rates on the loan. At a modest 3 per cent interest, the customer pays twice the initial loan in twenty-four years. But that is only the beginning of it.

> While neoclassical theory assumes that banks use customer deposits when extending loans, in practice, commercial banks generate loans by increasing a borrower's deposit account ex nihilo equal to the amount of the loan. The bank views the loan as an asset and that is offset by a deposit liability, while the borrower views the loan as a liability that is offset by a deposit asset. The process is that simple: there was no need for a savings deposit before extending the loan and banks will extend all the loans they deem profitable regardless of deposits. (Ament 2019: 9)

There have been limits to the amount of new loans that private banks may create. The government may require a certain level of deposit from the customer. The deposit varies as the government tries to encourage or discourage new mortgages – and affect the supply of new money into the economy. Separately from customer deposits on a new loan, historically, the banks needed to be able to deposit a small fraction percentage of the loan value with the Central Reserve Bank. Bank reserve requirements aimed to guarantor the value of any mortgage defaults (which usually average around 3 per cent of all mortgages). Within the parameters of reserve requirements – which were 10 per cent for a long time – the bank creates new money (as a double entry line in the accounts) on the basis of the estimated ability of the customer to pay it back. That means, a new line with the loan amount in the buying customer's bank account, and a new line with the loan deposited as a credit in the seller's bank account. This is called endogenous credit, as there is no government-backed currency but the banking system accepts each other's electronic transactions. The credit money can then start to circulate as further loans or further purchasing power. As the money moves around within the banking system, it provides the fractional reserve on further loans.

The question of the rate of interest seems to make a link with government policy and the real economy. But that link, too, is tenuous. At the end of each business day, the banks have to square their accounts. If they are in excess, they can lend money overnight, and if in debt, they need to borrow. Based on the need for these accounting transactions, the overnight cash rate (OCR) sets the value of money.

> Respective banks will exchange reserves to clear the payment. Reserves are created ex nihilo by the Central Bank in order to facilitate payments between intrabank customers and can only be accessed by commercial banks. At the end of each day, commercial banks net their incoming and outgoing reserve positions and borrow any owed reserves from either the Central Bank or the interbank lending market. (Ament 2019: 9)

Banks also lend and borrow from each other, at a rate set each morning, called the OCR, LIBOR (London Interbank offered rate) or SOFR rate (depending on country). The OCR rate is the cost of interbank lending, in other words, it is the cost of money. In this regard, banks argue, money is merely a commodity, like any other.

However, the money that is 'washed' each evening has nothing to do with credit creation. It only covers the actual physical cash that is extracted or deposited by customers when they interact with their savings accounts, business accounts or any other alteration in actual cash supply. Only government Treasury are legally entitled to create physical cash. (This expense is the reason banks are making it harder to extract cash, by closing ATM machines and branches.) The mortgages (including the customers deposit) are sent by electronic transaction, so they have no interaction with the daily wash. Nevertheless, in an attempt to preserve the fiction that deposits are backing the loans that banks make, the interest rate on the OCR is closely tied to the interest rates charged for mortgages and the interest rates given to deposits.

There is a compounding circularity of new ex nihilo money producing some inflation, and more profit, which combined with population growth, goes on to stimulate economic growth. This is the exponential growth spiral. In these circumstances, it is very difficult for banks to fail. However, because endogenous or ex nihilo credit has very little reserve backing and almost no relation to savings deposits, if the public lose faith in a particular bank, they will line up to withdraw all of their savings deposits, and the bank has no means to meet all of these claims. The banking fraternity is constrained by the need for a relatively stable economic situation that does not trigger public discontent with the banking system, or the entire edifice may collapse.

RESERVE REQUIREMENT

The American National Bank Act of 1863 introduced reserve requirement as a way for banks to honour each other's banknotes, and it was initially set at 25 per cent. Some countries have a capital requirement which serves the same purpose. Since 1863, the reserve requirement has been a tool of government to influence liquidity. After the major financial crash in 1929, the US Glass-Steagall Act of 1933 was introduced by Roosevelt as part of the New Deal. It was designed to separate retail domestic banking from riskier investment banking. It insisted on a security reserve of 10 per cent. The Central Bank regulated retail banks through setting interest rates and other measures to slow down or speed up the availability of money, inflation and the economy. Clinton overturned the Glass-Steagall Act in 1999 and replaced it with the Financial Services Modernization Act. The fractional reserve dropped to 3 per cent. This act allowed insurance, retail and investment banks to amalgamate, creating banks that were 'too big to fail'. The trading banks created new financial instruments that enabled them to 'bundle' mortgages and trade them on the stock market. Trading banks got a slice of the mortgage money-making machine but forgot that money supply needs to keep pace with overall market transactions unless runaway inflation topples the entire system. The rapid proliferation of new mortgages in the 2000s generated a housing bubble that was many times higher than the average annual wage. A natural ceiling to increasing house prices occurred, and people were paying high-interest rates on 105 per cent valuation of houses. Prices collapsed, triggering the 2008 global financial crisis (GFC). Government bailouts from public taxes saved the massive corporate banks instead of making significant changes in process or regulation. Governments have since stabilised confidence in banks by guaranteeing personal deposits up to a specified limit (e.g. prime minister and banker John Key introduced a government guarantee of up to eighty thousand for personal bank savings in New Zealand 2008).

Under fractional banking, the customer walks in with a deposit to buy a house and a good job. They promise to pay the capital loan back over several decades and guarantee their loan against the asset. The price of the house is deposited in the seller's bank account. That deposit is now available for the second bank to use as fractional reserve on any other ex nihilo credit it may create. Doorman (2015: 2) explains, 'Commercial banks create check book money whenever they grant a loan, simply by adding new deposit dollars in accounts on their books in exchange for a borrower's IOU'. If, for argument's sake, the seller leaves the money in their account, the bank can use it as part of their reserve and leverage it by 97 per cent (at 3 per cent reserve requirement). Or the seller uses the money to buy further assets, which circulates the money across further banks. At the third bank, 3 per cent must be set aside to the reserve, but 97 per cent is available for redeployment to leverage on more loans, and so on, and on, and on. The 'multiplier effect' has a natural diminishing point, that is many multiple times in excess of the original deposit and many multiple times the extra $900,000 that was created for loan against the million-dollar house. On top of that, is the interest rates on the loan that was initiated by the ability of the consumer to pay it. While $900,000 worth of new money has suddenly been introduced into circulation, no new productivity is associated with the original loan. So, the *double entry* of $900,000 (and interest) dilutes the pool of public money in circulation in the real economy. Without any overt permission from the social contract, the bank has appropriated value from the public. If this situation seems incoherent and unlikely, your disbelief is not isolated. The great Nobel winning chemist Frederick Soddy wrote on banking,

> To any business man a knowledge of the truth ought to be sufficient to condemn the system according to the canons of ordinary competitive business. Where else in the whole realm of human activity is it possible to create capital by an act of imagination and to derive from its supposed existence a perennial revenue, just as though it were real wealth put to productive use? (Soddy 1933: 148)

In 2018, US banks with less than 16 million had no reserve requirement, and between 16 to 122.3 million had a 3 per cent requirement. As motivation to stimulate the economy, in 2020 COVID regulations reduced the reserve to zero (Federal Reserve Bank 2020). During COVID-19, 100 per cent of the loan deposit was available for further lending, as none was required to sit in reserve to guarantee bank defaults. It seems unlikely that the reserve ratio will be reintroduced, although the commitment from banks to tie their interest rates to the OCR remains intact. In 2023, New Zealand required banks to double their inhouse cash reserves, as the expectation of a run on the banks is more acute with higher inflation and economic instability.

INFLATIONARY LEVERS

Banking officials are reluctant for the public to know how easy it is for private banks to create money. Christopher Kent, the assistant governor (financial markets) for the Reserve Bank of Australia, gave a speech called 'Money – Born of Credit?' in 2018, where he presented debt as loosely related to savings deposits. 'Money can be created, however, when financial intermediaries make loans. Accordingly, the concepts of money and credit are closely linked in the modern economy, albeit not one for one'. His speech promotes the myth that customer deposits are deployed by the banks as loans. In which case the bank itself (not the customers) are financial intermediaries. His speech obscures the lack of relationship between ex nihilo loans and the cascades of deposits that are generated from this new credit, on the one hand, and savings deposits that are accrued from the 'real economy' through hard work, on the other hand. There are, effectively, two economies: a massive one generated through abstract leverage of ex nihilo debt (sometimes called 'nominal' currency) and the 'real economy' (sometimes called the 'broad economy') which is transactional – where labour or products are exchanged in a like-for-like circulation of currency and value.

Soddy makes the important point that ex nihilo credit, generated out of thin air, devalues the total circulation of money. To compensate, to some degree at least, the new debt introduces a demand on overall growth to offset the devaluation, or inflation. This is the unspoken core of the demand for economic growth.

> Though by the creation of money and the inflation of the currency by bank credit the community as a whole are robbed of the wealth equivalent to the new creation, it must not be supposed that the banks ever had or claimed any legal title to the ownership of the money so created. They got the permanent use of it and the ownership of the interest it was issued for. (Soddy 1933: 144)

The interest is gathered off the original debt but also the fractional multiplication of subsequent debts through the banking system. Thus, hugely escalating the requirement for broad economic growth. The abstract debt economy sits 'on top' of the broad economy. It is justified by enabling capital expenditure that promotes future productivity 'growth'.

The dilution of the value of money as large amounts of new cash swamps the existing real economy is a very important problem for both the banks and governments. Galbraith (1975) details the first major banking crisis in France, in 1719, when the Banque Royale had offered millions in new credit notes to the government and borrowers with the 'backing of new gold mines in Louisiana'. Inflation improved prices, and newly affluent producers stimulated the economy. In the short term, France was thriving. However, as inflation continued to escalate, the public became suspicious that the gold rush in Louisiana would not meet the proliferation of credit notes. Once the public became aware that there was not enough gold in the bank to meet the value of the credit notes, there was a huge run on the bank and the economy collapsed.

Governments have several levers they utilise to slow down or speed up money creation and thus rates of inflation. They can raise the OCR rate, making interest rates more expensive and slowing down house sales. Governments often introduce rules about the level of deposit the mortgage customer must have, ranging from 40 per cent to slow down an overheated housing market, to –5 per cent to enable more housing volatility. The graph below (figure 3.1) shows the close relationship between credit creation and the value of cash and GDP transactions, or 'broad money'.

Figure 3.1 from the assistant governor of the Australian Federal Reserve shows how 'broad money' tracks with credit generation in the Australian economy more than forty years (Kent 2018). The slight variation is the impact of various factors on the rate of growth and liquidity in the system, from bullish housing market, unemployment rates, natural disasters, global trade exchange and levers applied by government to slow or rev up economic growth. The graph is the percentage of growth in the system, where Broad money (which is available in current accounts for 'transactional purposes and as a store of value' 'roughly doubling as a share of nominal GDP'). Over the same forty years, 'Growth of credit has been stronger still, roughly tripling as a share of nominal GDP, from 50 per cent in the early 1980s to over 150 per cent currently' (Kent 2018). Although the graph looks flat, it charts

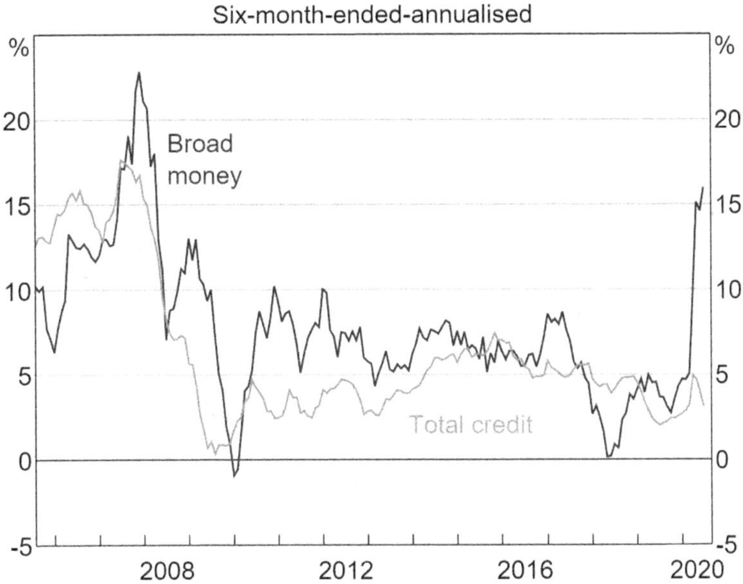

Figure 3.1 How 'Broad Money' Has Tracked with Credit Generation in the Australian Economy over Forty Years (Reserve Bank of Australia)

percentage increase in growth, so its overall impact is exponential increase in GDP and consumption. The change in credit ratio to 150 times the current total GDP demonstrates the 'marvellous' expectation that the continual rate of growth has transcended the 'bounds of the physically possible' (Soddy 1933).

Once the Glass-Steagall Act had been repealed in 1999, all around the world, the Residual Reserve has been reduced or replaced. Kent tries to present this as a fundamental change in the money multiplier effect. 'Traditional money multipliers' he says, 'had earlier required banks to hold a minimum share of their deposits as reserves with the Reserve Bank – have been replaced with modern prudential regulation and market discipline' (Kent 2018). These 'new' government levers include overnight cash interest rates and bonds.

If the world market was static, then this expansion in monetary quantity would have a considerable inflationary effect, and the value of money would fall dramatically, as happened in France in 1719, America in 1929, Nazi Germany during the 1940s, in Argentina in the 1990s and in Venezuela in the 2010s and 2020s. However, for most countries, monetary value stays relatively stable because multiple moving parts are growing, and so with some adjustments by governments, inflation can appear to be kept roughly stable. Governments tend to aim for 2 to 3 per cent growth in GDP and less than 2 to 3 per cent inflation. Through central reserve banks, the governments influence the OCR rate, which is the rate that banks have to borrow or deposit to square their transactions at the end of each day. Governments can leverage the amount of new money that banks create by setting the treasury rate higher or lower. An expensive rate raises interest rates and slows down house sales – slowing the introduction of new ex nihilo money into the economy. The value of inflation also shows up on the government bond rate. If the four-year bond rate is high, the market estimates inflation will move up. If its low, then money is flowing in the direction of the housing market or shares.

Under the hard work and savings deposit model, when banks are genuinely intermediaries between depositors and borrowers, banks would not make a great deal of profit. As Fred Soddy put it in 1926, for intermediary banking, interest rates are justified to pay for the cost of administering accounts. Knight and Simons came up with something similar in 1933, which they called the Chicago Plan (Tavlas 2020). Endogenous banking is altogether different. It stimulates the multiplier effect that enables the financial industry to charge interest many times over by leveraging the same ex nihilo deposit across multiple banks. More importantly, in regard to climate change, is the effect that the multiplier has on consumer growth and, thus, emissions growth (Noll 2023). Effectively, the endogenous multiplier scheme of bank credit forces the economy to grow, as shown in the graph above, to offset the demand made by nominal credit money on the broad economy. People do not merely pay back the capital over time; they pay back triple or more the quantity they first borrowed, and this money comes from labour and transactions – the real economy – making necessary overall economic growth to meet these new demands that have no original productivity exchange value.

The loose association between the volume of cash in the system of the broad transactional value of like for like, is why economists need to associate volume of money with GDP – gross domestic product. It gives them an estimate of real productivity through gross transactions. Kent is proud of Australia's economic growth since the 2008 GFC. 'The stock of (broad) money increased noticeably, rising by around $1.1 trillion dollars, from around 80 per cent of GDP in 2007 to around 115 per cent in 2018'. This is directly related to the bank bailout which allowed the continuation of invented debt instead of allowing a full (if accidental) collapse of the banks. The Bank for International Settlements show that Australian household debt continue to hover around its height of 123 per cent of GDP in 2018.

The global climate change fraternity is not oblivious to the impact of the financial system on emissions, but protection of the term economic 'growth' is still rampant. When Nick Stern released his famous review, *The Economics of Climate Change* (2006), he refrained from publicly critiquing growth and instead framed the green economy as new opportunities for green growth. Stern reiterated that in 2018. In the short term, this is very accurate. To meet net-zero goals, Robins (2020) argues the next two decades will require large sums of capital. But overall, the aim must be to reduce energy throughput. Energy accounting is beginning to include CO_{2e} pollution as a cost in the real economy. In 2026 the EU will collect a tariff called the Carbon Border Adjustment Mechanism on emission intensive imports such as steel and concrete with the aim that higher prices will lower consumerism. But flooding the market with new credit, swamps the sensitivity of the invisible hand to price. In other words, money devalues, becomes more prolific, and although prices go up, inflation means overall production expenses gets cheaper and cheaper. The tariffs will only work if they are inflation adjusted.

The role of consumerism in emissions that are produced overseas are being included in offshore scope 3 emissions accounting. As yet, there are no mechanisms or regulations for including detailed information on embedded or transport emissions on products. Changing consumer preferences is fundamental to neoliberal market theory but only a small aspect of reducing emissions. If the goal is to reduce energy throughput, then relying on *increased consumerism and disaster capitalism* (next chapter) as ways of growing GDP needs to become unpalatable.

BANKING REFORM

Since the 1990s, central banks have played an important role in levers for government objectives regarding inflation and employment rate. At the same time, they are positioned as 'apolitical'. Financial institutions have managed to avoid major reform after the 2008 GFC but a certain unease and better popular understanding about ex nihilo endogenous loans and credit derivatives (Mengle 2007) have not gone away. Nevertheless, discussion on 'structural change' at Central bank level continues to avoid the issue of ex nihilo endogenous banking (Buch 2022; Lane 2021; Buch and Dages 2018; Stevens 2003; Werner 2014). Climate denial no longer washes politically, and central banks are having to respond to a changing policy expectation and shareholder demand particularly from the Paris Accord (2016), pension funds, and insurance companies. 'A clear net-zero strategy will be crucial for Central banks to deliver on their mandated goal of safeguarding macrofinancial stability and ensuring policy coherence' (Robins et al. 2021). The demand for change is getting louder.

In the 2020s, the banking system is under the spotlight for the biggest structural review since 1863. The starting point of divesting from fossil fuel companies has begun with the Green member for Brighton, Caroline Lucas, suggesting to the European Union (just before Brexit) that futures trading to provide capital investment for fossil fuels is too risky and should be phased out, because in twenty years, they will be stranded assets. Similar demands for change are aiming at financial institutions. Organisations to encourage green investing are springing up, such as the Network for Greening the Financial System (NGFS), and superannuation funds have been under enormous pressure from subscribers to alter their investment portfolios to rule out fossil fuels. In 2021, the TCFD, or Task Force on Climate-Related Financial Disclosures, was set up, that encourages banks to voluntarily join an umbrella organisation that promises to check the moral credentials of investment opportunities. There is no doubt that huge investment in green infrastructure is urgent and necessary. On the face of it, central banks are enthusiastic about these initiatives.

However, the self-reflection of the central bankers does not go very deep. Lane (2021) explains that bankers need to accommodate the new dual dilemmas of digitalisation and climate change as risks and opportunities for further responsible financial activity. However, there is no analysis of how endogenous credit – that is entirely confined to electronic transmission – contributes to the problem.

> Digitalisation and the greening of the economy require significant levels of investment, both from the private sector and the public sector. In addition, public transfers and subsidies will be required to ensure an inclusive approach. The funding of digitalisation and the carbon transition poses significant challenges for the financial system, the tax system and the public finances. While these funding decisions will be heavily influenced by political economy considerations, the design of optimal funding schemes remains a high research priority.

The danger remains that 'net-zero' changes in finance will be limited to greener investment, leaving the fundamentals of banking intact. According to Mark Carney of the UN Special Envoy on Climate Action and Finance, current calculations suggest that 'the financial system as a whole is funding temperature increases of over three degrees centigrade' and there is 'a striking gap between what society wants and what the market values' (Carney 2020 in Robins et al. 2021).

Green investment does not solve the fundamental systemic problem of ex nihilo loans as a source of new money, that dilutes the value of currency unless offset by economic growth. Nor does green investment engage with the role of interest rates attached to loans which force triple or higher repayment from real economy work and like-for-like transactions. New money and interest forces growth into the system or families risk defaulting and losing their homes. Justification for the steady rise in house prices has been attached to 'market demand' and for those already in the market, the price of one will approximate the price of another. However, the ratio of house price against average income has been steadily going up, from three times the average income in the 1970s to ten times the average income in the 2020s. 'Market demand' is fuelled by global population growth and immigration but in my view, population is not the primary reason for rising house prices.

Excess money sloshing around in the system has to go somewhere. Although neoliberal theory is that markets will 'balance' between supply and demand, the multiple billions that flowed suddenly into economies as a result of COVID-19 showed a different story. Neoliberal market theory is a static or stable 'perfect' market with small variations in innovative profit generation. Keynes introduced a more dynamic understanding of money (1991/1936). My view is that money is like a fluid. Money is in flow. In the broad economy, when inflation is low, it subtly alters the like for like transactional value of money. That 2- to 3-per-cent-level failed to accommodate the enormous glut of monetary 'easing' generated during COVID-19 lockdowns. Governments printed some money to compensate workers for their loss of work. More importantly, the reduced OCR during COVID-19 dropped the cost of loans almost to zero, which allowed interest repayment to be very low and stimulated a big boom in the housing market. New house sales enable the creation of new *ex nihilo* loans. So fast housing turnover promotes enormous amounts of new ex nihilo money to enter public circulation. Post- COVID-19 inflation has subsequently overrun productivity growth to more than 7 per cent in most 'developed' countries.

As COVID-19 wound back, energy supply has become more obviously problematic. As oil prices rise, productivity and transport are more expensive. Governments and Central banks are altering the OCR or (EU and UK) LIBOR (soon to be revamped as the secured overnight financing rate (SOFR), to make money more expensive. This adds to interest on mortgages, dampening demand and slowing new money creation, high inflation alongside high mortgage rates strengthening the risk of mortgagee sales. The nearly nonexistent liquidity to back up bank loans becomes a pinch point in the finance sector. To avoid a run on the banks, the New Zealand government insisted banks double their on-site cash reserves in 2023.

FUTURE DERIVATIVES

Derivatives trading and betting on future commodity prices are two examples of sophisticated financial products that enable banks to make more money with lower theoretical risk. The International Bank of Settlements estimated that the derivatives markets were worth 20.7 trillion US dollars in 2015 (Bank of International Settlement (BIS) 2016). That dollar value is 1 per cent of the 'notional' value of the packages (as 1 per cent deposits get traded, but the full worth is one hundred times more). 'The gross market value of OTC derivatives – that is, the cost of replacing all outstanding contracts at current market prices – rose to $20.7 trillion at end-June 2016 from 14.5 trillion at end-2015' (BIS 2016). In contrast, global GDP – the total monetary value of all transactions – at the end of 2015, was 74 trillion (World Bank National Accounts Data 2022). The notional value of OTC derivatives was 598 trillion at the end of 2021 and went up to 632 trillion at the end of June 2022 (BIS 2022). In contrast, global GDP in 2021 was 96.51 trillion (World Bank National Accounts Data 2022). The massive disparity between 600-odd trillion in derivatives, and 96 trillion in actual transactions, is an indication of how alienated the finance sector is from material 'reality' despite a necessary relation between financial capital investment and the real economy. To give an idea of how variable the data is in derivatives trading, the Bank of International Settlements estimates 2022 nominal derivatives at 120 trillion dollars against real trade of 20.7 trillion (BIS 2023).

At present, there is no safe way to end this type of speculation. The systemic alienation of abstract ex nihilo credit and the derivatives that trade on it makes it impossible to reintroduce a connection to finite planetary resources. Debt and inflation speculation bear almost no relation to environmental concerns. As we saw during the 2008 financial collapse, the impact of financial system collapse is far reaching. Not only did house prices radically plunge, affecting millions of people, but with them the faith in derivatives and securities plunged, causing financial institutions that were 'too big to fail' into bankruptcy or, worse, enormous taxpayer bailouts. In the 'developed' world, taxpayer resources have been diverted from the welfare system – education, hospitals and so on – towards bailing out the banks. We have no alternative system for creating credit for large-scale projects outside of the current banking system, despite knowing that it is only a case of when, not if, it will fail again, with predictable impacts on working- and middle-class communities. These are catastrophic risks, not calculated risks.

GREEN CAPITAL VERSUS FINANCIAL RESTRUCTURE

The call for change is reaching conservative halls and institutions entrenched in long-held normative assumptions. The Financial Stability Board's Task Force on Climate-related Financial Disclosures (TCFD) argues that net zero is a vital component of all four pillars of the TCFD recommendations. The network NGFS and Robins et al. (2021) refer to Article 2.i.c of the 2016 Paris Agreement, which sets out the goal of 'making finance flows consistent with a pathway towards low greenhouse gas emissions and climate-resilient development' (Robins et al. 2021). They point out that even if central banks are obliged to be politically neutral, they are also obliged to meet the policy directions set by government.

The 2021 annual remit letter from government to the Bank of England's Financial Policy and Monetary Policy Committee is most prepared to consider deep-set change. They have made it clear that 'structural reform' is necessary 'to transition to an environmentally sustainable and resilient net-zero economy' (HM Treasury 2021a and b in Robins et al. 2021).

Other central banks are sitting on the fence, making a standard operating procedure (SOP) towards sustainability (Brundtland 1987) without requiring substantive change to their priorities. The Swedish Sveriges Riksbank is typical of these newly emerging goals, seeking 'sustainable growth and high employment, assuming that these targets are compatible with its other objectives' (Sveriges Riksbank 2020). In 2021, an executive on the Board of the European Central Bank (ECB), Frank Elderson, reminded the ECB of its obligation to 'to support the general economic policies in the [European] Union'. These policies include environmental protection and sustainable development. But these obligations still come second; 'without prejudice to the primary objective of price stability'. Likewise, academics like Robins et al. (2021), while quoting the above changes in banking rhetoric, are still claiming that regulation is unnecessary, confirming the Non-Binding Guidelines on Nonfinancial Reporting Update (NFRD) (NFRD European Commission 2019).

Matikainen et al. (2017) and Oustry et al. (2020) both argue that 'market neutrality' needs updating or liquidity will continue to contribute to the carbon crisis. This recommendation is nevertheless based on ridiculously low estimates of climate risk to the financial sector. Dietz estimates, 'the average global value at risk for global financial assets due to climate damages between 2015 and 2100 in a business-as-usual scenario to be 1.77 per cent, but reaching 16.86 per cent at the 99th percentile' (Dietz 2016, in Matikainen et al. 2017). Even with such low estimates of impact on the financial sector, the fear of stranded assets is starting to change market behaviour and banking regulations. Reducing exposure to the carbon intensive sector limits the risk of potential stranded assets. A net-zero goal by 2050 sends clear market signals, and 'a managed transition in order to ensure a scenario that minimises harm to the financial system and the wider economy in the future' (Chenet et al. 2019, in Robins et al. 2021).

At present, a lot of that managed transition is more talk than action. Modelling by the NGFS (2020a) retains business as usual (BAU) with an additional greenwash stance, by advocating a 2°C-pathway that relies on CO_{2e} carbon capture technology – that does not exist. But some central banks are making a genuine effort to consider banking at a structural level. Klaas Knot, the governor of De Nederlandische Bank claims that not only can central banks 'help to correct the carbon bias in capital markets', they can, more importantly, also 'redesign their monetary policy instruments to prevent such biases from occurring, and instead contribute to unlocking more green investments' (Knot 2021, in Robins et al. 2021).

There is still a lot of greenwash swirling amongst the banking fraternity, not least amongst academics. Nevertheless, key personages in central banking, in some of the oldest banking centres of the world, are beginning to make the case for change (Turner, 2015). At present this is aimed at green finance and not change to the ex nihilo endogenous banking system. Reconsidering how the endogenous monetary system works, and how banking is currently leveraged, is crucial. A new imaginary for a system that enables raising capital for big projects, without locking ex nihilo banking into the monetary system needs to be forged.

IMPACTS

People have forgotten about the necessity for debt to remain aligned to the carrying capacity of the ecosystem in modern times. An increasingly abstract system of nominal modern debt is exacerbating the alienation of people from ecology. In its very early inception in fifteenth-century Italy (De Roover 1963; Buch and Dages 2018), double-entry bookkeeping and loans ex nihilo were applied only to a tiny elite. Interest was defined as the sin of usury during this period. Gifts and allowances benefited the bankers and contributed to their personal wealth, along with the growth of the local economy. The exponential impact of the devaluation of coinage would not have affected the public very much because direct bartering was still the dominant form of exchange. Economic growth must have been slight enough that fertile Italian ecology was not significantly affected.

As time has gone on, the disparity between nominal credit and the broad economy of real exchange has grown ever wider. The International Monetary Fund (IMF) estimates global GDP was 88 trillion in 2020, 94 trillion in 2021 and 104 trillion in 2022. They estimate global nominal debt in 2020 as 226 trillion. Total debt was 2.5 times planetary gross exchange in goods and services in 2020, during COVID-19 (Gaspar et al. 2021). The most recent update for this book for global debt in the first quarter of 2023 is 305 trillion.

Governments do still print money, but the importance of ex nihilo credit to money supply is far more significant. 'Modern money is created when commercial banks issue loans to individuals and corporations at interest. This 'horizontal' money makes up approximately 97% of money in the UK and approximately 92% of money in the US (data from the Federal Reserve Bank of St. Louis)' (Ament 2019: 9).

Nominal debt is effectively borrowing from the common circulation of currency both now and in future. Even without the complication of CO_{2e} emissions, the risks embedded in the current financial system are extreme. The schism between the value of abstract financial instruments and the real economy is vast. It is an indicator of how alienated and abstract economics, and neoliberal culture, are from ecology and like for like transactions in the real world.

A slowing and aging population will eventually take the wind out of the housing market, which has underpinned the abstract debt economy for the entire lifetime of all of us alive. The Energy returns on investment (EROI) has radically reduced over the last century from one-one-hundredth, or one barrel input for one hundred extracted, to less than one-sixth (Hall et al. 2014; Brockway et al. 2019). Several rare earth minerals are running out (Stockholm Resilience Centre), soil fertility is decreasing (Shiva 1989), deforestation has shrunk most remaining virgin forests, and fish stocks are in rapid decline. All these issues show the physical limits of the planetary system. An aging population has fewer requirements for goods and greater requirements for services. Union busting, combined with the impact of robotics, IT and AI on the gig economy has reduced real income for many people in 'developed' countries (Measuring Worth 2022; Statista 2023). How the aging of the population will play out in relation to the continuously growing abstract generation of debt is difficult to determine. At present, debt seems to continue its exponential acceleration despite limits in the real economy. The consequences are exponential economic growth and accelerating global warming (Noll 2023).

The staggering data that James Hansen's (2023) team gathered (figure 1.1 and 1.2) shows how dangerous this exponential warming is becoming. Since 2010, the *rate* of global warming is escalating. Staying beneath the 1.5°C limit has become practically impossible. Already the escalation of climate impacts in fire, flooding and extinction levels are prolific.

Despite the anguish of so many people affected by climate disasters, the discourses in political theory remain oblivious to structural problems of exponential growth. Yet, for those who are looking, the correlation between economic growth and growth in CO_{2e} emissions is hard to miss (Le Quéré et al. 2020; Caporale et al. 2021; Mitić et al. 2022; Onofrei et al. 2022; Noll 2023). Finding a pathway towards a sophisticated, low-emissions global economy has begun. Hidden in the present normative structures are commitments to emissions growth that are not yet interrogated. Commitment to net-zero CO_{2e} emissions is a good start and a necessary transition. There are problems with the 'net'-zero rhetoric, including faith in 'discounting' (or putting less weight on the needs of future generations (Nordhaus 2018), and 'carbon reduction technologies' (Paris Agreement 2016; IEA 2021) that have not emerged at scale despite more than thirty years of expectation. The 'net'-zero approach leaves economic growth objectives intact, despite the dangerous correlation with emissions growth.

System balance through growth – or dynamic stability (Rosa 2013) – has guided government policy and endogenous ex nihilo banking since the Keynesian settlement in the 1950s. Pollution and CO_{2e} emissions were never factored in during the development of the financial system or modern economic theory (Keynes 1991/1936; 1933; Hayek 1960; Friedman 1970). Thermodynamics and energy throughput were ignored, and market theory is built on older Enlightenment principles of cause and effect and the rational individual. Growth was justified as an expanding pie, a rising tide, the multiplier effect that would benefit the working classes through access to prolific, cheap consumer goods. The rich could get richer if the poor benefit a little too. Decoupling emissions from growth has proven impossible. Accounting for emissions is an important component of change. However, the systemic factors of financial endogenous ex nihilo lending are the crucial component of forcing growth into the system. Until this changes, the correlation between economic growth and emissions growth will remain in force. We need to reimagine a better theory of capital, and debt, that returns banks to a genuine financial intermediary – without the powers to create money ex nihilo (Jackson and Dyson 2013). To avoid the growing emissions associated with economic growth, we need systemic change.

Given how banking has been an important policy lever since the 1990s, the volatility of the banking system is increasingly problematic. At the moment, many countries with dynamic housing markets have reached the average wage-income ceiling. While the spiralling feedback loop that inflation forces up wages will compensate for a little of the rapid rise in ratio of debt to annual income, it is unlikely to shift the cost of a house from ten times the average annual wage back to three times the average wage. The long growth curve has reached an indefinite bubble. In collaboration with CDO's or innovative financial products, there remains high potential for future market collapse. There was limited reform of the banking system after the 2008 GFC and the entire industry expects it to occur again. Sophisticated 'bundles' of debt are bought and sold amongst trading banks in an effort to share the risks of 'toxic' or default loans, but this spread of risk does not alter the overall volatility of exponential debt in the context of a relatively sticky transactional or 'real' economy.

Catastrophic risk is impossible to calculate, and at present we are heading into the perfect storm – as endogenous banking and the stock market transcend the relationship with the broad economy, the impact of continent-sized weather events escalate with accumulated greenhouse gases. The two things are firmly linked but, as of yet, have failed to show up in political discourse.

Clearly, while banks are allowed to create new ex nihilo money, diluting the value of existing circulation of like for like fiat currency, both inflation and growth stimulation are forced into the system. This makes it necessary to counteract inflation through economic growth. While these structural factors are in play, the reliance on economic growth has to remain in place, and the risks will continue to escalate. To shift to a new mode of generating the resources for substantial capital projects, while avoiding the problems of new money creation, will take fresh eyes. It will need to combine a thermodynamic approach to economic throughputs, with a more sophisticated understanding of how essentially integrated are the fields of economics, culture, epistemology and ecology.

Chapter 4

Calculation, Risk and Fear

Billions of ordinary people are facing an abrupt collision with the future. Control over nature, or even control over macroeconomic forces, seems to be very unreliable. Yet, the urge to calculate and control is very strong, especially according to Enlightenment rationality. Technical optimists are often called Prometheans, after the ancient Greek Titan, Prometheus, who gifted humankind the original technology of fire. Promethean optimists are sure that innovative technologies can solve all problems, once the scientific analysis of cause and effect has taken place. In the spirit of optimism, 'risk' has become commodified and encoded into the market since the 1980s. Disaster clean-up and rebuild contributes to an overall expansion in gross domestic product (GDP) (Waring 1990). But black swan events, such as COVID-19, or climate, or other 'unknown unknowns' seem to be increasing in scale as exponential growth consumes disproportionate resources from the limited planetary pool. Risk is tipping towards system collapse. Beck (1992) calls this the shift from the risk society to the society of fear.

> 'Risk society' means that we live in a world out of control. There is nothing certain but uncertainty. But let's go into details. The term 'risk' has two radically different meanings. It applies in the first place to a world governed entirely by the laws of probability, in which everything is measurable and calculable. But the word is also commonly used to refer to non-quantitative uncertainties, to 'risks that cannot be known'. When I speak about 'risk society', it is in this latter sense of manufactured uncertainties. These 'true' uncertainties, enforced by rapid technological innovations and accelerated societal responses, are creating a fundamentally new global risk landscape. In all these new uncertain risk technologies, we are separated from the possible end results by an ocean of not knowing. (Beck 1992, in Yates 2003)

To understand how the risk society has emerged, it is necessary to consider Enlightenment science, philosophy and politics, and how they have forged accelerating modern industrialisation. Incredibly dense and portable fossil fuel energy has metamorphosed the world of work in just two hundred years. Fossil-fuelled technologies intensified the capacity and wealth of modern humankind to an unprecedented scope and scale. Throughout this period of abundance, knowledge proliferated and the global population multiplied, growing from a few million to 8 billion in a very short time span. Mass manufacturing and global supply has transformed nearly every community on earth. The trajectory of 'progress' seems to make sense. Yet the utopian promise of promise has always been tempered by a dystopian shadow.

Industrial modernity is a colossal shift in scale and speed. It has extended storage capacity (Heidegger 1977) and accelerated the nodes of information and distance exchange (Stiegler 2018). Time has become compressed and rates of information flows are twenty-four/seven. The acceleration of information has enabled banks to invest overnight, speeding up endogenous banking yet further. Likewise, there is a share market always open, and speculators have a continuous opportunity to bet on minute changes in value trends, completely alienated from material constraints of the business in real time.

The calculation of risk is the premise upon which the insurance industry is based. Over the last forty years, risk assessments have permeated the practices of education, business and public events and programmes. If risks can be anticipated and calculated, then they can be enfolded into the measures of the market.

The level of risk present in this late stage of modernity is unprecedented. The Intergovernmental Panel on Climate Change (IPCC) has been warning against emitting more greenhouse gases into the atmosphere since 1988. Many politicians and largescale businesses have been unwilling or unable to sufficiently change practices to reduce overall global emissions. The impact of climate change is no longer possible to deny (IPCC 2021; Hansen et al. 2023; Noll 2023). Denial dominated the climate discourse for decades but now has weakened legitimacy with the public. Genuine commitment to zero emissions is rapidly emerging. However, the infrastructure in nearly all sectors – transport, the built environment, agriculture and especially finance – are still deeply embedded in a fossil fuel culture of continuous exponential growth.

The diverging association between the real economy and the abstract financial economy is a good indicator of the alienation of humanity from ecosystems. Modern industrial impacts on planetary ecological systems are still insulated from the modern concept of the 'world'. This chapter explores the consequences of alienation. The next chapter delves deeper into how alienation emerged. Alienation can be found in the fundamental premises of Enlightenment modernity. Far from accidental, the concepts of cause and effect, rationality and freedom are actively pursued in foundational modern philosophy. Descartes (1980/1641) and Kant (1992/1781) both tear the fabric of subjectivity from its relationship with the wider natural world.

The alienation of human beings from the ecosystem is the core of the Enlightenment presumption of scientific mastery over nature. Alienation is fundamental to the concept of progress. Alienation is also behind the mounting popularity in 'social constructivism'. This is where modernity constructs its own hyperreality (Baudrillard 1994, see next chapter). Promethean mastery and the simulation of control is particularly visible in Silicon Valley. The billionaires who are shaping modern culture from information technologies through to geo-engineering take seriously the idea that we are living in a computer simulation.

This late in the exponential growth of the modern system of extractive capitalism, mastery over nature is in the process of toppling. Monetarising risk, such as insurance companies, or the boom for a building industry after climate events, has become normal. But these industries are struggling to keep up with the level and frequency of climate calamity, in multiple sites at once. As I write, New Zealand has just had the three worst storms in its history, within weeks of one another. The storms and flooding dumped two to three metres of silt throughout the broad Esk Valley, annihilating dozens of roads and trainlines under landslides, and smashing thousands of houses. Eight hundred houses are 'red stickered' and too dangerous to live in. In Australia, there have been three years of frequent, unprecedented flooding, where towns, such as Lismore, have been two-stories deep in flood waters. As one-in-thirty-year flood events become more frequent, affected houses are uninsurable and, thus, unsaleable. New Zealand and Australia pale in comparison to Turkey, where three cities were smashed with major earthquakes in 2023. (Polar ice melt is changing ocean weight and shifting the pressure on tectonic plates.) This week the same region has vans and trucks tumbling in fast-flowing flood waters. Bangladesh and Pakistan have suffered similarly severe flooding this year. At the other end of the same scale, ferocious, very hot fires are impacting areas all over the world. In what Beck calls the society of fear (1992), environmental and pollution disasters have become so damaging that the risks of civil war and chaos are increasing. Consequently, the super-rich are buying remote islands or estates in distant lands like New Zealand, where they build bunkers. Or, in the case of Elon Musk, they are preparing to leave earth altogether by colonising Mars to 'make humanity interplanetary' (Junod 2012). Musk reports to Carroll (2013) that he aims to transport a colony of eighty thousand people to Mars.

> The lessons of history would suggest that civilisations move in cycles. You can track that back quite far – the Babylonians, the Sumerians, followed by

> the Egyptians, the Romans, China. We're obviously in a very upward cycle right now and hopefully that remains the case. But it may not. There could be some series of events that cause that technology level to decline. Given that this is the first time in 4.5bn years where it's been possible for humanity to extend life beyond Earth, it seems like we'd be wise to act while the window was open and not count on the fact it will be open a long time. (Musk in Carroll 2013)

The intrinsic paradox of mastery and disaster has been noted throughout the modern epoch. Literature on the calamitous acceleration of pace in knowledge, technology and consumerism includes Karl Marx's book on *Capital* (1887/1867), Nietzsche's *The Antichrist* (1985/1895), Oswald Spengler's books *The Decline of the West* (2002/1918) and *Man and Technics* (1932), Heidegger's text 'The Question Concerning Technology' (1977/1954), Alvin Toffler's *Future Shock* (1970) and Bernhard Stiegler's *The Neganthropene* (2018). Alienation is at the centre of this paradox. The impacts of modern industrialism strains eco-social connection and drive a philosophy of alienated individualism. Consumerism is exacerbating the dislocation of communities from their traditional territories, freeing them from localised ecological events. At the same time, consumerism intensifies reliance on the capitalist economy, where the only reliable source of subsistence is the human resource of your own skilled labour power (Marx 1887/1867). Accelerating technological progress has created rapidly changing, transient, alienated and dislocated populations. People now have to generate their own social 'tribes'. The promise of Keynesian growth, to furnish the working classes with the benefits of capitalism, are becoming tenuous. Srnicek and Williams (2015) describe the inherent tensions and despair of a precarious world:

> For all the glossy sheen of our technological era, we remain bound by an old and obsolete set of social relations. We continue to work long hours, commuting further, to perform tasks that feel increasingly meaningless. Our jobs have become more insecure, our pay has stagnated, and our debt has become overwhelming. We struggle to make ends meet, to put food on the table, to pay the rent or mortgage, and as we shuffle from job to job, we reminisce about pensions and struggle to find affordable childcare. Automation renders us unemployed and stagnant wages devastate the middle class, while corporate profits surge to new heights. The glimmers of a better future are trampled and forgotten under the pressures of an increasingly precarious and demanding world. And each day, we return to work as normal: exhausted, anxious, stressed and frustrated.
>
> Neoliberalism has held sway for decades, and social democracy exists largely as an object of nostalgia. As crises gather force and speed, politics withers and retreats. (Srnicek and Williams 2015: 2–3)

On the one hand, the calculation and control generated by modern science has extended and accelerated knowledge and utility of natural systems in unprecedented ways. Inversely, as techniques of exosomatic memory such as the printing press and the computer changes the way learning takes place, the proletarianization of work and knowledge is extending from the manufacturing sector through the knowledge economy (see chapter 9). The basis of modernity's paradox in the attempted calculation of risk and control emerges from Enlightenment science.

SCIENCE

The extent of scientific knowledge about the universe has developed rapidly. Telescopes from the time of Kepler and Galileo, through Hubble and now the James Webb Space Telescope (JWST) are probing the earliest evidence of the universe back 14.8 billion years, to shortly after the Big Bang. This golden age of understanding was transformed by Copernicus (1543) who decentred the flat earth from its position as the centre of the universe, to one more planet orbiting our central sun. Kepler (1995/1621) and then Newton (1687) unpacked the measurements that proved Copernicus correct. Newton's physics built a mechanical view of the universe, where forces act upon other forces, more commonly known as cause and effect.

The 'scientific revolution' seemed to enable humanity to discover everything there is to know, if we turn our attention to it (Weber 1948/1919). Extrapolating an argument about the essence of technology, new knowledge has resulted in new resources, waiting to be exploited. Heidegger calls the modern scientific revolution, the 'ordering of order' (1977: 17) which lends itself to illusions of mastery. This critique will be explored further, below.

Scientific knowledge from the last hundred years grows ever more sophisticated, and the 250-year-old Newtonian assumptions of a mechanical universe based on straight forward cause and effect are giving way to more paradoxical laws, such as thermodynamics, quantum physics and relativity. With this extension of the known universe, currently modelled as LamdaCDM, there is more recognition of evidence of what we do *not* know. Dark energy makes up 68 per cent and dark matter 27 per cent of the universe (Farnes 2018). While better comprehension is rapidly emerging, humanity understands less than 5 per cent of the universe. Black holes have such immense gravity fields that they swallow up matter and energy that nobody really understands. Humanity probes the evidence of the earliest known solar systems and the background universal radiation through JWST's telescope images. Physicists are imagining not just the beginning and end of our universe but speculating about the scope of possible multiverses, the role of black holes and quantum entanglement of their two-dimensional holographic horizon (Maldacena et al. 1997).

The philosophy of science has changed along with the expansion of knowledge. The discourse of cosmology has evolved natural laws based on atoms, forces, quantum field theory and, more recently, information. Two things are notable in these transitions: (1) cosmology from ancient Greek atomism to Newton's impact of forces acting upon other forces and (2) the current focus on relativity, and informatics as a substrate, are all inclusive of humanity as an intraconnected component of the universe as a whole. Nevertheless, important shifts have occurred in conceptualising the mechanics of the universe. From cause and effect, through the Copenhagen interpretation of quantum field theory, and the ability to inscribe data across many material substrates, concepts of the universe are changing the way modernity understands the capacity for social communication, technology and evolution.

Importantly, these amazing changes in modern technology and concepts of the laws of nature are more open about the dynamic unknown. However, outdated science and philosophy from the Enlightenment continues to inform politics and social norms, despite major developments in science that have rewritten the natural laws of physics and chemistry.

Modern society is based on an *Enlightenment* science of calculation and control. In the 1980s, calculation and control was extended to encompasses risk. The obsession with mastery informs Promethean technological optimism. But paradoxically, it feels like the progressive juggernaut of modernity cannot alter course and successfully adapt to the vulnerabilities of planetary finitude. The rigid premises of control are failing to keep risk at bay. The scale of calamity is disintegrating into the contemporary society of fear (Beck 1992).

In the following chapter, I will show how alienation of the individual from their ecological context generates hyperreality. The sort of hyperreality that justifies nominal credit being multiple times greater than the entire yearly aggregation of goods and services. I argue that sceptical Idealist philosophy progressively alienates people from the real towards the simulacra (Baudrillard 1994; Kant 1992/1781). It has disturbing consequences in the Anthropocene (Crutzen and Stoermer 2000; Crutzen 2002). Philosophical normativity as alienated individualism has implications for the future visions and aims of leadership. Alienation affects Keynesian and neoliberal governments. But it also affects the influential multibillionaires. Alienated abstraction justifies the tenuous relation between nominal ex nihilo debt and the real economy.

THE POLITICAL SCIENCE OF CALCULATION AND CONTROL

Modernity, as a civilisation, is often lauded for being unique and essentially different from any previous human society. Fossil-fuelled industrialisation has enabled human beings to plumb resources at a rate never seen before. In just two hundred years, the world has transformed from subsistence home economics to the 'wealth of nations' (Smith 1904/1776). The accuracy and extent of knowledge has expanded, new technologies have proliferated and the global population has thrived. Colonisation and land clearances, combined with industrialism and better techniques of global distribution, have imposed upon and changed nearly every culture on the planet. Of course, historical trade routes like the Silk Road have acted as channels of distribution for thousands of years. Modernity differs because information, transport and storage are on a completely different scale and speed (Heidegger 1977).

The famous sociologist, Weber (1948/1919) enthused about modern rationality and calculation as a mode of knowledge that enabled modern governance. Through science, it is possible to know everything – and utilise it to capitalist ends. With a dim sense of poignancy for a romantic era, now past, Weber believed that 'increasing intellectualisation and rationalisation' enables access to all realms of knowledge. Rationality indicates,

> an increased and general knowledge of the conditions under which one lives. It means . . . the knowledge or belief that if one but wished one could learn it at any time. Hence, it means that principally there are no mysterious incalculable forces that come into play, but rather that one can, in principle, master all things by calculation. This means that the world is disenchanted. one need no longer have recourse to magical means in order to master or implore the spirits, as did the savage, for whom such mysterious powers existed. Technical means and calculations perform the service. (Weber 1948/1919: 138)

Casually scathing about premodern societies (which is unpacked in chapters 5 and 6), Weber depicts modern, colonial, Promethean optimism about human mastery and control of natural forces as built on technology and rationality. Calculus and deductive rationality were developed by Enlightenment thinkers such as Bacon (1696/1597), Descartes (1980/1641) and Kant (1992/1781), in alignment with the scientific calculations of cause and effect in Kepler (1995/1621) and Newton (1687). The Enlightenment has reduced the opaque 'mystery' of nature, replacing it with the clarity of observation and mastery. Forces acting upon other forces are understood as a kind of 'mechanical' universe, where linear cascades of cause and effect, or as Kant puts it, a 'series of alterations' make predictable logical laws of nature. These laws underpin all events and relations. As Swer contends, 'science settled into its new role as motor of the industrial revolution' (2018: 12).

Kant (1992/1781) finds the logic of cause and effect so determining that he worries about the freedom of the individual to make his or her own decisions.

Kant's text, *The Critique of Pure Reason* (1992/1781), which is explored in the following chapter, sets out to find room for individual freedom in the context of a determinist Newtonian mechanical universe.

In 1776, Adam Smith (1904) transformed Enlightenment individual rationality into the basis of the invisible hand of the marketplace. Rational individuals each make self-interested decisions. An aggregation of all the rational individual utility maximising decisions combine into a perfect price point. The price balances the cost of extraction and production against the benefits of ownership and consumption. Progress stimulates new technological developments that reduce costs or add unique selling points. Innovative producers can generate more profit than their competitors. None of it makes sense if individuals make choices that are not rational and predictable.

Neoliberal market theory still rests on these seventeenth- and eighteenth-century assumptions underpinning science and philosophy. Adam Smith's *Wealth of Nations* (1904/1776) is a book of persuasion to push people from traditional ecological subsistence or 'home' economics and put more emphasis on producing and selling local products on a global marketplace. Hotly contested during the land clearances, global trade was promoted as an advantage. Globalisation alleviates the hazards of local catastrophe, because a community can buy in food and necessities during a disaster and sell more during the good years. Modernity promises to free the community from the uncertainties of local seasonal productivity. And at the same time, its frees the need for close awareness of natural seasonal cycles.

ALIENATION FROM 'NATURE'

Heidegger (1977) argues that the technological innovation that characterises the industrial revolution is essentially 'nothing technological'. Rational calculation enables the emergence of consumerism. Consumerism dictates the pace, type and distribution of production. Absolutely everything in the modern world is understood as a resource to be utilised on the market, including soil, water, atmosphere, animals, plants and humans. In earlier societies, the seasonal availability of food stuffs and energy defined when things were available and how far they could be traded. Fruit and vegetables were confined to local markets and ripened according to local conditions. But the invention of refrigeration and faster mass transport means that perishable materials can be transported globally and stored almost indefinitely, until consumer demand brings them to the table. The calculation and invention of modernity has shaped our lens of the world, so that now everything is understood as 'in reserve' – waiting for consumer demand to extract resources and put them into production. All elements of the world are understood, Heidegger argues, as a resource, waiting for utilisation – including human beings, who are now characterised as 'human capital'.

Paradoxically, the freedom offered by global markets has effectively alienated communities from their local ecological conditions. So much so, that urban communities are often unaware that there is flooding just north of the city, which will make local produce harder to find and more expensive. The global market obscures the impact of all such perils.

The alienation of communities from their local ecology is built on a dualism that elevates the importance of humanity over and above other species and ecosystems. The definition of one side relies on the disparagement of the other. Dualism proliferate – men versus women, culture versus nature, technology versus wilderness (Merchant 1980) and so on. The production of knowledge itself is alienated from its origin. An object presents an appearance that is experienced by human subjectivity. Through the phenomenological apprehension of these appearances, people make a representation in the mind about what the object might be (Kant 1992).

The cascades of cause and effect are a method for deductive logic to appraise, predict and model future effects. Statistics become incredibly important in the calculation of the truth of a particular appearance in relation to others. Dupay explains Heidegger's critique of technologies of calculation, mastery and alienation as the foundation for governance. Mastery and calculation has evolved from the eighteenth century to the contemporary information society.

> In the metaphysical conception of the world, Heidegger holds, everything that exists is a slave to the purposes of man; everything becomes an object of his will, fashionable as a function of his ends and desires. The value of things depends solely on their capacity to help man realize his essence, which is to achieve mastery over being. It thus becomes clear why technoscience, and cybernetics in particular, may be said to represent the completion of metaphysics. . . . Technoscience, insofar as it constructs mathematical models to better establish its mastery over the causal organization of the world, knows only calculating thought. Cybernetics is precisely that which calculates – computes – in order to govern. (Dupay 2011)

Artificial intelligence (AI) is the ultimate example of calculation and control. Large machine learning sifts gigatonnes of data to create self-learned algorithms that reproduce, model, estimate, target, advertise and amplify answers to human requests. As Heidegger put it in 1954, 'no prophecy is necessary to recognize that the sciences now establishing themselves will soon be determined and steered by the new fundamental science which is called cybernetics' (1973b/1954: 376).

The alienation of communities from local and planetary ecology through the discourse of anthropocentric calculation and mastery has become the modern default mode of understanding the world. It incorporates a future orientation with assumptions about progress and exponential growth. Until very recently, calculation and mastery were exclusively focused on a narrow analysis of costs versus benefits that ignored the impact of industrialism on the environment, on mental health, on communities and on the planetary ecosystem. The calculation and mastery inherent in governance is forged on Newtonian mechanics of cause and effect: a linear cascade of forces acting upon other forces. It ignores more recent and more complex scientific analyses of the 'laws of nature'. The hundred-year-old science of thermodynamics and entropy, and the theory of relativity and quantum physics, do not impact economics or policymaking. Alienation and mastery have largely ignored the 'externalities' of pollution and resource exhaustion. Colonial imperatives utilise the probabilities of risk to exert control and mastery over the future.

Beck (1992/1986) argues that early modern disasters were outside of human control. The forward-thinking constructivist approach to risk calculation is a monetarization and governmentality of risk (Yearly 2001). However, that calculating approach to risk is getting overturned by the extent and frequency of climate catastrophes. Human activity is breaching the predictable, calculable and controllable. With the improvement in technological calculation and mastery, disasters such as nuclear accidents and climate change are increasingly man-made. What is worst is that self-reflection and democratic voices also seem to have little traction in changing the course of monolithic modern capitalism.

PROMETHEAN OPTIMISM AND RISK

In the early Greek myth, the Titan, Prometheus, gifted the forgotten and unprepossessing humans the first gift of technology – that of fire. The association of fire with the emergence of culture is tight knit across all human societies (James 1989). Prometheus was optimistic that humanity deserved this powerful gift from the gods, a gift contested by the god of lightening, Zeus.

During modernity, the optimism of Prometheus about technology confirmed the faith in technological progress. In their neoliberal text, Davidson and Rees-Mogg bind 'Prometheus Unbound' with the 'rise of the sovereign individual' (1997: 7). Promethean technological optimism presumes that technological creativity emerges when market conditions require progress. Neoliberal theory proposes that the perfect market leaves little profit for producers, as rational consumers quickly establish the price point that covers costs but little else. So creative innovation, and new ways of generating efficiency, are necessary to create a higher price point against competition in the market. The neoliberal narrative is that this competitive tension *generates* innovation, creativity and progress. Under the theology of the invisible hand, neoliberalism expects the market to accommodate all conditions, including the pricing of risks. This is the conceptual background to emissions trading schemes.

The trouble is that while competition may function as a progressive force occasionally, more often than not, technology as a means of progress is overblown. Innovative technologies frequently meet a great deal of resistance. New technologies have been quashed by producers with a stronger existing market share, who exert ruthless measures to ensure monopoly. The electric car is a classic example. Invented in 1832 at about the same time as the petrol car and with much the same capacity and scope, electric vehicles (EVs) became practical by the 1870s. In 1900, there was six hundred electric cabs operating in New York City (Wilson 2023). Thirty-five thousand 'Detroit' electric vehicles were built between 1907 and 1939. However, the petrol Model T Ford was just $300 by 1923, and electric vehicles with expensive batteries were up to ten times more costly. Thus far, the theory of market competition holds. Despite cheap and portable oil, EVs kept getting reinvented and, by 1965, consumer advocate Ralph Nader complained in court that petrol companies were buying up and burying new EV patents. In the film *Who Killed the Electric Car* (2006) Paine describes the history of quashing EV development. The film focuses on the production and protests about the destruction of GM's 1990s EV1 electric car. All of which were recalled and destroyed after a successful trial period in California. Market competition did not generate successful electric cars. Petrol oligarchies tried to destroy them. Finally, after three decades of concern about the hazards of climate change, Elon Musk produced Tesla EVs from 2008 which changed the market for electric vehicles.

Promethean optimism assumes that technological solutions to risks will emerge as market opportunities. The sociologist Anthony Giddens admires the ways neoliberalism has digested the risks of mass production and circulation and converted it into an optimistic framework that can be monetarised.

> Traditional cultures didn't have a concept of risk because they didn't need one. Risk isn't the same as hazard or danger. Risk refers to hazards that are actively assessed in relation to future possibilities. It comes into wide usage only in a society that is future oriented – which sees the future precisely as a territory to be conquered or colonised. Risk presumes a society that actively tries to break away from its past - the prime characteristic, indeed, of modern industrial civilization. (Giddens 2002: 22, in Zinn 2016: 3)

Despite Giddens's optimistic framework of a forward-looking analysis of risk as a method of calculating a future-oriented modern society, risk *is* based on hazards. Risk analysis introduces a mode of calculation that makes it possible to profit further from catastrophe. But the short- and long-term psychosocial and physical impact of hazards on ecological and human communities are not alleviated by an improvement in GDP.

DISCIPLINE AND THE RISK SOCIETY

The risk society reveals changes in modernity's approach to discipline and the constitution of subjectivity. In *Discipline and Punishment* (1991/1975), Foucault describes how terrifying but infrequent feudal punishment was performed on a person who had already transgressed. The proof of criminality gave way in the eighteenth century to Bentham's (1995/1787) panopticon. Institutions including the prison, the school, the factory, the hospital, the shopping mall, and even the family, had become continuous surveillance regimes. Constant surveillance encourages people to voluntarily curtail freedom and discipline their own behaviour (Foucault 2010; 1990; Kelly 2015). Discipline anticipates and looks for indicators of potential transgression (nothing need actually have taken place). Deleuze (1990) argues that discipline has evolved further, into a society of control. Where, Deleuze writes, the 'walls of the institutions are breaking down in such a way that their disciplinary logics do not become ineffective but are rather generalized in fluid forms across the social field' (in Hardt 1998: 139). Modernity expects transgressions and monitors for it throughout society. Foucault (1991/1977) shows how societies of discipline put emphasis on the individual, not only as a cypher of market choices but also as a fundamental target of self-discipline for social control. Discipline is both embodied and educational. The individual is the primary element of the Enlightened philosophy of mind.

Risk became part of policy, legislation and financial discourse during the 1980s. Now, community events or endeavours need to have a risk analysis before getting approval. Foucault's (1991/1977) insights into surveillance and Deleuze's (1990) control society are useful analyses. Discipline aims at potential behaviour – not necessarily what has happened but the likelihood of what might happen. Risk analysis attempts to anticipate problems and solicit promises of responsible behaviour. Effectively risk analysis makes a bureaucratic barrier to 'risky behaviour' – to the freedom of children to take school trips, groups to convene in public. Or in the case of the UK's prevent policing policy, to prohibit universities from allowing established protest groups like Amnesty International or Greenpeace on campus. The discourse of risk entails the politics of the surveillance and control society (Foucault 1991/1977; Deleuze 1990; Hardt 1998).

Risk analysis is not simply an attempt to save lives – in fact, that often has little to do with it. Neoliberal theory monetarises everything. The balance of the market can put a price on actions, events and resources. Risk has been calculated and monetarised by the insurance industry and the political classes. They estimate statistical likelihood of disaster and make an educated guess about the profitability of a development or event. Marilyn Waring (1990) alerted readers that risk and GDP as a measure of national economic activity enables the concept of an environmental disaster to be seen as an 'opportunity' to increase economic growth. Perversely, cleaning up the *Exxon Valdez*, a huge oil tanker that ran aground on the West Coast of North America causing thousands of miles of oil pollution, enhanced that year's GDP in Canada and the United States. Disaster capitalism improves the indicators of economic growth. Risks and productivity are all included in the discourse of the market as a universal method. Market universalism incorporates all elements of human behaviour and all potential resources in the (re)distribution of market transactions.

According to classical and neoliberal economic theory, competition between producers encourages technological innovation. Innovation will address any problems with resource depletion, labour shortages and other issues. Lately, as climate change is introduced into market discourse, there have been ongoing calls for innovative 'carbon capture' technologies to address the problem of emissions and the climate crisis. Net-zero policy relies on this Promethean expectation that carbon capture will solve climate change. After some thirty years, technology capable of capturing carbon to any scale remains nonexistent. This is not the first time that the market rhetoric of technological innovation, as a crucial component of market theory, has failed to emerge as the theory indicates.

In both the risk society and the society of fear (Beck 1992), an assumption of technological innovation and mastery prevails. In early modernity, the scale of exponential growth and pollution was still relatively small. By the 1980s, the impact of pollution and risk was prevalent and costly. Risk analysis provides safety measures and standards that aim to lower the chance of accidents. Since then, the acceleration of exponential economic growth and exponential pollution has become so extreme, ignoring it is impossible (Steffen et al. 2015; Ripple et al. 2020; Higgs 2014; Heidegger 1989; 1990; Irwin 2019b). Late modern society is becoming aware of the dissonance between ungovernable but localised accident, global pollution and economic risk. There is greater expertise in modelling scientific statistical evidence about such risks. And there is also more awareness of the total loss of control, the 'unknown unknowns'.

The society of risk remains embedded in Enlightenment assumptions, so well-articulated by Weber (1948/1919). By turning attention to a problem, modern science and technology can find the cause and thus the solution. However, as capitalism has grown exponentially, it has ignored the impacts of resource exhaustion, pollution and ecological devastation throughout. The era of small-scale risk is becoming displaced with planetary-scale risk. Beck (1992) points out that the rational calculations of risk are giving way to greater uncertainty. Legrange and Lotz Siska describe the shift from Weber's optimistic view of the control society to the risk society and, ultimately, the society of fear:

> Ulrich Beck argued that modernity can be broken into two epochs; industrial society (or traditional modernity with its fundamental belief in certitude and progress) and the risk society (where modern industrial stability has broken down at a variety of levels, including the individual subject, precarious, casualisation of labour, financial collapse, and arguably, climate change). In the risk society, certainty no longer governs knowledge, and 'new' science is able to expose the risks and consequences of its own activities. (Legrange and Lotz Siska in Irwin 2010a)

The forward-looking calculation of risks signals a future-oriented modern society for Giddens. This assumes that risk is predictable and can be incorporated into structures of governance. However, risk is based on *increasingly uncontrollable* hazards. In the first instance, risk analysis makes use of information and data to avoid dangers to health and property. In the second case, risk analysis introduces a mode of calculation that makes it possible to profit further from catastrophe. Risk aversion is the first element of calculating a risk analysis, and profiting from disaster is the second phase. The third phase is the ungovernable society of fear.

THE SOVEREIGN INDIVIDUAL

Perhaps the Promethean example, *par excellence*, is the billionaire preppers who combine technological optimism with awareness of dystopian despair. The society of risk (Beck 1992) brought environmental hazards within the sphere of the neoliberal market. The conceptual basis for the widely shared neoliberal ethos of preparing and profiteering from risk is articulated in a book written by James Dale Davidson and William Rees-Mogg (1997) called *The Sovereign Individual: How to Survive and Thrive During the Collapse of the Welfare State*. O'Connell (2018) describes how rich and elite 'Founders' are encouraged 'not simply to evade the apocalypse, but to prosper from it'. Rees-Mogg is a Tory peer in the United Kingdom whose son was one of the key promoters of Brexit. They point to neoliberal New Zealand as the ideal place for prepping for Armageddon 'with its rich resources and clean air, away from the chaos and ecological devastation gripping the rest of the world' (quoted in O'Connell 2018). Foreshadowing the internet's capacity for working at home, during the end times, companies can still operate from anywhere. As international entities, Davidson and Rees-Mogg advocate avoidance of interference from democratic governments. Thus established, they recommend, 'you mined the moon for its ore and other resources, before moving on to colonise Mars'. This strange coalescence of leadership and doomism is showing up already. Sam Alton, the developer of ChatGPT, is 'rapidly building out an economy within Silicon Valley that seems intended to essentially supplant Silicon Valley – a guild of hyper-capitalist entrepreneurs who will help one another fix the broken world' (Friend 2016). Tegmark wrote an entire book on AI and the benign dictatorship of Silicon Valley programmers (Tegmark 2017; Irwin and White 2019). Many preppers, like PayPal partner Peter Thiel, are building safe rooms and luxury bunkers in remote locations. Elon Musk's major passion is SpaceX embarking on the prepper's dream of 'fleeing a dying planet Earth for privately owned colonies on Mars' (O'Connell 2018; Carroll 2013).

Peter Theil, Elon Musk and the rest of the preppers (O'Connell 2018) are exemplars *par excellence* of rational individual utility maximising. These men demonstrate the successful individual to which the rest of us should aspire. Individualism is swept into the system via Adam Smith's narrative of the invisible hand. The invisible hand, a gesture to God's hand which has guided kings for centuries, could guide the market, through the gathering together and balancing of aggregated individual consumer choices (Smith 1904/1776). For Adam Smith, individual freedom that is so important to Christian morality is encompassed by rational choice. Individuals make the most self-serving decisions regarding consumer opportunities, which are behaviourally predictable, making policy to channel behaviour relatively easy. Smith explains that,

> The annual revenue of every society is always precisely equal to the exchangeable value of the whole annual produce of its industry, or rather is precisely the same thing with that exchangeable value. As every individual, therefore, endeavours as much as he can both to employ his capital in the support of domestic industry, and so to direct that industry that its produce may be of the greatest value; every individual necessarily labours to render the annual revenue of the society as great as he can. He generally, indeed, neither intends to promote the public interest, nor knows how much he is promoting it. By preferring the support of domestic to that of foreign industry, he intends only his own security; and by directing that industry in such a manner as its produce may be of the greatest value, he intends only his own gain, and he is in this, as in many

> other cases, led by an invisible hand to promote an end which was no part of his intention. Nor is it always the worse for the society that it was no part of it. By pursuing his own interest he frequently promotes that of the society more effectually than when he really intends to promote it. (Smith 1904/1776: 349)

Massumi (2018: 6) points out the paradox between a rhetoric of fair distribution with the built-in condition of rationality, with 'utility maximising' greed. It is this acceptance, even reverence, for greed that justifies the extraordinary narcissism outlined by Davidson and Rees-Mogg (1997). The narcissism is updated thirty years later in Tegmark's (2017) book *Life 3.0: Being Human in the Age of Artificial Intelligence*. Tegmark's version of the future advocates a small group of computer programmers who quietly instruct a very powerful AI to manipulate democracy and the stock market through bot written newspaper articles and fake news. By siphoning off tiny percentages of the profits this manipulation generates, for public services, Tegmark assumes that the masses will be content with this arrangement (especially as they are unaware of it). The benign dictatorship can continue accelerating the role of cybernetics, machine learning and robotics indefinitely (Irwin and White 2019).

These narratives of Promethean technological optimism are blissfully ignorant of the impact of extractive capitalism on natural habitats. There is no awareness that local communities, and especially indigenous communities, retain a relationship to their territories. Alienation and technological acceleration have been repeated across the planet through successive land clearances. But remnant locals, passionate environmentalists, and the larger frequency and scale of climate disaster events is bringing the travesty of alienation back into view (Heidegger 1977; 1999a; Irwin and White 2019). Tegmark (2017) assumes that controlling massive machine learning is the ultimate calculation and control. His narrow computer science discipline means he has no concept of climate change, macroeconomics, finance and the volatile disproportional discrepancy between the broad economy and financial mechanisms, as outlined in chapter 3. The belief is that machine learning will analyse enough data to have complete control.

The neoliberal narrative of the market still dominates proposals to cure modernity of climate change. Market solutions ought to solve the entropy of mass production and economic growth. Consumer decisions about the quality of the products they consume is assigned the moral carrier for improving emissions. Household consumption should influence the renewable energy decisions of producers. Green consumer choices make sense according to market rhetoric. However, consumerism utterly fails to address the intrinsic infrastructure, both material and financial, that forces individuals to continue high-entropy lifestyles.

Governments insist that urban households remain on the national electricity grid, often provide no or low tariffs to buy back solar energy and remain reliant on national coal supply. Countries such as the United States have transitioned to more renewable energy and reduced their fixed energy emissions. But they still grow an economy that distributes embedded coal emissions in cheap, shoddy plastic goods. China's emissions continue to grow despite building more efficient coal-burning power stations and some renewable generators.

Extractive economics rules the manufacturing of electric vehicles as much as it does conventional petrol ones. Enormous amounts of fossil fuel–powered steel creates the chassis and body of almost all cars. Then there is the plastic, lithium and other elements that make up other car components. We have currently a transitional approach to emissions accounting. The transition to renewable energy is genuinely underway (see chapter 8 on thermodynamics). But Jevons paradox indicates that energy efficiencies are absorbed and over-reached by accelerating energy requirements in the extractive economy. An introductory period to acclimatise to the depth of change is necessary, but it will not be adequate to reduce the accumulation of emissions building up in the atmosphere. The continuous accumulation of emissions are pushing the climate to ever more volatile conditions (Hansen et al. 2023).

As a mode of organising trade and distribution, modernity has the highest entropy market that has ever existed. But up until very recently the impacts of entropy, and specifically waste and greenhouse gas emissions, have been ignored. The neoliberal theory of the perfect market accounts for the costs of production in balance with the distribution of benefits to consumers via the aggregated price point. There is no theory of finance. Up until very recently, there was no acknowledgement of pollution, which has been simply externalised into the global commons. And there is no comprehension of the manipulative influence of advertising on the consumer choices of apparently rational individuals (Irwin 2014). These anthropocentric factors underpin the dissonance between calculation, mastery, risk, hazards and a looming sense of calamity and Armageddon.

THE SOCIETY OF FEAR

In *The Risk Society* (1992/1986), Beck writes that environmental and social jeopardy has been hidden under jargon or ignored. Increasingly, people are embedded in a society of fear. Beck writes, 'It is the increasingly global risks of modernization *per se*', risks that 'induce systematic and often *irreversible* harm, generally remain *invisible*, are based on *causal interpretations*, and thus initially only exist in terms of the (scientific or anti-scientific) *knowledge* about them' (Beck 1992: 22–23, italics in the original).

The society of fear has outstripped the discourse of calculated risks that permeated modernity since the 1980s. Disasters had been localised and relatively small in scale. The climate crisis has seen the scale of turbulence change from focussed ecological epicentres, to large-scale storms that cover an entire continent. Harsh continental-size storms, for example, as some years the weakened polar vortex unleashes intensely cold conditions across Europe and North America.

Higher levels of evaporation and drought are challenging long established ecosystems. I was on an island in Croatia holidaying with my son in 2017 when we noticed the lovely smell of the beautiful old trees, sweltering and sweating sticky, oily sap in the unbearable heat. We forgot to get off the bus and ended up on the coast instead of the walking track at the top of the hill. By the time we returned, the forest had spontaneously combusted. The fire had already swept over the hilltops as far as we could see – all that sticky tinder dry wood in ferociously hot conditions. Forgetting all thoughts of walking, I was very worried the bus would not make it through the road, surrounded on either side by fire. Yet another scale again, was the forest fires in Australia in the summer of 2019–20 that burnt 24.3 million hectares of trees and 3 billion animals (Van Eedon et al. 2020). These kinds of numbers are so horrifyingly high it is impossible to comprehend. This level of existential threat is beyond the capacity of fire rescue, emergency services, local government or insurance companies. Whether or not the clean-up and rebuild stimulates the economy, species that are becoming extinct cannot be re-accommodated into the standing reserve of capitalist resources. They are gone.

Their loss is a tangible grief. If each of us is a rational individual, replaceable by any other faceless rational individual – as a consumer, a worker, a member of a cultural group, or gender identity, or even as a member of the collective – then the point of existence is called into question. Who cares if I exist, if the 'I'-ness is so easily replaceable? Rational subjectivity has been alienated from ecological objects through the processes of modernity and 'freedom'. It feels progressively harder to make sense of human abundance when our mode of inhabiting the planet is so destructive. The silence of forests, the lack of insects and moths, the disappearance of lizards and small mammals. I even miss the flies. It is hard to sit with the knowledge of the disappeared. It creates a deep upwelling of desolation and loneliness. I feel this existential threat compounds a kind of facelessness that operates within neoliberal theory.

Regardless of economic growth, the ability of insurance companies or national governments to accommodate these calamities is disintegrating. In

2020, the largest global forest fire was replaced in Australia by extremely widespread 'once in a hundred year' flooding. The *La Niña* ocean circulation dominated for an unprecedented three years in a row in the Pacific Ocean. In New South Wales (NSW), Australian insurance premiums have gone from $2,000 per year to $14,000 per year in places not yet hit by flooding. Towns such as Lismore, that have had repeated 'hundred-year' floods in the last three years, are now uninsurable – and unsaleable.

The earth is currently, on average, 1.3°C higher than preindustrial temperature (Hansen et al. 2023). Depending on which of the seven-plus greenhouse gases, it takes between twelve and more than fifty thousand years for greenhouse gas emissions to be reabsorbed from the atmosphere (Centre for Sustainable Systems 2022). That means that for the last two hundred years, modern emissions have continued to *accumulate*. For millennia, there was a steady average 280 parts per million volume (ppmv) of CO_{2e} in the atmosphere before industrialisation. In 2023, as I write this, it is 419 ppmv CO_{2e} and the *rate* of accumulation is increasing (Hansen 2023; Noll 2023). By final edits in 2024, it has made an unprecedented jump to 427 ppmv. Kevin Anderson suggested there is a budget of 570 gigatonnes (Gt) for the entire century, if the planet is to remain within the parameters of 1.5°C average over preindustrial temperatures. At current rates of 34 Gts/year (Noll 2023), this budget will be used up by the end of the decade.

Modelling devised by climate scientists and used by the Intergovernmental Panel on Climate Change (IPCC) have consistently underestimated the impact of CO_{2e} from 'business as usual' (BAU) in the speed and depth of ecological changes. Land-based glacier ice melt, Arctic and Antarctic ice melt, sea level rise, deforestation and unforeseen feedbacks are very complex interrelated systems. Feedbacks such as the subarctic thawing of tundra which is releasing methane pingoes are major players but difficult to predict (Serov et al. 2015). There have multiple subtle feedbacks in the intraconnected, symbiotic ecosystem that make modelling simplistic and difficult. The Anthropocene names the shift to an unstable climatic epoch after twelve thousand years of placid Holocene (Crutzen 2002; Irwin 2010b). Multiple tipping points have already been exceeded. In 2005, the ocean's carbon sink, which had absorbed the continuous accumulation of CO_{2e} in the atmosphere, became so warm it started to release CO_{2e} for the first time. Hansen suggests that at 2°C warming, there is a very high probability that sea levels will rise, as has happened in the past at similar temperatures, by six to nine metres (Hansen et al. 2016: 3762). Many of these feedbacks are irreversible and because more and more CO_{2e} is accumulating in the atmosphere, feedbacks, turbulence and calamity continue to escalate (Hansen et al. 2023).

By the end of his life, Beck (1992) saw that climate change is stimulating a metamorphosis in geographic, ecological and sociopolitical organisation.

Climate change has no respect for the boundaries of the nation-state; the political-economic unit that Adam Smith argued was the key unit of modern society in *The Wealth of Nations* (1904/1776). Countries that have contributed very little carbon emissions, such as Bhutan, Bangladesh, Pakistan or the Pacific Islands, are at the frontline of coping with climate impacts. The melting of mountain permanent ice caps is jeopardising clean water supply. Storms, flooding and sea level rise are beginning to displace millions of people. UNICEF (United Nations International Children's Emergency Fund 2023) just released the first study on climate displacement and regards 135 million people a very low estimate. The new geological vulnerability of coastal property – where the bulk of human population reside – is beginning to reshape the viability of whole cities and regions, with a cascade of consequences on inland areas having to accommodate climate refugees. Meanwhile, the aging and falling global birth rates are changing the status of young immigrants, who are needed to prop up the workforce in developed nations. Erstwhile normative power dynamics during the modern era are in a period of dynamic change.

> Change brings a characteristic future of modernity into focus, namely permanent transformation, while basic concepts and the certainties that support them remain constant. Metamorphosis, by contrast, destabilizes these certainties of modern society. It shifts the focus to 'being in the world' and 'seeing the world', to events and processes which are unintended, which generally go unnoticed, which prevail beyond the domains of politics and democracy as side effects of radical technical and economic modernization. They trigger a fundamental shock, a sea change which explodes the anthropological constants of our previous existence and understanding of the world. Metamorphosis in this sense means simply that what was unthinkable yesterday is real and possible today. (Beck 2016)

Part II

TIME AND TECHNOLOGY

Chapter 5

Alienation and the Simulacra

The *driver* of continuous greenhouse gas emissions is exponential economic growth. The *cause* of CO_{2e} emissions is more obviously fossil fuels. Alienation has enabled the schism between economic growth and finite planetary conditions to emerge. The current disparity between ex nihilo–generated nominal money is 250% more than the realities of fiat exchange in actual goods and services – and continuously increasing (Gaspar et al. 2023). Derivatives trading is even more out of kilter with the real world – there are 120 trillion dollars' worth of nominal derivatives and 20.7 trillion of actual goods they are traded against (BIS 2023). That schism between the 'nominal' and reality is only possible because of the alienation of culture from nature. This chapter explores the roots of philosophical alienation and how it results in the belief that we are living in a simulacrum of reality.

Modern economic theory is built on an aggregation of rational individual behaviour. Assumptions about behaviour are based on Enlightenment philosophy of subjectivity. To fully understand the emergence of alienation, it is necessary to take a deep dive into the philosophy of two seminal contributors – briefly of Descartes (1980/1641) and, with more depth, Kant (1992/1781). What emerges is an interesting and important focus on scepticism. Scepticism is not the only focus of their philosophy but, like rationality, assumes the character of human essentialism. The process of sceptical self-reflection interjects a conceptual gap between subject and the object under consideration. That sceptical gap is the key moment of alienation between subject and object.

> I suppose therefore that all things I see are illusions; I believe that nothing has ever existed of everything my lying memory tells me. I think I have no senses. I believe that body, shape, extension, motion, location are functions. What is

there then that can be taken as true? Perhaps only this one thing, that nothing at all is certain. (Descartes 1980)

Both rationality and scepticism are incredibly useful characteristics of humanity. However, the escalation of these two key factors to transcendental proportions – as occurs in Kant – lay the foundations for the imposition of human mastery. They also forge a contrived approach to truth and knowledge claims. The interaction between an object or event, and its meaning and linguistic representation, are increasingly tenuous. Truth merely holds explanatory power as a sort of foible of the human mind. Nietzsche (1989) writes that the human mind needs to impose a sense of regularity and predictability over random events, even when there is no permanence, relationship or regularity between them. Taken to its extreme, this type of conceptual imaginary can be completely open, with no ties to the empirical world at all. It requires no connection to actual events, objects or what is commonly understood as 'reality'.

Unpacking Kant (1992/1781) gives a new understanding for the method and aims of 'constructivism'. Kant also gives light to the narrative, popular amongst the information technology (IT) fraternity, that this world is a simulacrum, generated by sophisticated artificial intelligence (AI). Both cases are a more extreme version of the normative alienation built into regular economic assumptions. Economics has a tendency to inventive, abstract 'nominalism' which I will argue emerges from Kant's concept of the transcendental aesthetic. Alienation is only possible because of a constructed belief in the world as a 'transcendental' nominal simulacra. Pollution, ecological destruction, social alienation, local land clearances, resource exhaustion and climate change are disregarded because they do not fit the hypothesis of the world as progressive. They are inadvertent side effects of alienation – externalised and forgotten – that is built into social and economic theory. This level of alienation is important to understand because (despite scopes 1, 2 and 3) it continues to inform extractive business practices today. Just as it informed colonialism a few hundred years ago.

Alienation combines with the paradoxical tension of mastery, risk and the society of fear examined by Beck (1992). Alienation is a key component of the continuation of economic growth. They constitute the architecture of the visions of the future in IT and the Silicon Valley multibillionaires that control such vast proportions of the earth's wealth. Alienation and transcendental constructivism are key factors that conceptually support the edifice of nominal credit.

Heidegger and Baudrillard (1994) both offer important insights into the feeling of stuckness that permeates the modern worldview. Ideas from 250-year-old Enlightenment philosophy and science have informed neoliberal market conceptual underpinnings. These normative assumptions saturate policy and business models. Recognition of contemporary science, that emerged one hundred years ago, offers remarkably different concepts of natural laws. These are the 'new' laws of thermodynamics, gravity, relativity and quantum mechanics. These ideas extrapolate upon the Copernican Revolution by decentring the human subject. Inadvertently, contemporary science reinvigorates the legitimacy of indigenous understanding of the intraconnection of all things. Relationships rather than alienation needs to permeate the economics of the future.

The mechanical universe was a controversial concept in the Christian medieval context. By the time Kant was writing in the late 1700s, it had become the normative paradigm. Forces acting upon other forces, or cause and effect, was the latest, cutting-edge science when Kant wrote his book *The Critique of Pure Reason* (1992/1781). Kant found the determinism of the mechanistic universe both compelling and overbearing. He worried that individual freedom would be subsumed in the cascades of cause and effect. Building on Descartes's sceptical individual, Kant refashions the free will of transcendental individualism, despite the determinism of cause and effect. In doing so, he elevates the Christian view of individual free will to an essential cypher of philosophy. Later, this view is promoted via Adam Smith (1904/1776), as an essential cypher of economics.

CAUSE AND EFFECT

The alienation of ecological nature is justified philosophically through the discourse of idealist scepticism. Objects cannot be understood intrinsically but are represented in the mind based on phenomenological perception of appearance. Deductive logic is informed by the appearance of empirical evidence. Logic enables the categorization of an object and forges the basis of rationality. This alienation between subject and object creates a normative basis for economics. Consumer decisions and production processes can ignore the environment. Econometrics about capital, labour, transport, insurance, energy and so on is 'balanced' against consumer benefits. But the costs of energy pollution and waste are 'externalised' from the calculations. The impact accumulating in the environment is ignored. Rapid, intense fossil fuel combustion produces entropy in the form of carbon dioxide, methane, sulphur dioxide, nitrous oxide and a few other greenhouse gases. According to the second principle of thermodynamics, these externalities are now impacting the planet. The entropy generates climate change and immense plastic pollution (another by-product of fossil fuel). We are currently in one of the most rapid extinction events since the dinosaurs (WWF 2022; IPBES 2019; Goulson 2019; Tong et al. 2019). In a dyad between Hell and

Heaven, Armageddon and progress are tightly tied in the worlding of the Enlightenment.

As fossil fuels became better understood, chemists began to develop the theory of thermodynamics. The ideas on chemistry culminated in Clausius (1856) who articulated the two laws: (1) nothing is ever lost and (2) for every increase in order there is also an increase in disorder (Xue and Guo 2019). Thermodynamics (like relativity and quantum physics) are very different from Newton's physics. They are more complex and diverse than forces acting upon other forces with a linear logic of cause and effect.

Cause and effect, forces acting upon other forces or the 'series of alterations', as Kant describes it, are crucial to understand the logic of the modern world. The principle of calculation (Weber 1948) and rationality underpins the Enlightenment. However, in this mechanical universe, the determination of destiny unfolds with its own logic. Even free will can be understood as caught up in this series of alterations. Kant was particularly concerned about the lack of room for freedom, especially given the need for free moral choices in the Christian framework of good and evil. If even evil doings can be explained by cause and effect, then humanity is no more significant than a rock or a tree. Sceptical idealism reaches back to the ancient Greek ideas about epistemology and introduces the importance of doubt into regimes of truth (Descartes 1980/1641). Scepticism can be applied to empirical relations of cause and effect by composing a 'meta' analysis of truth claims.

PLATO AND APPEARANCE

In Plato's cave (427 to 347 BC/1965), prisoners are chained with their back to the cave entrance and can see only the shadows playing on the cave wall. When one of the prisoners is released and turns towards the light, they are blinded by the direct sunlight and find the experience of direct truth impossible to contemplate. Plato's allegory of the cave is that humanity functions best on the representation of truth rather than the direct experience of it. Those few that even begin to directly apprehend the actual are vulnerable to violent blindness. They also suffer an emotional and intellectual falling away of all conventionally understood 'ground' that they have in common with other ordinary people in society (who continue to be imprisoned by that conventional representation). Plato makes the important point that narrative concepts 'world' a community and its understanding of things, and that our interpretation of truth is often based on representations.

Plato drafted the basis of sceptical epistemology, later taken up in extreme ways by Descartes and Kant. The methodological process of reaching a representational norm is carefully considered by Baudrillard (1994).

BAUDRILLARD AND THE SIMULACRA

Baudrillard's book *Simulacra and Simulations* (1994/1981) is a meta-analysis of Western epistemology as a process of mimesis and simulacra. Baudrillard explains a methodological evolution that enables the legitimacy of representations to progress from the copy to the hyperreal. He begins his book with an interesting quote from Ecclesiastes, 'the truth which conceals that there is none' (1994: 1). Baudrillard argues that there is a progressive course of alienation in the fundamental principles of philosophy. It starts with mimicry of the real, and turns towards the simulacra itself as a concept of reality. As alienation becomes more complete, the simulacra eventually become its own 'hyperreality', replacing entirely any connection to the real. Humphreys et al. (2022) describe Baudrillard's startling argument of the hyperreal, 'the simulacrum could replace reality to such an extent that it became the truth in and of itself, completely divorced from any concretizable anchors' (Humphreys et al. 2022).

Baudrillard (1994/1981: 6) describes the steps of progressively alienated representation. It undergoes advancing stages that end up with the simulacrum itself standing in as the hyperreal.

1. It is the reflection of a basic reality.
2. It masks and perverts a basic reality.
3. It masks the absence of a basic reality.
4. It bears no relation to any reality whatever: it is its own pure simulacrum.

Initially, truth mimics the real, as a reflection of empirical reality. Next, is the value of the counterfeit, which remembers and perverts the real. Importantly, this second stage has a focus on the material conduit of the truth, such as the medium of language or the medium of paint, or the medium of time, of space or of the mind itself. The third stage of the counterfeit is a dissimulation where reality is concentrated in the representation, shifting emphasis to the subject's interpretation rather than the real. In the medium of language, for example, the speech acts that gesture towards a correlation between an object and its name get coded and over-coded in the interactions of meanings across the system of signs. The language, especially in writing, refers to nuances in meaning with reference to other meaning-signs, not in reference to an object itself. As the validation of the simulacra progresses, it reorients the truth from the real to the subjective experience, within a system that connects a society of individuals, not in relation to reality or intraconnection with each other but via the medium of a system of signs (de Saussure 1966). Finally, the world is constituted by the hyperreal.

Baudrillard makes an important analysis of the movement from mimicry to simulacra. The simulacra are not merely a copy. Simulation involves parody, perversion or the loosening of reference to reality; it involves *dissimulation*.

> To dissimulate is to feign not to have what one has. To simulate is to feign to have what one hasn't. One implies a presence, the other an absence. But the matter is more complicated, since to simulate is not simply to feign: 'Someone who feigns an illness can simply go to bed and pretend he is ill. Someone who simulates an illness produces in himself some of the symptoms' (Littre). Thus, feigning or dissimulating leaves the reality principle intact: the difference is always clear, it is only masked; whereas simulation threatens the difference between 'true' and 'false'. (Baudrillard 1994: 366)

Some slippage from representation to its real correlate takes place in Baudrillard's third stage, gradually legitimising the 'hyperreal'.

> The material thing becomes the concept and the name is the sound image; these both together make up the sign. The concept is the signified and the sound image is the signifier. Language precedes the material spoken word and it exists in our consciousness. We understand 'to precede' as not to mean to exist prior in time, but prior in constitution. Through language we start to see the spread of simulacra through the loss of the material referent and its replacement with the concept. The connection between land and language is lost and language becomes self-referential. (In Humphreys et al. 2022)

Baudrillard explains this passage from the real, to the copy, and then to the medium of language as a progressive deterioration of integration with eco-reality.

> In this passage to a space whose curvature is no longer that of the real, nor of truth, the age of simulation thus begins with a liquidation of all referentials – worse: by their artificial resurrection in systems of signs, which are a more ductile material than meaning, in that they lend themselves to all systems of equivalence, all binary oppositions and all combinatory algebra. It is no longer a question of imitation, nor of reduplication, nor even of parody. It is rather a question of substituting signs of the real for the real itself; that is, an operation to deter every real process by its operational double, a metastable, programmatic, perfect descriptive machine which provides all the signs of the real and short-circuits all its vicissitudes. (Baudrillard 1994: 366)

The real doubles back from its association with an actual object, through the interconnected system of signs, to the self-referential interpretation of the subject, almost entirely dislocated from reality at all. For this shift from the medium of truth as a system of signs, to the hyperreal, Humphreys et al.

(2022) use the example of the King James Bible. It is written in the common vernacular, emphasising each personal relationship with God. That contrasts with the whole of Being, or universal, as the Creator God of the Latin text in the Catholic Church. In the penultimate stage of the hyperreal, Baudrillard claims that objects 'have no meaning beyond the dimensions of the industrial simulacrum' (Baudrillard 2001: 140, in Humphreys et al. 2022).

These claims are audacious, and sound implausible. Closer examination of Kant's *Transcendental Aesthetic* (1992/1781) shows how the philosophical justification of alienation from reality – even reality understood simplistically as cause and effect – has taken place. Kant tracks through all four stages. He firmly establishes the phenomenological experience of a real thing but reaffirms its personal interpretation. In Baudrillard's second stage which focuses on the *medium* of reality, Kant utilises the medium of logic, time and space to mediate representations from subjective to universal and, thus, systematic. Then comes Baudrillard's final two stages of dissimulation and hyperreality. Kant perverts the relationship of the correlation between the thing in itself and subjective experience with the noumenon – a hyperreality that is independent of cause and effect. This is how he creates a space for individual free will.

SCEPTICAL IDEALISM

Slowing down and carefully examining how these four phases from representation of real objects to full blown hyperreality is necessary to fully appreciate what is happening in modernity at large.

At first glance, Descartes's commitment to deductive reason and universal truth is polar opposite to the mimicry and manipulated simulacra described by Baudrillard (1994). Descartes is very anxious to avoid the false testimony of his body or the inadequate normative judgement of his peers. So, he sets out a methodology of sceptical doubt that informs rationality to ascertain the indisputable 'truth' about objects in the real world. Descartes framed this dislocation of truth from reality in terms of the struggle between God and the devil. It results in his sceptical rejection of bodily experience, and complete reliance on rational deduction.

> I will regard the heavens, the air, the earth, colors, shapes, sounds, and all external things as nothing but the bedeviling hoaxes of my dreams, with which [the devil] lays snares for my credulity. I will regard myself as not having hands, or eyes, or flesh, or blood, or any senses, but as nevertheless falsely believing that I possess all these things. I will remain resolute and steadfast in this meditation, and even if it is not within my power to know anything true, it certainly is within my power to take care resolutely to withhold my assent

> to what is false, lest this deceiver, however powerful, however clever he may be, have any effect on me. (1980: 62–63)

Descartes carefully separates the subject from the object under observation. Even the medium of his own body is apprehended sceptically, as insufficient empirical evidence of an object. As nothing about the appearance of the object can be taken for granted, Descartes makes the case for rational deduction to ensure the object is what it appears.

For Descartes (1980/1641), the universal categorisation that classifies objects through the medium of deductive logic enables each rational individual to reach the same conclusion. This creates a replicable system of rationality that erects the representation of the appearance of an object as its truth. Descartes background as a mathematician informed his approach. Grünbein (2010) describes Descartes explanation of poetry, in letters to the Countess Palatine in 1649 in mathematical terms. Descartes brings all intellectual capacities into line with the geometry of reason, including intuition and poetics. Grünbein explains how Descartes characterises poetic inclination, 'from the geometry of conic sections, to the planetary orbits of the astronomers, to the geometer's art of measurement made possible by the earth's magnetism and the deviations of the quivering compass needle from the horizontal, in a word: *inclination*' (2010).

Kant is aware of the criticism of Descartes's caricature of scepticism. If dreaming, or the devil, might intervene in the interpretation of an object, Descartes makes the alienation of subject from object so complete that empirical reality might as well not exist. Kant redressed this alienation by arguing that the rational mind needs something to work on. The mind needs access to an appearance of something to begin to speculate about its essential characteristics. Empirical reality is not completely lost, but it is virtually inaccessible. The mind can only comprehend the appearance through phenomenological experience, and this is irreducibly interpreted subjectively by the individual. Empirical evidence gives only sketchy insight into the inner workings of the thing itself.

Kant judiciously advances the concept of representation, from Descartes's scepticism towards truth, to the medium of comprehension. The medium of comprehension maintains a connection with empirical reality, which is not as abrupt as Descartes's rational focus on the representation as truth itself. For Kant, the representation is always personal and subjective, and what is closer to truth is the medium of comprehension itself. Objectivity is achieved through rational analysis of representations of the thing in itself. The medium of by which knowledge is possible are the a priori conditions of time, space and rational subjectivity. All three are necessary to comprehend any phenomenological or empirical information about a thing.

At the same time, Kant is highly concerned that the Newtonian mechanical universe is too determinist. He worries that free will is lost in the destiny of forces acting upon other forces. Free will being an essential component of both Christian accounts of good and evil and, also, the free choice of consumers in the neoliberal marketplace, where individual rational decisions aggregate according to the invisible hand (Smith 1904/1776). Free will is essential to create the perfect market. Individual rational choice is aggregated utility maximising decisions inadvertently lower producer prices to the lowest possible and enable fair distribution. Without the free choice of rational individuals, market theory makes no sense. To rescue the meaningfulness of free will, Kant insists on separating empirical nature with its 'series of alterations' from an abstract essential reality that is a kind of prime mover of those empirical natural things, the thing in itself and the noumenon. What follows is a deeper dive into how Kant arrives at a position that validates the subjective and manipulates yet further the real into the realm of the simulacra and the hyperreal.

KANT

The Critique of Pure Reason (1992/1781) is a meticulous examination of how real things impart phenomenological sensations on an individual's senses. Real things stimulate an 'intuition' of the 'appearance' of an object or thing. This is akin to the first stage of Baudrillard's (1994) simulacra – the appearance mimics the real object but becomes 'pure' representation. Building on the connection to real things, Kant makes the second and third element of Baudrillard's process of simulacra. Kant masks reality with a focus on the a priori conditions of time, space and subjective mind. Then, with a duplicitous and perverse shift to the universal thing-in-itself, he masks an entire absence of reality, completing the task of hyperreality with the realm of the noumenal.

Kant's famous text, *The Critique of Pure Reason* (1781, revised 1787), is where the argument for alienation is made. It is a key component of shifting attention from the correlation between representations (or signs) of the 'real' to the second phase of Baudrillard's process of simulacra; the medium through which interpretation is understood. Kant argues that subjectivity, time and space are three separate but related a priori (prior) conditions – a medium necessary for comprehending the appearance of something. The three a priori conditions are the most real part of Kant's method. '(I)t is proved that space and time are only forms of sensible intuition, and therefore only conditions of the existence of the things as appearances' (Kant 1992: 115).

Kant is not oblivious to developments in science. He claims the a priori is his philosophical 'Copernican Revolution'. Copernicus, who wrote

Six Books Concerning the Revolutions of the Heavenly Orbs (1543), had upturned the flat earth model and made the case for a heliocentric, or sun-centred system, that the earth orbits. Copernicus explained day and night is not the movement of the sun but rather the earth rotating around its own axis once per day. Later, Kepler wrote *Astronomia Nova* (1609) that made the solar centrist argument more acceptable to Christianity by arguing the revolution of the earth around the sun is a result of God's will. Time and space could be associated with the physics of the mechanical universe. By the time Kant published his text in 1781, Isaac Newton's book *Principia* (1687), which formulated the theory of universal gravity, had already well established the 'mechanical universe' where forces act upon other forces in a cascade of cause and effect.

Kant accepted the logic of God's determinism. But the mechanical universe signifies a problem, for in a fatalistic world, free will is not possible. Ironically, it is the logic of cause and effect that both confirms the perfection and logos of the God-given universe but, paradoxically, threatens free will. Kant is wary that the determinism of forces acting upon other forces discovered by Copernicus, Kepler, Galileo, Newton and others will destroy the capacity for human free will.

Firstly, Kant establishes that real things are needed for the phenomenological apprehension of the appearance of a thing. He carefully elucidates the material conditions needed a priori for those truths to be ascertained. Kant goes on to argue that the appearance of sensible intuitions offers the stimulus for the rational individual to deduce an entirely abstract, universal 'thing in itself' that is beyond the empirical world. This 'pure intuition' is the introduction of the simulacra. The dualist separation between the thing in itself as an essential abstract and universal category from the everyday experience of empirical things is necessary, according to Kant. Otherwise humanity will be just another cog in the machinery of the Newtonian universe.

> I call all representations pure (in the transcendental sense) in which nothing is to be encountered that belongs to sensation. Accordingly the pure form of sensible intuitions in general is to be encountered in the mind *a priori*, wherein all of the manifold of appearances is intuited in certain relations. This pure form of sensibility itself is also called pure intuition. (1992: 156)

Refocusing on the individual subject confirms the Cartesian sceptical individual and deductive logic. Or in Kant's case, augmented by sensible intuitions, to a self-referencing system that revolves around a 'pure' intuition. By positioning human subjectivity as the centre of pure intuition, Kant makes an individual a prime mover (much like God) that is not swept up in the fatalism of empirical cause and effect.

Pure intuition is the basis for Kant's 'mature' transcendental aesthetic. Descartes was ferociously critical of the evidence of his own senses (as dreams, tricks, demonic illusions). Kant presents his scepticism in a mode both more mundane and, ultimately, in even deeper alienation from the real. Objectivity is difficult to discern when phenomenological experience is necessarily personal. Personal intuition of objective phenomena show up in necessarily subjective ways making scepticism of phenomenology necessary. Each person experiences empirical phenomena differently. 'Things like colors, taste, etc., are correctly considered not as qualities of things but as mere alterations of our subject, which can even be different in different people' (1992: 161). Because of these subjective variations between people, understanding the universal essence, a thing in itself, is not attainable through any empirical path. The thing in itself can only be conceptualised through pure intuition or rational deduction. There is no observed appearance of a thing in itself because it does not exist in the empirical realm of cause and effect. The thing in itself is a simulacrum of real things.

> For in this case that which is originally itself only appearance, e.g., a rose, counts in an empirical sense as a thing in itself, which yet can appear different to every eye in regard to color. The transcendental concept of appearances in space, on the contrary, is a critical reminder that absolutely nothing that is intuited in space is a thing in itself, and that space is not a form that is proper to anything in itself, but rather that objects in themselves are not known to us at all, and that what we call outer objects are nothing other than mere representations of our sensibility, whose form is space, but whose true correlate, i.e., the thing in itself, is not and cannot be cognized through them, but is also never asked after in experience. (Kant 1992: 161–62)

This is the route Kant takes to invert empirical objects and events and subvert them into subjective constructs. The a priori conditions of subjectivity and space are necessary to apprehend things. But the real is located in the a priori, not the thing – or even its representation – making possible a hyperreality of the essential thing 'in itself'.

ARISTOTLE, BEING AND BEINGS

Kant did not arrive at this pure reason without precedence. Aristotle's influence over eighteenth-century European thought was immense. In *The Metaphysics* (Aristotle 350 BC/1972), Aristotle argues that the essence of a species is only known over the entirety of its existence, across the totality of historical time and places. Long before the discovery of DNA, Aristotle (384–322 BC/1972) argues that each individual example of a species is vulnerable to historical accidents. These accidents damage the potential held in the essence of the seed or infant. Thus, the category of 'oak tree' might be expected to spread its branches evenly as it grows up but historical accidents such as lightning strike could take out a chunk of the tree and the scar would obstruct an observer of this particular tree from understanding the perfect essence of all oaks. Only from the universal standpoint encompassing all locations across the entire existence of oaks, after collating every single historical example, can the universal essence of an oak be truly understood. The universal *transcends* the pattern of multiple individual examples, staying firmly secured to *all* real cases but not bound by the real events that impact any one of them.

Kant, in *extremis*, wants to sever the universal from the everyday. The universal noumenon is unknowable from an empirical standpoint. Aristotle (384–322 BC/1972) does not really explore incomplete knowledge of the universal essence of a category, but it would be plausible that over a long period someone has multiple (statistical) examples and thus a partially complete (and more 'objective') understanding of the essence of oak. For Kant, the universal essence, or 'pure concept of the understanding', is completely unknowable for human subjectivity, other than as a 'boundary condition' for human intuition. He explains that the only way to arrive at 'pure concepts of the understanding' is through deductions 'which apply to objects of intuition in general a priori, as there were logical functions of all possible judgments'. The thing in itself is not knowable empirically but can be transcendentally deduced via the conduits of logic, time and space. The a priori medium of truth presents a boundary condition, 'for the understanding is completely exhausted and its capacity entirely measured by these functions. Following Aristotle we will call these concepts categories, for our aim is basically identical with his although very distant from it in execution' (Kant 1992: 212).

Aristotle (384–322 BC/1972) regarded each historical iteration of a thing as participating in the universal essence. Each one is vulnerable to accidents which impinge on the perfect expression of its categorical essence. Kant has avoided the participation of each thing in the universal category and turned towards the a priori as the carrier of what is real. The intuition of a thing is bound to haecceity, but the representation (carried by the a priori of subjective rationality) is elevated through logic into a signifier of categorical truth: the thing itself.

BEYOND REALITY: THE NOUMENON

If that sounds to you like a lot of complex words with poor definition, I agree. It is difficult to explain Kant's method, because despite his fame as a master of reason, his methodology scarcely makes sense. Initially, the appearance has to have some correlation with the real thing. But almost immediately, the really real thing is in the realm of the noumenon, outside of the a priori conditions of reason, time or space and outside and independent of sensibility.

> This was the result of the entire Transcendental Aesthetic, and it also follows naturally from the concept of an appearance in general that something must correspond to it which is not in itself appearance, for appearance can be nothing for itself and outside of our kind of representation; thus, if there is not to be a constant circle, the word 'appearance' must already indicate a relation to something the immediate representation of which is, to be sure, sensible, but which in itself, without this constitution of our sensibility (on which the form of our intuition is grounded), must be something, i.e., an object independent of sensibility. (Kant 1992: 348)

It is difficult to understand why Kant is so interested to make such a complicated argument, unless we remember that he is struggling against being trapped in the interrelated, determinist mechanical universe of cause and effect. It turns out that individuals need not know much about these hyperreal universal categories. Just knowing they exist outside the cause and effect of time and space is enough. Kant makes a case for the noumenal as a 'boundary condition' (1992: 350) where the subject is aware of the abstract thing in itself through deductive logic and 'pure understanding' from the appearances of sensible objects. 'Now from this arises the concept of a noumenon, which, however, is not at all positive and does not signify a determinate cognition of any sort of thing, but rather only the thinking of something in general, in which I abstract from all form of sensible intuition' (Kant 1992: 348–49)

It is enough to surmise the noumenon exist. Making a case for the role of the noumenon enables him to make a parallel case for the free will of the subject, a mover and shaker in their subjective interpretation of the world. The pure understanding enables the individual to be free, and not merely a pawn in the determinism of the mechanical universe.

Kant's gap between the intuition of an appearance of a real thing through the a priori of time, space and subjectivity is the first stage of alienation. The transcendental aesthetic, as a rationally deduced representation of a thing in itself, is the second stage of alienation. The process makes most sense through Baudrillard's synopsis. Baudrillard argues that the progress of the simulacra goes from straight forward mimicry, that addresses the real, with a subsequent turn towards the materials of the message, that then creates a type of counterfeit or dissimulation of the real into the hyperreal simulacra (Baudrillard 1994: 366).

> The concept of a noumenon, i.e., of a thing that is not to be thought of as an object of the senses but rather as a thing in itself (solely through a pure understanding), is not at all contradictory; for one cannot assert of sensibility that it is the only possible kind of intuition. Further, this concept is necessary in order not to extend sensible intuition to things in themselves, and thus to limit the objective validity of sensible cognition (for the other things, to which sensibility does not reach, are called noumena just in order to indicate that those cognitions cannot extend their domain to everything that the understanding thinks). In the end, however, we have no insight into the possibility of such noumena, and the domain outside of the sphere of appearances is empty (for us). (Kant 1992: 350)

This is the path that brings the alienation of the real to its hyperreal culmination. Concepts are not limited to empirical sensibility. The real object somehow correlates to the appearance of things. Then, through a counterfeit process of the transcendental logic, the truth is attributed to an a-historical thing in itself. Like Aristotle (384–322 BC/1972) – but unlike him – Kant produces universal categories. For Kant, the realm of cause and effect is severed between the universal category and its empirical iteration and replaced with the hyperreal which now bears no relation to real things.

A SERIES OF ALTERATIONS AND AN UNMOVED MOVER

The question of Kant's need to create a two-tiered conceptual apparatus – empirical and transcendental, emerges from his attempt to synthesize or reconcile Descartes, Christian theology, Aristotelean universal categories and the physics of Copernicus and Newton that had arrived at a mechanical universe of cause and effect.

> The world of sense, as the whole of all appearances, at the same time contains a series of alterations. For without these, even the temporal series, as a condition of the possibility of the world of sense, would not be given to us. Every alteration, however, stands under its condition, which precedes it in time, and under which it is necessary. (Kant 1992: 490)

The world of sense refers to the philosophy of science, and especially the mechanical universe, which emerges from the mathematics of Newton (1687). Newton's work is in turn based on the insights of Kepler (1995/1621), Galileo (1632) and Copernicus (1531), who realise that the earth is not the centre of the universe, as has been normative across many millennia. The decentring of the earth met with a lot of resistance, particularly in religious dogma. Copernicus avoided Church censure by publishing on his deathbed, but Galileo spent his final decades under Ecclesial house arrest for side-lining the role of God and the Seven Days of Creation of the Earth (Leveillee 2011). Evolution was not originally included in the explication of God's creation.

The resolution of the conundrum of a decentred universe has been to emphasise the first origin of cause and effect. God can be the prime mover that initiates the universe, whether in the Garden of Eden or the Big Bang. The theological argument is that there needs to be an origin, with a supreme being who initiates the world. Kant makes the case for humanity to reflect God's supreme free will, in contrast to an object caught in the mechanical universe conditioned by forces acting upon other forces.

> Now every conditioned that is given presupposes, in respect of its existence, a complete series of conditions up to the unconditioned, which alone is absolutely necessary. Thus there must exist something absolutely necessary, if an alteration exists as its consequence. This necessary being itself, however, belongs to the world of sense. For supposing it is outside it, then the series of alterations in the world would derive from it, without this necessary cause itself belonging to the world of sense. Now this is impossible. For since the beginning of a time-series can be determined only through what precedes it in time, the supreme condition of the beginning of a series of changes must exist in the time when the series was not yet (for the beginning is an existence, preceded by a time in which the thing that begins still was not). (Kant 1992: 490)

During the Protestant Reformation, God was repositioned from his role as the universal figure of everything – the universal prime mover, to a more personal God intimately invested in the goings-on, in the life of each supplicant. Kant makes a similar step. To achieve this remarkable transformation, the relationship of the a priori is fundamentally necessary. Time, subjectivity and space function as the medium through which all apprehension of empirical series of alterations can occur. But the transcendental individual can intuitively apprehend pure understanding of the noumenon. With this dissimulation, Kant positions each individual subject in a very privileged position, almost as a prime mover themselves.

The subject could now be in the position to initiate a series of alterations in the real world of cause and effect. But this, too, is troublesome because it would create anarchy in the evolution of the empirical world of objects, as multiple free decisions to intervene would disrupt the universal laws set in place by God. Kant concludes that individual subjectivity might achieve freedom, but it will not be in the world of cause and effect; rather, it will be in the realm of the noumenon.

> Moreover, even if a transcendental faculty of freedom is conceded in order to begin alterations in the world, then this faculty would in any case have to be

> outside the world (although it always remains a bold presumption to assume an object outside the sum total of all possible intuitions, which cannot be given in any possible perception). Yet it can never be permitted to ascribe such a faculty to substances in the world itself, because then the connection of appearances necessarily determining one another in accordance with universal laws, which one calls nature, and with it the mark of empirical truth, which distinguishes experience from dreaming, would largely disappear. For alongside such a lawless faculty of freedom, nature could hardly be thought any longer, because the laws of the latter would be ceaselessly modified by the former, and this would render the play of appearances, which in accordance with mere nature would be regular and uniform, confused and disconnected. (Kant 1992: 489)

Thus, Kant is more circumspect in his proposition of the individual's capacity for free will. The trio of conditions that form the a priori, dissimulate the real from the empirical to the noumenal, and the noumenal over and above the empirical. The transcendental aesthetic takes the empirical objects in space and funnels it through subjective time, and rationality, to emerge as the noumenal free will. In this way, the priori conduits of time, space and subjectivity are more significant than the series of alterations that make up the fundamental laws of nature. This 'transcendental aesthetic' is what Kant calls his Copernican Revolution.

KANT'S A PRIORI AND COPERNICUS'S HELIOSYSTEM

To achieve the status of free will as outlined above, Kant made the case for the a priori – time as a subjective, inner experience. 'Time is nothing other than the form of inner sense' (Kant 1992: 163). Kant argues rather than an independent historical 'river' (Heraclitus 540–480 BC/2001) that each individual is thrown into, time is an 'intuition of our self and our inner state'. This inversion of time from outward historical flow that is like a container to subjective experience is the philosophical inspiration that underlies Kant's 'Copernican Revolution'. Kant goes on, 'time cannot be a determination of outer appearances; it belongs neither to a shape or a position, etc., but on the contrary determines the relation of representations in our inner state' (Kant 1992: 163).

For Kant (1992/1781), time is even more closely tied to subjectivity than space, which is the other a priori vehicle for natural objects to appear. Instead of the flow of the series of alterations, time is subjective. This is a dissimulation of pure transcendental understanding. It removes time from the determinism of the series of alterations that form the Newtonian laws of nature. It helps to situate the free will of subjective reasoning in the realm of the hyperreal noumenon (a universal realm, rather than historical).

Subjective time forges what Kant regards as his philosophical 'Copernican Revolution'.

Kant claims his transcendental aesthetic is revolutionary, comparing his 'anthropocentric procedure in philosophy and Copernicus's heliocentric revolution in astronomy' (1992: n715). I am not as enthusiastic. On a positive note, Kant's inspirational characterisation of a priori time, space and subjectivity remains a fertile foundation for considering individuation (which is very different from individualism) (Siegal 2022, see chapter 5). Individuation is a psychological term for the specific and distinct character personality and identity that is not merged, but in relation and intraconnected with others. Universal individualism on the other hand, has everyone with the same universal rational characteristics. Personality is uninteresting to the philosophy of universalism.

The transcendental aesthetic is a brilliant dissimulation that 'frees' any and every individual rational will into a hyperreal noumenal sphere outside the empirical world and relations of cause and effect. The key to the disparity at the centre of Kant's Copernican Revolution is to differentiate subjective transcendental alienation from decentred intraconnection.

COPERNICUS

Importantly, and in contrast to the decentring of Copernicus, Kant's achievement reprioritises anthropocentrism. Humanity retains its status at the top of the great chain of being. Kant lost touch with the cosmological integration that Copernicus and subsequent astronomers like Kepler, Galileo and Newton traced with their science. Against prevailing norms, Copernicus inverted the geocentric system to the heliocentric solar system. At the time Copernicus wrote *De Revolutionibus Orbium Coelestium*, or *Six Books Concerning the Revolutions of the Heavenly Orbs* (1543), the flat earth theory dominated Christendom. But millennia earlier, ancient Greeks and, earlier still, in Mesopotamia and ancient Egypt people understood the cosmos as an earth-centric sphere, where the five known planets and the sun revolved around the earth.

Copernicus and other physicists of the sixteenth and seventeenth century were astronomers and astrologists, continuing the ancient traditions that emerged from Axial cultures. They related the movements of the heavenly bodies directly to fate, destiny and the politics of the day (Westman 2022). Their age old premise was an onto-epistemology of interconnection between the cosmos, destiny and events in the eco-social and political sphere (Steele 2000a, see chapters 7 and 9). Kant successfully liberated the individual from the stars of their fate.

Copernicus developed his radically decentred astronomical geometry that challenged the flat earth with influence from earlier physicists. Of fame were Aristarchus of Samos (276 to 194–5 BC) who had calculated the circumference of the earth, and according to Archimedes (287 to 212 BC/2011), suggested the heliocentric solar system. Undiscovered in Copernicus's time, the Antikythera mechanism (approximately 206 BC) had elaborate mechanics to show the erratic pathways of the planets. It suggests the Greeks already knew that the planetary orbits were not circular (see chapter 7). But Ptolemy's later *Almagest* (AD 150 est.), which dominated the normal understanding of the cosmos for the next thousand years, was an earth-centric almanac which described the circular motions of the stars and planets. Hypatia (around AD 400) wrote a commentary on Ptolemy's *Almagest.* She also developed the mathematics of cones which underlies the mathematics necessary to determine that orbits are elliptical not circular. For her pagan philosophical views, Hypatia was stoned to death by early Christians. Copernicus delayed publishing his heliocentric view of the solar system for thirty-six years until on his death bed in 1543.

Christianity was fiercely averse to the heliocentric change because earth is the centre of God's creation. The heliocentric system was a substantial break with the previous norms. Copernicus had the intuition that the planets revolved around the sun – which resolved some urgent anomalies in the earth-centric model. Copernicus's intuition placed the five known planets in the correct order, but he retained the belief in a circular rather than elliptical orbit. The Church outlawed Copernicus's books in 1616. The same year, Galileo was declared a heretic for further elaborating on the heliocentric solar system. Galileo was threatened with torture until he recanted. He was again charged with heresy in 1632, when he was subsequently confined to house arrest for decades until his death. In the *Epitome of Copernican Astronomy*, Kepler (1995/1621) proved Copernicus's intuitive heliocentric system. Kepler resolved the mathematics of erratic pathways by carefully measuring the conical shadows of astronomical bodies and establishing their orbits were not circular, but elliptical. Kepler regarded Copernicus's insight as divine revelation and called his physics of the elliptical orbits, God's harmony (Westman 2022).

Copernicus had understood that it made better sense to recognise all angles and data from the position of earth. He also made the daring step of decentring the normal model of the cosmos, shifting the central locus from earth to the sun. Copernicus made the case that the earth and other planets orbited the sun, and the earth made a daily rotation around its own axis. Kant is eloquent and comprehensive in his footnote on the inversion of anthropocentricism. 'Copernicus, who, when he did not make good progress in the explanation of the celestial motions if he assumed that the entire celestial host revolves around the observer, tried to see if he might not have greater success if he made the observer revolve and left the stars at rest' (1992: 110). In an important footnote, Kant writes an acute synopsis of Copernicus's model and methodology.

> All apprehended change of place is due to movement either of the observed object or of the observer, or to differences in movements that are occurring simultaneously in both. For if the observed object and the observer are moving in the same direction with equal velocity, no motion will be detected. Now it is from the earth that we visually apprehend the revolution of the heavens. If, then, any movement is ascribed to the earth, that motion will generate the appearance of itself in all things which are external to it, though as occurring in the opposite direction, as if everything were passing across the earth. This will be especially true of the daily revolution. For it seems to seize upon the whole universe, and indeed upon everything that is around the earth, though not the earth itself. . . . As the heavens, which contain and cover everything, are the common locus of things, it is not at all evident why it should be to the containing rather than to the contained, to the located rather than to the locating, that motion is ascribed (Nicolaus Copernicus [1473–1543], *De revolutionibus orbium coelestium* [Nuremberg, I S43] 1: 5). (Kant 1992: n715)

Copernicus understands that viewing the solar system from the central perspective of the sun is impossible, and that the relative orbits necessarily need to be measured from earth. He could still *intuitively* articulate the heliocentric solar system in all its relative relationships of position and motion. The new hypothesis means the planets are accurately ordered around the sun, explaining numerous anomalies that the geocentric model could not accommodate. Decentred but irreducibly located, Copernicus's hypothesis requires making multipart calculations about the comparative velocity and direction of celestial bodies in relation to each other. While it has to be the location of our perspective, the earth is merely one cosmic object orbiting around the sun, amongst others. The celestial relations were later better calculated by Kepler and then further mathematically analysed by Newton. Kant seemed to understand the general principles but refused to condone the philosophical consequences.

Despite understanding the fundamentals of Copernicus, which he articulates in footnotes, Kant consolidates the anthropocentric Enlightenment sceptical tradition. Subjects can only apprehend an appearance of alienated objects. He admired Copernicus for his revolutionary intuition that inverted the geocentric norms and letting the decentred observer (on earth) revolve around the sun. Then immediately, Kant goes on to lay claim to intuition rather than science.

> Now in metaphysics we can try in a similar way regarding the intuition of objects. If intuition has to conform to the constitution of the objects, then I do

> not see how we can know anything of them a priori; but if the object (as an object of the senses) conforms to the constitution of our faculty of intuition, then I can very well represent this possibility to myself. (1992: 110)

Kant's proposition does not decentre the 'I' relative to other objects. On the contrary, Kant succeeds in subverting all objects to the sceptical apprehension of the anthropocentric, individual, rational subject. The apprehension of the appearance of a sensible object as a thing in itself becomes completely dependent on the rationality or 'synthetic' understanding of subjective comprehension.

The anthropocentric perspective of decentred individuals could be understood as relative, within the medium of space and of time. But Kant steadfastly stuck to the principles of the Enlightenment, that situates the rational individual as a superior being, at the centre of conscious comprehension of the known universe. Kant maintains the emphasis on 'appearance' which upholds the alienation between subject and object mooted by Descartes.

It is difficult to achieve this tortuous recentring of individual subjectivity and anthropocentricism in the face of radically decentred cosmology. Kant focuses on the material proximity of subjective apprehension, via a priori space and time, and then alienates further the representation of the appearance of any empirical object or thing by insisting synthetic judgement is necessary to ascertain the thing itself. Synthetic judgement is a step beneath the 'transcendental' 'pure understanding'.

Initially, Kant makes an attempt to legitimise the intrinsic reality of objects. Objects generate an appearance which can be ascertained by the subject, through the medium of a priori conditions, subjectivity, time and space, Kant ultimately undermines the connection to real things. He prioritises the freedom of the will which resides in the noumenal rather than the Copernican/Newtonian universe of forces acting upon other forces in a 'series of alterations'. Genuine transcendental reality, Kant argues, is in the sphere of the noumenon. The noumenon is completely outside empirical evidence, beyond time and space, and at the outer boundaries of the knowable. The transcendental aesthetic exceeds the mechanical universe of cause and effect, and situates truth outside the boundaries of historical time.

Ultimately, Descartes's schism between subject and object, culture and nature, is confirmed and deepened (Merchant 1980). In this fashion, Kant (1992/1781) takes the original reflection of the shadowy representations on the wall of the cave in Plato (427–347 BC/1965) and then exceeds it. In Plato's cave, the truth, as the sun, is too fierce to confront, and most people are reduced to believing the shadows on the wall of the cave are normative representations of reality. It is tempting to position the absolute truth of the light of the sun in the schema of the noumenon, but more accurately, the noumenal is a hyperreality, completely absorbed in a self-referential system of logos, with no connection left with empirical truths. Through the process of dissimulation, focusing on the a priori conduit and then perverting truth from appearances to rational synthesis and from objects 'given to us' to the thing in itself deriving from the hyperreal, noumenal realm. Kant fails to notice that he remains shackled in the cave, back to the sun, absorbed in shadow play.

PROGRESS OF THE SIMULACRA

Kant's so-called Copernican Revolution is an important step in the progress of the simulacra. Baudrillard (1994) outlines the stages from intimate absorption in a co-relation with the truth of the ecosystem to alienated hyperreality. The first step is the copy of reality that Kant makes when the appearance of the empirical object is mimicked by representation. The second stage is the focus on the materiality of the message, rather than the message itself, as Kant does with the a priori of time, space and rational subjectivity (that removes focus from the 'appearance' of the nature of objects). Thirdly, the tricky dissimulation of the thing in itself from empirical examples to the deductive logic that comprehends the rational universal essence of a thing. The 'thing in itself' is poised at a delicate point of abstraction that it is no longer an empirical example of the object, nor yet lost in the unknowable noumenon. Finally, is a complete immersion in the simulacra. The hyperreal noumenon is the fourth stage, established through pure understanding of the transcendental aesthetic. The alienation of individual subjectivity from empirical eco-reality is complete when there is a shift in language systems from gesture and correlation to empirical things, to the internal circulation of meaning in the system of logos. Hyperreality occurs when meaning depends on other signifiers of meaning (de Saussure 2011) rather than any direct relation to empirical objects.

The simulacra has become so normalised that it no longer has to hide behind a dissembling claim to reality. In the world of Silicon Valley, the development of layers of computer languages from binary code through to normal language commands and machine learning are intrinsically self-referential. Computers are incapable of gesturing towards a rock and correlating it with the empirical thing which is 'rock'. The capacity of machine learning, at its best, can make use of enormous quantities of data, available on the internet and in government or enterprise databanks, but these are its confines. Machine learning is forced to be entirely dependent on the prevailing system of signs, with all 'thinking' as a copy or duplication of always-already existent material within the entirety of the system. Artificial machine learning can amplify and select, but it is never generative in an original way (Bender et al. 2021). Like

Baudrillard's fourth stage of the hyper-real, all information within the system circulates within the system of signs and has no connection to objects, or the real. It is completely alienated from ecology. Nevertheless, it is a powerful tool and while it is constructed entirely of inputted network data, it is also profoundly influencing how people think (Stiegler 2018).

AI AND THE SIMULACRA

The final stage of the simulacra as the hyperreal, can be seen in action where the language and amplification of artificial intelligence dominates civil understanding and activity. Complete absorption in the simulacra is best outlined in popular culture by the film *The Matrix* (Wachowski and Wachowski 1999). Keanu Reeves's character, Neo, is woken from his inert life in a gigantic pod farm wired into the AI simulacra called *The Matrix*. The director of the *Future of Humanity Institute* at Oxford University, Nick Bostrom, seems to take this storyline very seriously. Without referencing the film directly, Bostrom suggests, with the inventive help of Bayesian probability curves, that something like *The Matrix* is our reality. He considers that we might currently be living in a 'historical' simulacrum that is inhabited by trillions of people – presumably wired into their pods – spread across the galaxy.

> Simulating the entire universe down to the quantum level is obviously infeasible, unless radically new physics is discovered. But in order to get a realistic simulation of human experience, much less is needed – only whatever is required to ensure that the simulated humans, interacting in normal human ways with their simulated environment, don't notice any irregularities. (2003a: 5)

The anthropocentric alienation of Enlightened universal discourse sets the scene for farfetched narratives that are having important influence on the billionaire class. It influences their technological and business interests and their orientation towards the future.

The rational individual that underpins market theory is completely alienated from empirical objects: from ecology. Due to the noumenon – such as the invisible hand – the hyperreal has become more normalised than the resource boundaries of the physical planet (Higgs 2014; 2023, Stockholm Resilience Centre nd). Together with the elitist assumptions that successful individuals deserve extreme wealth (as members of the Heavenly elect), the hyperreal is taken deadly seriously by Silicon Valley. Elon Musk, Peter Theil and Sam Altman have funded Bostrom's Future of Humanity Institute at Oxford University by tens of millions of pounds. Űber-rich tech billionaires have forged a curious approach to prepping for Armageddon, where they both take seriously the crises of climate change, drought, food and water shortage, extinctions, disruptions, poverty, refugees, homelessness and civil unrest and, at the same time, wear it lightly enough to continue to make money out of catastrophe. The unprecedented disparity between rich and poor means that more resources are held in the hands of fewer people than the height of feudalism (Piketty 2022). For these few, the society of fear (Beck 1992) includes significant moments of financial opportunity. At the same time, billionaires offer no leadership towards a viably sustainable future. They are committed to disaster capitalism. That means they seem to have fundamentally given up on genuine sustainability within planetary boundaries.

Bostrom's simulacra (2003a) justifies the idea that exhausting earth is irrelevant: firstly, there are plenty of other planets to push out the wild west frontier and, secondly, if this is a simulacrum and it all goes wrong, just push delete and reboot the programme. Bostrom's matrix gives the über-rich a secular rationale for treasuring the heavenly other worlds, at the expense of earth (Nietzsche 1974; 1982; 1985; 1989; 1909).

Bostrom's simulacra gives credibility to ongoing exponential population growth. Growth underpins continuing exponential resource exploitation and economic growth, and correlating exponential increase in greenhouse gas emissions (Noll 2023; Hansen et al. 2023). Bostrom argues that humanity needs a *huge* population to generate the finances and technological innovation needed to populate the Virgo Supercluster. He believes the future population of biological humans will reach 10^{38}. It is hard to grasp such a massive number but for comparison, today as I write, it is 8 billion, or 8,000,000,000. Bostrom's speculation is for 100,000,000,000,000,000,000,000,000,000, 000,000,000 people (2003b: 1). Why Bostrom comes up with this specific number is unclear. He is convinced that the trillions of future humans will produce 'historical' computer simulations, probably millions of them. The energy required for so much computer power will be easy to obtain with the solar energy from millions of stars. The technical difficulty of developing a completely believable historical narrative – like the one he speculates we live in right now – is difficult but not unachievable.

> On the surface of Earth, macroscopic objects in inhabited areas may need to be continuously simulated, but microscopic phenomena could likely be filled in ad hoc. What you see through an electron microscope needs to look unsuspicious, but you usually have no way of confirming its coherence with unobserved parts of the microscopic world. Exceptions arise when we deliberately design systems to harness unobserved microscopic phenomena that operate in accordance with known principles to get results that we are able to independently verify. The paradigmatic case of this is a computer. The simulation may therefore need to include a continuous representation of computers down to the level of individual

> logic elements. This presents no problem, since our current computing power is negligible by posthuman standards. (2003a: 5)

Bostrom thinks that the programmers will be required to 'see into' a person's mind, to keep the simulation as real as possible. This is all feasible because the speed of the human brain is limited, whereas with entire solar systems at the disposal of computing power, the speed and number of computer operations are very large indeed. Bostrom has a pretty specific number in mind; 'we can use ~10^{33}–10^{36} operations as a rough estimate' (2003a: 5). With a God's-eye view,

> a posthuman simulator would have enough computing power to keep track of the detailed belief-states in all human brains at all times. Therefore, when it saw that a human was about to make an observation of the microscopic world, it could fill in sufficient detail in the simulation in the appropriate domain on an as-needed basis. Should any error occur, the director could easily edit the states of any brains that have become aware of an anomaly before it spoils the simulation. Alternatively, the director could skip back a few seconds and rerun the simulation in a way that avoids the problem. (2003a: 5)

Like Tegmark's (2017) book *Life 3.0: Being Human in the Age of Artificial Intelligence*, Bostrom (both enthusiastically read by the Silicon Valley billionaires) seems to believe in the ideas of a benign dictatorship. This is where a small handful of tech engineers direct a monster AI programme that runs the world through subtle online nudges. The AI is capable of writing news, adverts and music, and influences the various bubbles in the population to vote for friendly politicians. These already existent AI amplifications enable Tegmark's programmers to control the resources and financial flows of earth. Davidson and Rees-Mogg (1997) argue that it is possible to keep over-utilising earth's resources during the Age of Information while prepping for civil war and other disasters. Davidson and Rees-Mogg advocate for 'Founders' who are an elite set of millionaires, to leverage 'the age of information'. They regard New Zealand as a suitably sophisticated, benign and neoliberal country for these 'Founders' to hide out from the civil unrest that climate change, resource exhaustion, pollution and overpopulation will undoubtedly produce. Prepping while continuing to run their huge corporations online; making coin while the world burns.

Elon Musk concurs. As he said in an interview with Rothman (2016), 'I've had so many simulation discussions it's crazy'. Bostrom (2003a) is convincing that present reality, and even the humans within it, are in fact, a huge, sophisticated simulation.

> Suppose that these simulated people are conscious (as they would be if the simulations were sufficiently fine-grained and if a certain quite widely accepted

> position in the philosophy of mind is correct). Then it could be the case that the vast majority of minds like ours do not belong to the original gender but rather to people simulated by the advanced descendants of an original race. It is then possible to argue that, if this were the case, we would be rational to think that we are likely among the simulated minds rather than among the original biological ones. (2003a: 1)

Musk and other Silicon Valley billionaires share the sentiment – Musk 'suggested' that the development of simulations 'indistinguishable from reality' was inevitable. The likelihood that we are living in 'base reality', he concludes, is just 'one in billions' (Rothman 2016).

Nietzsche (1974; 1982; 1985) warned of this narcissistic replacement of God with human mastery in the 1880s. Nietzsche (1974) argued that monotheism began the discourse of devaluing life on planet earth by placing the highest value on an 'otherworldly' heaven, which is situated beyond the planetary surface. With the 'death of God' (Nietzsche 1974; 1982; Heidegger 1977), humanity has taken over the role of mastery and continued to devolve value away from living systems towards the hyperreal.

Bostrom makes explicit his disregard for climate change and earth's ecosystem in a footnote in 2013, where he claims that the earth is not sensitive to carbon dioxide, given that five hundred thousand years ago, during the Cambrian period, CO_{2e} reached 410 ppmv, not vastly different from 426 ppmv today. He seems to have no idea that this level of atmospheric CO_{2e} caused species with exoskeletons and vertebrates such as trilobites to dissolve in the highly acid ocean and become extinct (Cutting 2021). Some organisms did survive and evolve during the Cambrian, but not species such as our own. Bostrom really could not care less about ecological impacts because he has convinced himself that we all live in a computer simulation (Bostrom 2003a; Rothman 2016). Peter Theil, Musk's PayPal cofounder, concurs, 'People are spending way too much time thinking about climate change, way too little thinking about AI' (in Waters 2014). Musk himself is focussed on taking humanity 'interplanetary' with his company SpaceX, beginning the colonisation of the Virgo Supercluster by jumping off Mars (Carroll 2013; Bostrom 2003b).

> There's no rush in the sense that humanity's doom is imminent; I don't think the end is nigh. But I do think we face some small risk of calamitous events. It's sort of like why you buy car or life insurance. It's not because you think you'll die tomorrow, but because you might. (Musk in Carroll 2013)

The impetus driving the billionaires, and their pet philosopher of technology, Nick Bostrom, is a firm belief in Enlightenment rationality. Their science is firmly based in Newtonian physics of linear cause and effect that is easily leveraged by the Masters of the Universe.

At the same time, these men, like the rest of us, are living in the society of fear. They feel the need to prep for disaster with luxury bunkers at multiple sites. They remain within the modern technological enframing of thought, bankrupt of any genuine methods or ideas to lift humanity from the material extraction, pollution or emissions growth that tumbles us all towards an uncertain future.

THE ABSTRACTION OF MONEY

Baudrillard (1994) argues that the representation, or simulacra, of real events and objects is gradually transformed from mimicking the real to progressively more and more abstract and alienated copies of the copy. The sceptical Enlightenment proposes representation is the best comprehension human subjects can have about natural objects, or each other. Gradually, reality shifts from environmental objects to the context and medium of representation. These conditions of knowledge consist of time, space and rational deduction (Kant 1992/1781). The conditions of representation become more important than the content of the representations themselves. Progress towards a constructed reality alienates modern society from ecological reality.

Impacts of the alienation of modernity are everywhere. I arrived in Australia in December 2019. Weeks later, enormous forest fires broke out in Australia, killing more than 3 billion animals that summer. The scale of it is impossible to comprehend. Unlike normal Australian forest fires, this one was far hotter and swept through multiple states, including rainforest as well as burn-adapted forest. Scorched land, burnt fur, desperate animals with nowhere to escape. At 1.3°C above preindustrial levels, the atmosphere carries 10 per cent more water vapour (Coumou and Rahmstorf 2012), escalating the turbulence and severity of evaporation, wind, drought and also storms, powerful rain and flooding. Climate change, ubiquitous plastic waste, incredibly high levels of extinction, ocean warming, ice melt, flooding, desertification, deforestation, intense storms and food security are all adding up to a global existential threat (IPBES 2019; WWF 2022). While specific civilisations have been annihilated in the past, because of war, famine or other climate conditions (Diamond 2004), we have never faced this level of annihilation on a *planetary* scale before. The earth is sending us signals that our mode of being, and social organisation is detrimental and unhealthy. Modernity is a form of trauma for the earth, for other species and for humankind.

NOMINAL MONEY

The progressive trajectory towards the simulacra is guiding the ethics of some multibillionaires. The concept of simulacra and alienation also converge in the construction of money. As outlined in chapter 3, since the 1400s, banking has utilised double-entry bookkeeping and the creation of new credit ex nihilo – out of nothing (De Roover 1963; Buch 2022). This is a perfect example of the simulation of value, as it tends towards the simulacra, which I would argue is realised most completely in Bitcoin and the stock market. These 'reservoirs' of value storage are almost completely alienated from the real world. But because everything is actually connected, when the stock market crashes, the real-world impacts are significant. New credit creates vast quantities of new money, and stimulates a spiral of economic growth that meets the interest and repayment demands of ex nihilo credit generation. That economic growth is directly correlated to continuous accumulation of CO_{2e} emissions (IPCC 2021; Hansen 2023; Noll 2023). Given that nominal credit has reached three times the value of all global transactions in 2022 (International Monetary Fund (IMF) 2022), and continues to escalate, it is vital to begin to comprehend how money has become so alienated from function. There are two main approaches to money – one is that it is a neutral token of exchange and the other is that money is imbued with social relationships and fetishises consumerism. The idea of money as neutral or as a conduit of power have an important commonality: they understand money as an abstract cypher, which lends itself to the dissimulation of the finance economy from fiat exchange.

Early in his book, *Money Whence it Came – Where it Went*, Galbraith (1975) exclaims, 'The study of money, above all other fields in economics, is the one in which complexity is used to disguise truth or to evade truth, not to reveal it' (1975: 4–5). Former British financial regulator Lord Adair Turner (2013) has been forthcoming about the process of banks generating money out of thin air. 'Banks do not, as too many textbooks still suggest, take deposits of existing money from savers and lend it out to borrowers: they create credit and money ex nihilo – extending a loan to the borrower and simultaneously crediting the borrower's money account'. The question is, How did this simulation of value become legitimised? And what are its impacts?

Money is a stand-in for exchange. But multiple economists make a distinction between a direct value attributed to money, and the standards of exchange. Ament describes money this way – 'money is a medium of exchange that represents underlying commodity values, though without any value itself, and allows economic agents to store value in time and space – to lubricate the gears of trade' (2019: 5). Adam Smith wrote 'The powers attributed to money are not inherent in money, but derive from its functions as a means of exchange' (Tynan 2017: 215).

Ament tracked the history of economic attitudes to money:

> Samuelson wrote that 'money is a neutral veil that obscures barter transactions just below the surface' and argues, were it not for the inconvenience of barter, 'money . . . might have little function' (Samuelson 481). Schumpeter echoes Mill in writing that money is a 'technical device that has been adopted in order to facilitate transactions [between] goods and services' (Schumpeter 277). Wheelan writes that 'in theory money is not even necessary'. (Ament 2019: 5).

Marx considered money as a powerful symbol that enables purchase of commodities. Most importantly the purchase of labour as a commodity. 'From the Marxist perspective, all commodities, including money, are congealed labour power' (Tynan 2017: 215; Marx 1887). Thus, social relations are embedded in monetary exchange. Keynes was a chartalist, as was Locke, which means they associate money with the power of the state. Appaduri (1986) makes a case for money as the fetishisation of commodity exchange. Marx (1887), Simmel (1978/1907) and Marcel Mauss (1954) all explore modes of commodification, even in the gift exchange.

Georg Simmel (1978/1907), in *The Philosophy of Money*, has a different angle on the social value of exchange in money, believing value is 'never an inherent property of objects' (73). It is the desire, sacrifice, and difficulty of acquisition that is 'the unique constitutive element of value, of which scarcity is only the external manifestation, its objectification in the form of quantity' (100).

Money as a 'neutral veil' separates price from value, making money an abstract cypher that stands in for smooth bartering across different products or services and different places, cultures and circumstances. It is a plausible explanation of the bartering role of fiat money. But it also makes money an abstract concept, and far easier to enter the process of simulacra, increasingly immaterial and alienated from actual things and events. The abstraction of money justifies the structure of loans ex nihilo in the financial industry. Ultimately, it means there is nothing surprising about the massive disparity between real transactions for goods and services (the fiat economy) and the amount of nominal value attributed to stocks and shares (Bank of International Settlements 2022).

Tynan shows the historical trajectory of money as it loses contact with the gold standard and becomes increasingly abstract.

> The history of money since the establishment of the Bank of England in the seventeenth century has been an increasing abstraction from the material commodity (gold or silver) it was supposed to represent. What an ontology of money needs to address today is this dematerialisation, this spiritualisation, represented by credit. The ontological question of money involves an investigation into

> its credo. Since the so-called Nixon Shock of 1971, the dollar has operated as a 'floating' currency, unanchored from any metallic standard and therefore not convertible into anything beyond itself. The modern development of fiat money and credit money, in which monetary value becomes entirely symbolic or 'virtual', grounds the power of money in its value as pure sign operating without referent. A banknote is a promise to pay that can ever be fulfilled only by another promise to pay, in an infinite regress, a chain of signifiers. (Tynan 2017: 215)

Ament (2019) makes a related argument. The exchange value of fiat money is not a direct token, or 'numeraire' as Walrus described the 'symbolic representation of existing commodity values' (Walrus in Ament 2019: 6).

> Simmel escapes the logical trap that a direct comparison numeraire presents by arguing that measures need not, and most often do not, exhibit the same quality as the objects they measure (Simmel 1978/1907: 131–32). Thus, money, while acting as a 'stable pole' (121) of relative commodity values, does not represent the relation between those commodities, as a numeraire does, but rather, points to their relative relation to a third quantity (Simmel 1978/1907: 146). This quantity is an abstract value that constitutes a social relation and is starkly different from a 'direct comparison' [ibid.] of two quantities. In other words, abstract value allows for comparison, but is not the comparison itself. (Ament 2019: 7)

Ament then claims that 'commodity money has never existed and banks are not commodity lenders' (9). Credit-based money supply constantly pushes a requirement for economic growth into the system, to compensate for price inflation and to meet interest rates and loan repayment expectations. This is how the systemic expectation of economic growth has become normative. 'When money is understood as a credit-based unit of account, the model cannot reach equilibrium. Far from Schumpeter's argument that money does not affect the economic process (Schumpeter: 277), money itself is the economic process' (Ament 2019: 8). This is a continuous over-supply problem rather than the balance of supply and demand theorised through the invisible hand.

It is Keynes (1991/1936) that manages the sleight of hand that turns dissimulation into a simulation. The non-money of ex nihilo credit becomes pseudo money, until it is properly commodified through exchange. Keynes teases out the way the state leverages the creation of private ex nihilo credit that stimulates and simulates circulating money. This is a moment, or better yet, a movement, when ex nihilo credit 'manages' to cross over and enter fiat money circulation. Reminiscent of Kant's thing in itself, poised between the medium of representation and the pure intuition of the noumenon, the institutional line of credit is 'managed'. The state declares the value of currency through a variety of means, including the gold standard, the LIBOR or OCR, long-term bond offers and managing currency speculation. Keynes poises the objective standard of money defined by the state on one side, and the dynamic commodity market of fiat exchange on the other. 'Managed money' or credit created through private banks or government both balance between the poles of State defined value and market value.

> Commodity money and managed money are alike in that they are related to an objective standard of value. Managed money and fiat money are alike in that they are representative or paper money, having relatively little or no intrinsic value apart from the law or practice of the State. Moreover, managed money is, in a sense, the most generalised form of money – which may be considered to degenerate into commodity money on the one side when the managing authority holds against it 100 per cent of the objective standard, so that it is in effect a warehouse warrant, and into fiat money on the other side when it loses its objective standard. (Keynes 1991/1936: 6)

Keynes is an early advocate of the money 'multiplier' that stimulates the economy, by easing the money supply and allowing prices to rise, which stimulates profit and more spending – thus, growing the economy. He is oblivious to the impacts of economic growth on resource exhaustion, pollution, climate change, or the extraction of value from the public commons to the banks.

Tynan notes the abstraction and alienation of money as it progresses further towards simulacra.

> This *semiotic* transformation of the global economy – unprecedented in world history – is what facilitated the hegemony of financial capital in the latter half of the twentieth century and precipitated the economic crises that have menaced Europe and the United States since the subprime mortgage crash of 2008. This transformation should not be thought of as a purely economic affair but as a crucial moment in the history of speculative consciousness.
>
> Deleuze, for example, tells us that the 'power of language' derives not from its ability to refer to an empirical world outside language but 'to speak infinitely of or about words themselves'. Thought, likewise, has 'no object other than the power of reiteration in indefinite regress (I think that I think that I think. . .)'. Thought frees the cogito from empirical reality and language frees the word from the referent, just as money frees value from material commodities. (Tynan 2017: 215)

Money involves all the layers of Baudrillard's four stages of the simulacra. It represents the value of commodities in fiat exchange; it simulates value across disparate asset classes, it produces itself as a medium of value, and the medium has more reality than the value it represents. Finally, as nominal credit loses all relation to the real economy, the simulacra of money has transcended the first stages of mimicry and representation and forged its own semiotic system.

Money simulates value. Importantly, the abstraction of money into a semiotics of nominal credit is a deepening of uneven power relations, not a democratisation of them. The driver of ex nihilo credit of economic growth enables inflation to extract value from the labour commodity of the working classes and store it in the value reservoir of the stock market, Bitcoin, fine arts and the housing market. This has become a generational storage of value. 'The dematerialisation of money into a pure sign, far from democratising access to it, goes hand in hand with the intensification of money as an instrument of social power' (Tynan 2017: 215).

Most importantly of all, is that the abstract ex nihilo credit creation continuously drives too much money supply into the macroeconomic system. It jeopardises price stability and forces economic growth to counteract new demands for interest rates and servicing of the ex nihilo credit. This escalating spiral stimulates economic growth, which, in turn, stimulates increasing CO_{2e} emissions, which accumulate yet more in the atmosphere. Addressing climate change is impossible if we do not address ex nihilo credit creation.

Chapter 6

Technology Shapes the Way People Think

Heidegger is crucial for understanding the double bind of modernity and why it's been so hard to escape, even when science, and United Nations (UN) policy, such as the COP21 Paris Agreement and the Sustainability Goals, are trying to combat the existential threat of global pollution and ecological devastation. Heidegger's astute understanding of the philosophical underpinnings of modernism helps us to perceive how we got into this mess, even if he remained caught up in the prejudices of his time. This chapter engages with three important philosophers of technology: Heidegger, Spengler and Jünger. These three German, mid-twentieth-century philosophers initiated much of the discussion about understanding the parameters of modernity and why we remain so stuck in the modern paradigm of consumer growth. Heidegger is a vital leaping-off point in the discovery of a better, cohesive direction for the future.

The problems of climate change appear intrinsic to modernity. But greenhouse gas emissions are not intrinsic to all human societies – not even sophisticated, luxurious societies such as ancient Greece or the Roman Empire. So, the question is, What is essentially modern, that enshrines the motivation for fossil fuel use and consumer growth in gross domestic product (GDP)? The structural imperative to growth embedded in fractional banking that I outlined in chapters 2 and 3 is a large part of the issue. However, to contemplate changing the banking system or intervening in the stock market there is an overwhelming sense that the system is too big, too entrenched, to be altered. Nobody, certainly not isolated 'rational individuals', could intervene in the megalith of finance and global economics to such an extent, no matter how logical the argument or how firm the correlations between economic growth and emissions growth. This is why there proliferates small twittering around the edges, such as green consumer choices, going vegan, an emissions trading scheme, 'net' zero or the TNDF advocating an ethics of investment that criticises companies not sufficiently green. These endeavours are great, but as the continuous growth in global emissions demonstrates, they are entirely insufficient. Global direct primary energy consumption has increased from less than 6,000 Terawatt hours in 1800 to over 150,000 TWh's in 2021 (Our World in Data). Despite great advances in hydro, wind, and solar, the overall consumption of coal, oil, and gas continues to rise. The greatest philosopher of technology is Heidegger, who identified the 'essential' problem of modern technology in the 1950s, after the Second World War, and subsequently pursued a path to exceed or divert the entrenched determinism of modernity. This chapter explores Heidegger's ideas to further develop deep-set challenges to the modern technological enframing of the horizon of disclosure.

THE TECHNOLOGICAL HORIZON OF DISCLOSURE

Modernity is a unique culture in the history of the world, and it signals a profound technological shift from earlier civilizations. Technology can no longer be understood as handcrafted. Instead, modern technology has become mass produced and has taken over every aspect of the globe. Heidegger (1977) argues that the technological horizon of disclosure frames everything in terms of efficiency and storage. Each element of nature is in 'standing reserve', waiting to be consumed by global capitalism.

Heidegger developed his concept of the modern horizon of disclosure partly from a reading of Jünger's 1930 (1993) article called 'Total Mobilisation' in *Krieg und Krieger* (*War and Warrior*) and Jünger's 1932 book, *Der Arbeiter* (*The Worker*). In 'Total Mobilisation' (1993), Jünger makes the dystopian case that Germany is entering the same industrial modern world as Great Britain, which loses touch with the fragrance, peace and tranquillity of a tightly knit, somewhat romanticised community (Zimmerman 1990; Blok 2011; Neaman 2019). Nevertheless, Jünger's critique of modernity is powerful. Total mobilisation, he says, is 'an act which, as if through a single grasp of the control panel, conveys the extensively branched and densely veined power supply of modern life towards the great current of martial energy' (Jünger, cited in Zimmerman 1990: 126–27). Both Heidegger and Jünger have been influenced by Oswald Spengler's earlier book *Der Untergang des Abendlandes* (*The Decline of the West*) (1918) which traced the 'morphology' of the evolution of civilisations, from their birth, their heyday and then the rigidification of ideas during their decline (Spengler 2002; Hermans 2017). Spengler and the romanticism of *Kultur* over *Zivilisation* lent a feeling of historical destiny to the 'total mobilisation' of the technological horizon of disclosure.

Crucially, Heidegger (1977) argues that absolutely everyone is co-opted into this technological horizon of understanding. Other modes of knowing are forgotten. An untouched forest is no longer a virgin ecosystem but, rather, waiting to be exploited as wood, or generating soil so that it can be covered in palm plantations once cleared. A river is no longer a water channel with its own ecosystem but a source of hydropower, or a tourist destination. People also are understood only as 'human resource' as both labour power and consumers in the overall machinery of modernity (Heidegger 2012). Even rejecting consumerism, and living in an ancient traditional black house on a remote peninsula unattainable by road, is a *recognition* and *inversion* of standing reserve, which does not shift the dominant dynamic. Very importantly, Heidegger argues that we are not in control of technology. Rather, like bees gathering honey and inadvertently pollinating the flowers, we are the vectors of technological reproduction (Heidegger 1999a: 218; Spengler 1932). The tiniest subatomic particle, the largest star system, are all potentially exploitable by the technological horizon. This determinism is all encompassing. Heidegger, himself, spent most of his career trying to understand the philosophy underpinning modernity and exploring ways of overcoming its deterministic forces.

ORDER AND THE DISCLOSURE OF WORLDING

Modern technology is quantifiably different, according to Heidegger (1977), from all earlier periods, because of a shift from traditional craftwork to globalised production and storage. Efficiency and standing reserve characterise the essence of the new mode of modern technology. Heidegger regards the essence of modern technology as shaping our epistemology. Heidegger argues that we cannot think outside the 'box' of modern technology, which reduces all aspects of things to resources. These resources may not yet be in use; they may be waiting in reserve for the call from consumerism. Whether waiting in reserve or already partaking in the machinery of production, everything is part of the modern consumerist horizon of disclosure.

For Heidegger, the modern technological enframing of the world into standing reserve is an impoverishment of thinking. 'Enframing does not simply endanger man in his relationship to himself and to everything that is. As a destining, it banishes man into that kind of revealing which is an ordering. Where this ordering holds sway, it drives out every other possibility of revealing' (Heidegger 1977: 27). Stiegler (2018) later calls this the proletarianisation of thought. Far from exerting free will over our own actions, or mastery over others, Heidegger regards the ordering of order (Heidegger 1977: 17) as passively, and unconsciously submitting to the machinery of modern consumerism. The emphasis on calculation and order reduces meaning making to a utilitarian protocol that enhances productivity and reduces the ability to hear and apprehend other ways of knowing. 'The human is thereby an employee of requisitioning. Humans are thus, individually and in masses, assigned into this. The human is now the one ordered in, by, and for the requisitioning' (Heidegger 2012: 29). Mastery over nature and mastery of rational ordering as an indicator of capitalist success is an assumption – as Weber assumed in 1919, but it is also an illusion.

One of the most disquieting things about the shift from earlier technology to modern technology is that the pace of production is no longer at the rate of seasonal cycles, or shaped by the flow of the river or the force of the wind. Instead, the demands of consumption dictate the rate of production. Heidegger (1977) famously used the river as an example. The corn was ground at the pace of the season's harvest and the volume of water in the river as it flowed past the river mill. But now, the river is dammed, and a hydroelectric power station produces power at the rate of consumer demand. Human mastery over nature requires the ordering of the world. The pacing of consumerism is exacerbated with just-in-time (JIT) production, which only produces consumer goods at the rate that supermarkets and other stores can sell them. Only very severe natural calamities, such as earthquakes, drought, landslides disrupt consumerism – but they are localised. It hasn't been until the major COVID-19 pandemic that, for the first time, there has been a global disruption of the pace of consumerism. The failures help to illuminate the significance of smooth organisation of the world as we know it. For example, the river that had earlier had a cyclic relation to grinding corn, and now has consistent electro power generation. The failure of a disrupted electricity supply affirms the 'necessity' for continuous energy within the current world rather than a change in the tempo of production and consumption. Most people have no idea that electricity is flow and very difficult to store, and that supply and demand rarely match up. They are used to simply turning on the light switch and gaining instantaneous light. Heidegger argues that most of the time we are so deeply embedded in our technological world, that it simply unfolds without conscious thought. This 'forgetting' of the source of power, and its parameters, are fundamental to the way things 'world', and it is only when something fails that we notice how it functions, and whether we need to alter the tool, or the assumptions that are worlding. The pace of consumerism is entirely alienated from the pace of the seasons, or the pace of sustainable resource extraction. The enframing of our world view by modern technology, the pacing of modernity and a better conception of 'truth' and 'failure' are key concepts to perceiving the problems of the technological horizon of disclosure.

CALCULUS, MASTERY AND TRUTH

On the face of it, modern science and technology have positioned humanity as the apex species on the planet, in control of vast swathes of nature and seeking to extend that dominion further into space and deeper under the ocean. That is certainly the assumption of the sociologist Weber and those people who have embraced the Enlightenment paradigm of modern control.

As explored in chapter 4, in a vital way, modern humanity is *not* in control of technology. Heidegger (1977) is very clear that the scientific experiment, the ethos of calculation, is not the 'essence' of technology. Instead, science is the imposition of order on natural events. For Heidegger, ordering the order of resources is an anonymous and trivial occupation that deprives deeper meaning to both person and objects of understanding. But this role as the orderer is not something that humanity sets out to do. The modern human is co-opted into the role of orderer as much as nature has been co-opted as a 'fundamental piece of inventory of the technological standing reserve – and nothing else' (Heidegger 2012: 41). Humanity is a pawn in the machinery of standing reserve, as much as the river is a potential source of energy or the forest is a potential source of timber. The role of humanity in the standing reserve of consumerism is to set order in place, to demand utility at the behest of the machinery of capitalism. In 'The Question Concerning Technology', Heidegger explains this is 'the gathering together that belongs to that setting-upon which sets upon man and puts him in position to reveal the real, in the mode of ordering, as standing-reserve' (Heidegger 1977: 24).

The illusion of mastery is itself the 'supreme danger'. Heidegger's complex analysis situates humankind outside the typical Enlightenment view of the subject who knows about the object of knowledge. He writes that the truth (as unconcealment) 'does not concern man as object, but exclusively as standing-reserve, and man in the midst of this objectlessness is nothing but the orderer of standing-reserve (27).

Weber (1948/1919) smugly attested to the disenchantment of modernity with the advent of scientific knowledge, as though it offered up analysis and truth as soon as scientists focus attention on an issue. But for Heidegger, this excavation of facts as though they equal comprehension and truth is not how understanding emerges. Knowledge is a kind of revealing of that which is already there. One has to be attuned to what might be revealed (Guignon 2001). The difference between calculating and revealing is profound. Calculation demands or 'challenges' knowledge upon active examination, categorisation and accumulation of facts. Whereas allowing something to reveal itself is a more meditative and reflective process, where the pace of opening up and gathering is informed by the event, as much as the person seeking to know.

> Who accomplishes the challenging setting-upon through which what we call the real is revealed as standing-reserve? Obviously, man. To what extent is man capable of such a revealing? Man can indeed conceive, fashion, and carry through this or that in one way or another. But man does not have control over unconcealment itself, in which at any given time the real shows itself or withdraws. (Heidegger 1977: 18)

The calculation of cause and effect that underpins modernity gives the illusion of mastery. Technology has, somewhere along the way, been assigned to 'culture' and removed from 'nature', as though the two categories were in opposition. As a result, Heidegger is of the opinion that modern technology alienates humanity from deeper truths, older ways of knowing, that are less about projecting the regime of order and resources and more about a self-revealing that allows comprehension rather than succumbing to the demands of the consumerist imperative.

THINGS WORLD US

In *Being and Time* (1962/1927), Heidegger makes the case that tools form the ubiquitous, background 'worlding' of the modern age: the architecture; the infrastructure; the roads; the pipes; the sewers and water taps; the groomed gardens; the straight lines; the style and existence of buildings; the materials used in walls, ceilings, windows and furnishings; the shape of the chair; the use of a table, of cutlery and, more recently, of lighting and electricity; the surveillance cameras; the computers; phones; sound systems; TVs and so on, are all what he calls 'ready-to-hand' and shape our horizon of disclosure. One of these items in an alien context makes no sense. Together they 'world us'.

The Enlightenment case for a rational individual subject that observes separate objects is decentred by Heidegger (1973a; 1999b). Far from a separation of subject from object, Heidegger makes a phenomenological argument that the ready-to-hand (*das Zuhandene*) technological objects actually shape our sensory experience of the world. When I peer through glasses, they alter my focal point. Microscopes, earphones and smooth surfaces all shape our sensory expectations about the world we inhabit. The habitual movements of driving, writing, typing, operating an oven, a phone, a crane – all of these skillsets become deeply engrained in our bodies and shape our metabolism, our musculature and our unconscious expertise (Dreyfus and Hall 1992; Dreyfus 1998; Dreyfus and Rabinow 1983; Guignon 2001). Telescopes and microscopes, and enormous machines like the Hadron Collider, enable us to see further, to see more minutely, than ever before. This has changed what we know about the environment and cosmology and helps to shape the way we think.

Heidegger's argument that technology shapes our phenomenological experience and interpretation of the world is a radical philosophical move, in the face of Enlightenment idealism, where the rational thinking subject is at a remove from inert objects. For Heidegger, the tools have a form of agency, the world has agency, the earth has agency and the authentic subject, whom he calls *Dasein*, has agency (Heidegger 1982b; 1999b). *Dasein*'s agency is, most of all, to appreciate and notice the mode of Being that is exhibited in things around them (Heidegger 1949; 1969b). In stark contrast to Descartes's universal, abstract, rational individual subject, Heidegger is resituating subjects in deep interconnection with the historical, technological environment. We are at work, not at rationality (1999b).

> *The ready-to-hand is not grasped theoretically at all, nor is it itself the sort of thing that circumspection takes proximally as a circumspective theme. The peculiarity of what is proximally ready-to-hand is that, in order to be ready-to-hand, it must, as it were, withdraw in order to be ready-to-hand quite authentically. That with which our everyday dealings proximally dwell is not the tools themselves. On the contrary, that with which we concern ourselves primarily is the work – that which is to be produced at the time; and this is accordingly ready-to-hand too. The work bears with it that referential totality within which the equipment is encountered.* (Heidegger 1962: 99; italics in the original)

The ready-at-hand constitutes the world, and people occupy it without true consideration most of the time. Heidegger (1977) notices the way modern technology enframes the ways of knowing. He calls this the technological *Gestell*, which is best translated, according to Julian Young (2002), as the technological horizon of disclosure.

Its only on rare occasions that considering what kind of world we live in and its consequences for other species – the poor, the 'third world' or ecology – might bring the normative mode of doing things into perspective. Ready-to-hand tools are mostly taken for granted, unconscious or forgotten, and only show up when we learn to operate them for the first time, or when they fail. Heidegger (1962) says that we only register the ready-to-hand when it has broken for some reason. The failure makes us notice how useful it was when it worked and there is now a gap to be filled. Its utility value might be aesthetic, metabolic, symbolic or energetic. The failure shows up the way that tools contribute to the continuous emergence of the current world (Harman 2010).

FAILURE AND DISCLOSURE

Failure is crucial to understanding Heidegger's concept of truth, or *aletheia. Aletheia* is a kind of withdrawing, or obscuring multiple ways of understanding a thing, in service to the current dominant way of thinking. The forgotten aspects are not wrong, nor do they disappear entirely, but they are hidden, concealed in the beings, without showing up in the current horizon of disclosure. Heidegger often talks about what has been forgotten, and he prioritises questioning because the question forges the context for a possible answer to emerge. If we do not ask the right question, there is no way for the truth to reveal itself (Heidegger 1973a/1953; 1999b; Guignon 2001; Young 2002). When there is a fault in the ready-to-hand, we begin to ask questions, and this allows hidden aspects of the thing to be revealed. This illumination or 'clearing' happens with active agency on all parts; *the tool leans in*, and the person pays attention – enabling fresh modes of understanding to unfold. Without failure, the tools just continue to operate without being noticed. The computer does not freeze, the electricity does not go off, the kettle boils, the coffee is fresh, the milk has not turned. We inhabit the room without thinking about the fabrication of four walls rather than five, or the ethos of 'efficiency' which four walls exhibits rather than the elegance of a pentagon. While supply chains roll smoothly, people do not notice how the metabolism of consumerism is controlling the pace of what it is to be modern, violently wrenching materials from the earth, as 'standing reserve' (Heidegger 1977).

To gain most from the illumination of the process of failure, we need to contemplate truth in the context of the lifespan or, more so, the span of the historical epoch. Heidegger explored this in dialogue with his predecessor Friedrich Nietzsche and two contemporaries Oswald Spengler and Karl Jünger.

EXCEEDING THE MODERN ENFRAMING OF THOUGHT

Three thinkers – Spengler, Jünger and Heidegger – try to understand and overcome the determinism of the modern horizon of disclosure. After surviving the fascism of Germany during World War II, Heidegger and his friend Karl Jünger embarked on an important discussion about nihilism and the looming fear of nuclear devastation (Heidegger and Jünger 2016; Heidegger unpublished). Oswald Spengler importantly analysed the lifespan of civilisations in *The Decline of the West* (2002/1918) and then the essence of technology in *Man and Technics* (1932), which upend the concept of progress.

They each regard modern technology and the calculation and accounting of all facets of life as inherently totalitarian and dangerous.

In a less well-known discourse, Heidegger develops a theory of 'two beginnings' (1999a). He derives this idea partly from Spengler (2002/1918; 1932) and from a reading of Nietzsche (1989) and Kant (1992/1781). Nietzsche talks about how decisions being made in the present are in the context of a life-as-a-whole, from birth to death (Irwin 2010b; Irwin 2015a; Nietzsche 1989). Heidegger took this idea of life-as-a-whole and articulates it in relation to a civilisation rather than an individual (Heidegger 1982a; 1999a; Irwin 2010b). Decisions we make now are illuminated by the two 'beginnings'; the outset in early philosophy from ancient Greek philosophy (and potentially other Axial civilisations), and from the prospective end point of the civilisation (for Heidegger and Jünger this was the nuclear Cold War, in our case, climate change).

Heidegger (1977) was concerned that the enframing determinism of the technological *Gestell* obscured more important ways of being aware and alive. But the epistemological shaping of our understanding of all things, from the perspective of business-as-usual (BAU), is incredibly difficult to overcome. He thought that breaking through the normative epistemology of the technological horizon of disclosure only came from a couple of sources, failure, the two beginnings and poetics (Irwin 2015b). A 'failure' in a tool shows us in a more originary way its contextual usage. It reminds us of how the tool demonstrates the world. The failure of a tool either reinforces how the world shows up, or is a pivot point for change, as the world continues to evolve, becoming renewed. The first beginning at its origin, or *Ürsprung*, is a mode of understanding that emerges for the first time. The origin is when people are really exploring the impact of this new way of doing things – so the profound shift in the world is most clearly explored at its origin. The second 'beginning' is how current behaviours tend towards death or finitude. The fear of death causes reflection on the present. Poetics taps into deeper ways of encountering things in the world, which might have been forgotten but still exist in an underground, unconscious way and still have potency.

Heidegger conceives of birth and death in a more 'originary' way (1999a). Birth and death bookend the finite existence of someone's life. Heidegger proposes that authentic decision-making in the moment requires contemplating one's life as-a-whole from origin to finitude. He expands this process to the historical epoch. To authentically comprehend the historical context and make valuable decisions, people need to explore the current dilemma in the context of the birth and death, the origin and finitude of the civilisation as-a-whole. Each person can contemplate their own origin and their own potential death as two 'beginnings' which refracts back on present conditions – shining new light on why we arrived where we are and what the consequences might be if we continue. In the scope of modern civilisation, climate change poses an existential threat to many species and ecosystems, bringing about the potential collapse of modernity. Realistically recognising the finitude posed by climate change means we need to reconsider entrenched normative values in the present moment. The finitude of climate change is a 'new beginning' to current business-as-usual.

Heidegger's analysis relies considerably on Kant's a priori concept of time, space and subjectivity (Wood 1993). In this chapter, we focus on the two beginnings as methods for contemplating modernity as an epoch and the potential for a transvaluation of values beyond pollution, ecological degradation and consumerism.

TWO BEGINNINGS

Kant's philosophy of time and subjectivity in the preface of *The Critique of Pure Reason* (1992/1781) profoundly transformed our conception of time. Kant's insight was to reinterpret the 'external' flow of time as objective and linear. Once the time has flowed past, it is gone, never to be retrieved. Kant took the inverse position. Time, space and subjectivity all occur as a priori conditions, and none exist without the other two. Subjectivity, space and time all emerge as one piece, although each has irreducible properties, none of which can come into existence without the presence of the other two. While time flows, it can only flow 'through' the awareness of the subject and the extension of the place. Space has extension but that extension only exists in relation to time and subjectivity. Kant's philosophy of a priori time/space/subjectivity makes time and space relative to subjective interpretation. In other words, our relationship to time, to both history and to the future, is always 'through' the subjective now. Kant's conceptualisation of a priori mediation of knowledge is problematic (see previous chapter), but it has been incredibly influential, not least upon Heidegger (1985). We see this commemoration of 'now-ness' in many walks of modern life, from the self-care of meditation, to the short term, instant gratification of consumerism, or to banking where interest rates are 'calculated daily'.

Kant's reification of the intuition of individual subjects into the logic of the transcendental aesthetic, has a similar political outcome to Descartes's distinction of the mind from the body – 'I think, therefore I am' – which results in a 'hard' individualism (see previous chapter). Along with rationalism and the categorical imperative, individualism creates foundational philosophy for classical and neoliberal economics. Despite Kant's intuitive fundamentals of time, space and subjectivity, time continues to be understood by the Enlightenment as linear, cause and effect and, indeed, progressive, even teleological.

These assumptions underpin modern political theory and political economy. That progressive approach justifies the Keynesian belief in economic growth.

Heidegger maintains the fundamental elements of Kant's a priori conditions, where time, space and subjectivity emerge commensurately. However, Heidegger tries to overcome Kant's isolated individualism in two ways: through a philosophy of language (1982b) and by expanding the concept of subjectivity from individual to the entire community (1999b), an entire civilisation and an entire epoch. The 'lifespan' still plays an enormous significance, and the structure of a priori conditions holds, but the subjective mind is not reduced to individual rationalism but to technological cultural-language communities. The methodological structure of the two beginnings derives from this expanded concept of *Dasein* where subjectivity metaphorically encompasses the historical context of a civilisation as-a-whole (Heidegger 1999a; 1982b; Heidegger and Jünger 2016).

The two beginnings of birth and death, represent the early origins of modernity, and the looming finitude of civilisation at its death. For the birth of the modern epoch Heidegger (1975a; 1973a; 1969a; 1969b) points to ancient Greek thought, and to its end he indicates the nihilism of the technological *Gestell*, with its looming threat of nuclear war and the death of civilisation (Heidegger 1999a).

These are not lightweight ideas, and it takes emotional courage to engage – especially with looming finitude. For those beginning to learn about climate change, there can be a tendency towards head-in-the-sand withdrawal, or hedonistic fatalism or aggressive rejection. Climate denial is waning as more people are getting used to a legitimated, evidence-based discourse on climate. Hedonistic fatalism seems to permeate a lot of richer communities. It is easier for these people to consider the end of life entirely than to consider change to the status quo, which has served them and their own families so well. Rage is also evident, especially in fictional movies; some Armageddon narratives and the BBC seem to be taking an anti-activist stance lately, where climate activists are positioned as murderers (e.g. *Shetland*, second season). These early stages of engagement are sympathetic nervous responses: fight, flight or freeze. Heidegger is taking for granted that the parasympathetic response is online, that authentic consideration is in operation when he outlines the methodology of the 'two beginnings'.

CIVILISATION AS EPOCH

In their 1950s conversation about nihilism, Heidegger and Jünger discuss the 'second beginning' (Heidegger and Jünger 2016; Heidegger unpublished; Zimmerman 1990; Irwin 2010b; Herman 2017). In his earlier work,

Heidegger had sought to understand the philosophical assumptions of modernity by going back to its earliest genealogical roots in the *Ürsprung* of democratic ideals in ancient Greece (Heidegger 1998b; 1985; 1982a; 1982b; 1979; 1977; 1973a; 1970; 1969a; 1969b; 1968a; 1968b; 1962). In this later work, with Jünger, Heidegger focuses on the *end* of modernity, in a calamity such as nuclear devastation, or now, some seventy years later, in climate change and ecological crises like enormous forest fires.

In this development of a new methodology, Heidegger looks at a life 'as a whole', from inception to death, in order to authentically make decisions in the present (1999a). In his early work, he called this *Dasein* which focuses on the individual subject (1962/1927). He expanded this thesis of the life as-a-whole, of the individual subject, to think about civilization, as-a-whole, from its beginnings to its ending. In the case of modernity, Heidegger goes back to Ancient Greece for insight into the beginnings of democracy and philosophy. He considers the ending of modernity with the looming devastation of nuclear war, and even in the 1950 to the 1970s, is crucially aware of the risk to the environment. These two bookends – the birth and the end of a life as-a-whole, or a civilisation as-a-whole – refract back on decisions that are being made in the present moment. Just as Kant has time/subjectivity/space as a priori conditions, Heidegger folds the life as-a-whole through the lens of subjectivity/time/space awareness of the Present Moment. This large view inhibits the 'now' from becoming an ill-considered, hedonistic phenomenological experience and introduces the widest historical moral concerns. It intrajects the systemic processes by examining how they began and how we anticipate they will continue to unfold. It is a refractive method that takes the insights from the beginning and insights from looming finitude to 'begin' the task of examining why we accept normative discourse that shapes this moment, and produces or amplifies the potential future.

The 'death' of modernity presents a line in the sand which we cannot go over (Heidegger and Jünger 2016). There is no recovery from nuclear devastation; the radioactive isotopes pollute the area for hundreds of thousands of years. Or to carry this forward, the radical reduction in habitat caused by capitalist production, rising emissions and global warming is making unprecedented numbers of wild species extinct. There is no resurrection or rebirth from extinction. This end point is final. While it is true that we are all finite, and that all lives, all epochs, even all planets, eventually come to an end, contemplating our endpoint can help us to consider the healthy and wise decisions in the present and how to extend healthy conditions for a longer lifespan. The Holocene would have gradually cooled over the next ten thousand years, maintaining the pleasant stable conditions that have allowed homo sapiens to thrive, except that vast CO_{2e} combustion is drastically foreclosing that stability and transforming the climate into the turbulent Anthropocene.

Understanding early genealogical roots helps to unpack why we recognize and normalise things the way we do. Considering the lifespan as a whole, the horizon of the ending and the birth, or *Ürsprung*, helps to unpack the cultural norms we currently exist in. Heidegger (1999a) calls this refractive and reflective process the two beginnings.

The discussion with Jünger revolved around the line of nihilism, 'Over the Line' (1999a). Jünger wrote a *Festschrift* to Heidegger, in which he took a Nietzschean stance on nihilism (Jünger 1993; Zimmerman 1990). Nihilism is a loss of meaningfulness and requires a transformation of values at the culmination of modernity (Nietzsche 1909). The nihilism of fascist violence, surveillance and the 'ordering of order' (Jünger 1993/1930; Heidegger 1977: 17), could trigger a deep-set reaction and revaluation of values that overcomes modernity and transforms it into something more communally and ecologically oriented. In contrast, Heidegger's argument in his 1955 *Festschrift* to Jünger 'Über das Linie' which translates as 'Over the Line' or 'Concerning, the Line' (1999a). Heidegger argues that there is no 'after' modernity. Nihilism is forgetting to ask the question of Being (Heidegger 1973a; 1982a; Irwin 2003). People can only transform from within the present moment of modernity by contemplating its two beginnings: the origin and the finitude.

The looming finitude of humanity, whether nuclear or climate change, seems irrevocable, and yet at the same time is caused by modern practices. The will to death, or *thanatos*, described so eloquently in Ancient Greece, is the end of the world – that is, nuclear devastation or, in our case, the mass extinction period of late modernity and climate change (Hansen 2016; WWF 2022; IPBES 2019; Goulson 2019; Tong et al. 2019). In 'Concerning, the Line' (1999a), Heidegger's focus was on the line, itself, rather than a transvaluation of values *overcoming* the line. His reasoning is an atheist one. Heidegger had grown up in a deeply religious Catholic household with enormous iconography and reference to an afterlife. The text 'concerning, the line' strongly rejects the notion of an afterlife and, instead, presents a methodology where the finite ending of death has no modifying elements to soften it. The finitude of death is a refractive limit that brings the exuberance of life into focus. Death is nonbeing. The end of a civilization is a form of nonbeing because the conceptual framework that enabled that civilization is dead. Nonbeing is the necessary counterpoint to the sacred valuation of being itself. For Heidegger (1982a), the transvaluation of all values happens by contemplating the finality of finitude. 'Perhaps' Heidegger wrote to Jünger, 'the zero-line will emerge suddenly before us in the form of a planetary catastrophe' (Heidegger 1999a: 297).

For the most part, Heidegger regards the ending as creeping up on us. It is not a sudden zero-line but a zone, and we are mired in it with little recognition, for the most part. The zone of nihilism is as forgotten as the multiple layers of meaningful Being.

> The essential possibilities of nihilism can be pondered only if we think back toward its essence. I say 'back' because the essence of nihilism is not already its end. With the consummation of nihilism there first begins the final phase of nihilism. The zone of this end-phase, because it is pervaded by a condition of normality that sets in, is presumably unusually broad. For this reason the zero-line, where the consummation becomes an end, is not yet visible at all in the end. (Heidegger 1999a: 297)

Recognising the finitude of nihilism, particularly the existential threat of climate change, allows us to fully engage with the emotions it evokes (Bright and Eames 2021); first fear and denial, then acceptance and either inauthentic apathy or a transforming, life-affirming concept of what is truly meaningful and what is superfluous.

To get there, however, Heidegger does not take a Buddhist route through this reflective methodology. Nor does he apply a Husserlian focus on phenomenology. It is not a mediation on the body or the breath, or a quieting of the mind. The ending, as a zone of nihilism that draws closer to the line of finitude, is important for revaluing values *now*. For Heidegger, there is no afterwards where transvaluation can occur. In the text *Contributions to Philosophy (From Enowning)*, Heidegger argues that transvaluation emerging in the present is the *event* (1999b). That event has to be understood in the context of the whole – the lifespan of the subjective person experiencing it – and the lifespan of the epoch – from its inception in the originating beginning and the finitude of its looming ending. The event is not the same as the present moment, measured in a second or a millisecond. It is not the Buddhist meditation on nowness. Meditation has already been co-opted into the machinery of capitalism, as accelerating technology and greater work demands require people to manage their well-being and health to keep their place as viable human capital (Foucault 2010). For Heidegger, the event is evolving in the context of the civilisation as-a-whole.

The way I see it, the event emerges 'now', contextualised by multiple layers of evolving time. It is possible to take a more seamless approach to the Kantian a priori conditions of time, space and subjectivity, like Einstein's theory of relativity, so that time/space includes the subject in the weft of everything rather than segregating the experience through an isolated individual (see previous chapter). The event recentres and decentres subjectivity at the same time – not as an enlargement of *Dasein*, in the way that Heidegger evokes, but as a singularity in a relative field. This is a more defractive reading (Barad 2007). The weft of relativity takes hold at a range of different paces or speeds, all with a tug on the event that is presently unfolding. They operate in a coalescing, a collision, an entanglement with super large objects and super minute ones. They are all these things together, while each has its own pace and trajectory. It involves the body's metabolism, the transient breath – which is both self-regulating and exchanging 'me' with the atmosphere – with other creatures and plants and dust and oxygen. The emerging event is the tool functionality of a smooth, unstriated world (Deleuze and Guattari 1999). It is the normative background of consumerism and, at the same time, it is the failures of specific items (the underfunded, overburdened health system, erratic Wi-Fi and the condensing of space/time, surveillance, the crooked chair, monoculture palm oil trees replacing eons of diverse forest, the 44,000-year oscillation of the planet, sudden planetary overheating, the expansion of the universe, the event horizon) the failures of concepts to remain cogent (the concept of progress, economic growth, status as financial success), the failure of the epoch to stave off its finitude (nuclear devastation, climate change, extinctions) the end of the world – that already occurred for indigenous people everywhere (Stein et al. 2022), the end in extinction, the end of modernity.

By looking at the modern epoch from its beginning to its ending, modernity is no longer 'universal' or determining but, instead, is a historical period. All historical events have beginnings, middles and ends (Spengler 1918/ 2002). This contextualises the determinism of the modern technological *Gestell*, or horizon of disclosure (Young 2002). Modernity has been epitomised by a faith in the universal nature of science, linear cause and effect, the ordering of order (Heidegger 1977: 17) and modern concepts of distribution and individual, rational subjectivity. This feel of modern culture is a-historical, as it expanded, colonised and took over older, unique cultures all over the world. Modernity replaced their particular, peculiar modes of understanding the world and, more recently, co-opted these ancient ways of worlding into fixed, outdated, quaint, tourist attractions rather than living, dynamic modes of culture. Heidegger himself was not quite able to perceive the consequences of this historicization of the enframing of modernity and continued to valorise modernity over and above other more ancient cultures – unless they had direct precedence for the modern epoch, as he argued in regard to the democracy and philosophy of ancient Greece. This may be partly due to his methodology of the two beginnings, that a civilisational epoch is similar to an enlarged personal epoch. Nobody 'else' can change the person's habits but they, themselves. Nobody outside can make the decision to stop addictive practices. The change in fuel and metabolism has to be an internal decision-making process. Heidegger's only concession to these 'other' ancient cultures is the poetic – those forgotten knowledges that still form a hidden part of our truth about the world, although this cultural modern epistemology has obliterated their mode of meaningfully generating the world.

Aletheia

Aletheia is an ancient Greek word for 'truth'. Unlike other terms, such as *artrekes*, *adolos*, *ortos*, *apseudos*, *etymos* and others, *aletheia* means the whole truth, not a counterpoint to falsehood or accuracy. The core of the word is *lethe*, which means 'hidden', which Heidegger (1973a) sometimes phrases as concealed or forgotten. The 'a' at the beginning of *aletheia* means 'not', so its literal translation is 'not-hidden', or 'revealed'.

As mentioned earlier, according to Heidegger (1975; 1973a), there are three modes of helping *a-letheia* be revealed: failure, origins and poetics. Failure helps the forgotten background norms to show up, as a vital cog in the machinery of the system as a whole. In the originary and creative phase, Heidegger thinks essential meaning is clearer to understand, and *poesis* enables an intuitive apprehension of things as they offer to be known. Heidegger builds his concept of truth and epistemology around this core concept of 'revealing'. Yet it is important not to confuse revealing, or *a-letheia*, with the Enlightenment, even though both sometimes talk of things coming to light.

Enlightenment epistemology is a more mechanistic, 'essential' view of the universe. Enlightenment thinking assumes there is one essential truth underlying and constituting the elemental character of a thing. Whereas Heidegger, in similar fashion to Spengler, regards the 'essential truth' to be framed and made possible by the basic cultural and technological framework of the historical epoch. Spengler (2002/1918) argues, for example, that the ancient Greeks essentially believed in a finite universe, where the cycles of the planets revolved about the earth and were all within a spherical boundary of the heavens. Occasional evidence to contradict this earth-centric cosmology was heretical, and various mathematicians, including the renowned physicist Hypatia, were put to death for espousing a solar-centric rather than earth-centric universe. Spengler (2002/1918; 1932) argues that modernity is essentially about the infinite. So, we are constantly seeking smaller and smaller or larger and larger units of knowledge, with no 'bottom' to the pyramid. For Heidegger, it is modern technology that has made this view not only possible but also inescapable. The telescope and the microscope have irrevocably changed the phenomenological experience of human perception.

But while the telescope has revealed greater and greater detail of the infinite heavens, it has also obscured older, ancient Greek (and ancient Mesopotamian) ways of regulating culture, economics and politics through an earth-centric lens. The tempo of culture has shifted from a seasonal, finely-tuned-to-the-environment pace of sustainable reproduction to a view of infinite, technological progress and a mining of 'resources'.

THE FIRST BEGINNING

Heidegger's methodology of the two beginnings is useful for clearly distinguishing the normative functioning of discourse within an epoch. He focusses on the 'first beginning' – the leaping-off point of modernity in *Early Greek Thinking* (1975). For Heidegger, this involves looking closely at Parmenides, Heraclitus (540–480 BC/2001), Aristotle (384–322 BC/1972), Plato (427–347 BC/1965) and other 'great' thinkers of the Classical Greek tradition. He does not look much further than his classical education suggests – although his reading of these great thinkers is, in many ways, unique and important. Heidegger (1999b) derives a very dynamic understanding of Being and becoming. Truth as *alethia* is a close interaction between the agency of things which present themselves to a person for active apprehension. Agency operates for both parties – in contrast to the passive observer – the observed dyad suggested by Descartes. This will be more important in chapter 10 on quantum mechanics and taxonomy.

According to Stiegler (2018), Heidegger tends to limit himself to ancient Greek *philosophy* rather than mythology (this seems an overstatement; however, Heidegger did not specifically look at Prometheus that I am aware). This suggests that Heidegger misses a crucial element of Greek thought in the myth of Prometheus and his brother, Epimetheus, and the gift of fire. Stiegler (2018) describes their narrative as a 'double' failure, which brings the origins of culture itself into the technological and extends the concept of enframing from modernity to all cultures. I argue that both Stiegler and Heidegger miss the women in the origin myths – Gaia, Demeter, Persephone – the goddesses of fecundity and the cycles of death and rebirth (Irwin 2019a). These goddesses signify the core of the forgotten cycles of interconnection and (in) consistency of messy aliveness (see chapters 7, 9 and 10).

Neoliberalism is the exemplar, *par excellence*, of the modern technological horizon of disclosure in political parlance. The invisible hand of market balance is touted as universal by neoliberal thinkers. Market balance removes the necessity for political lobbying or action. The depoliticization of people has had far-reaching consequences. With its emphasis on universal rationality, identity has been alienated from history, culture, class, gender and sexuality (Benson 2020). The historical embodiment of consciousness almost disappeared from the political scene once neoliberalism dominated political and policy narrative. Communities were dissected into a-historical, universal, individuals, who each make self-interested, rational consumer decisions. In turn, these decisions can be aggregated by policymakers to anticipate future demand and ensure continuity of supply and distribution (see previous chapter on Kant). Emotional embodiment was an important feminist trope before neoliberalism dominated public discourse (Merchant 1980; Kristeva 1982; Irigaray 1985; Grosz 1994; Braidotti 2013). Both feminism and embodiment disappeared for several decades and have only recently re-emerged into public discourse.

From its origins in fire, the Promethean promise of technological innovation and the development of human culture has reached its current apotheosis in information technology (IT). For Heidegger, the culmination of metaphysics as the ordering of order is in cybernetics (Dupay 2011; Dicks 2011; Heidegger 2012; 1993a: 434–35; 1993b: 108). Technological progress was touted as reducing the need for work, giving more time and leisure to all classes. Far from saving us from the accelerating pace of modernity, Stiegler (2018) argues that, more recently, the miniaturisation of technology has begun to accelerate in pace and spatial distance is collapsing as technology moves into a new phase: quantum computing, nanotechnology and robotics. In an ironic double bind, as technology continues to proliferate using the human vector of creativity, human culture and the enframing of thought is becoming more proscribed. Late modern technology has accelerated to such a rate that no one is capable of keeping up with all the new developments. The epistemological impact of this new enframing is a sort of mass ADHD (attention deficit hyperactivity disorder), or attention deficit, where we are all too often in a zombie state, passively absorbing media rather than actively researching or thinking for ourselves about events in the world. Algorithms are an example – right-wing think tanks have been hiring IT specialists, such as Cambridge Analytica, to use big data to analyse millions of Facebook entries and discern who is vulnerable to 'fake news' (Cadwalladr and Graham-Harrison 2018). They have created sophisticated media campaigns designed to swing voters to the right through targeting vulnerable people to potentially vote for policies such as Brexit. The bubble effect of general algorithms gave left-wing voters little access or awareness of these campaigns, and so there was little push back. The sites that were targeted, such as Facebook, have long been emphasising their rights to 'free' internet from regulation, where they had no obligation to fact-check or censor material, even when it defamed people or blatantly lied. The passive acceptance of media in the zombie state of fast capitalism has allowed influencers pervasive effect (Irwin and White 2022).

The originary thesis of democracy, as a political method that put into action the 'general will' of the people (Rousseau 1892/1762), has been seriously damaged by cybernetics. Tied together with the zombie state of passive absorption in the algorithmic bubble, the early democratic rhetoric of 'freedom' has shifted from political will to passive consumerism. Free will used to mean political resistance to feudal totalitarianism. The original conception of free will is a negative freedom from being subjugated by a monarch. That evolved into the concept of neoliberal freedom *from* regulation, which is reinterpreted as freedom *to* shop.

The acceleration of technology has 'disoriented' our sense of self. It further deepens the 'forgetting' that Heidegger refers to frequently (1973a; 1999b), where we forget to notice any modes of Being outside the technological horizon of disclosure. It is a double failure – the recognition of the CO_{2e} calamity as a failure and the failure to recognise the depth of the crisis in the system. It is a double failure which attempts to ignore the systemic collapse signalled by the society of fear (Beck 1992). The IPCC has modelled the calamity of business as usual (BAU), but modernity remains committed to the intact system of BAU, with minor addendums that substantively change little, such as renewable energy and recycling. Like the frog that fails to jump out of the pot and is slowly boiled alive if the temperature increases very gradually, we are failing to notice the failure of capitalism – in climate change, deforestation, widespread ecological destruction, extinction at unprecedented levels, water contamination and the exhaustion of the water table, loss of soil, loss of insects because of poisons and monoculture, loss of healthy reefs as the ocean gets too hot and the list goes on (Hansen 2016; WWF 2022; IPBES 2019; Goulson 2019; Tong et al. 2019). The alienation of consumerism from the origins of productivity in the natural cycle is so complete that we fail to notice these failures. It is a different example of the doubling of failure but just as defining as the one noted by Stiegler (2018) in the origins of technology and culture, with the gift of fire. Neither Heidegger nor Jünger or Stiegler have moved far from Spengler's succinct pessimism in *Man and Technics: A Contribution to a Philosophy of Life*.

> Man was, and is, too shallow and cowardly to endure the fact of the mortality of everything living. He wraps it up in rose-coloured progress-optimism, he heaps upon it the flowers of literature, he crawls behind the shelter of ideals so as not to see anything. But impermanence, the birth and the passing, is the form of all that is actual – from the stars, whose destiny is for us incalculable, right down to the ephemeral concourses on our planet. The life of the individual – whether this be animal or plant or man – is as perishable as that of peoples of Cultures. Every creation is foredoomed to decay, every thought, every discovery, every deed to oblivion. Here, there, and everywhere we are sensible of grandly fated courses of history that have vanished. Ruins of the 'have-been' works of dead Cultures lie all about us. The hybris of Prometheus, who thrust his hand into the heavens in order to make the divine powers subject to man, carries with it his fall. (1932: 13–14)

THE SECOND BEGINNING

The second beginning refracts back from the looming calamity that signals the ending of civilisation as we know it. In the postwar discussion between Heidegger and Jünger (2016), they are both worried about the atrocities of industrial warfare and the escalating nuclear Cold War. Both have quite different approaches to nihilism and the form that emerging values take in the proximity to finitude (Irwin 2003).

Jünger (1993) was exploring Nietzsche's concepts of the transvaluation of values, and of the 'superman' or *Übermensch* in *Thus Spoke Zarathustra* (1982). Jünger speculated that stepping over the line of nihilism would completely transform the conceptual apparatus, our relation to technology, even the physiological parameters of what shows up as 'human'. The frightening spectre of nuclear devastation and the more insidious surveillance and fascism of Nazi industrialism clearly shows the threshold of nihilism. In response, our awareness of the threshold of possibility becomes crystal clear. The topic of looming finitude is very traumatic, and it stimulates the fight, flight or freeze response.

In a novel published during the Second World War, in 1939, Jünger wrote about the importance of peacefulness, that quality so often neglected and forgotten in periods of nonwarfare.

> Sweeter still becomes the memory of our years by moon and sun when their end has been in the abyss of fear. Only then do we realise that for us mortals even this is great good-fortune – to live our lives in our little communities under a peaceful roof, with pleasing discourse and with loving greeting at morning and at night. Alas! always too late do we grasp that, if it offered no more than this, our horn of plenty brimmed with riches. (Jünger 1947: 7)

Capitalism's progressively frantic activity makes these moments of peacefulness more difficult to establish and more difficult to protect.

The demands of zombie capitalism have been exacerbated by the internet. Stiegler (2016; 2018) argues that accelerating speeds of IT connectivity and the obscure black box of machine learning are dominating knowledge exchange and cultural production as never before. Technology is enframing our thinking by stereotyping search engine responses and duplicating and reinforcing like-minded assumptions, while reducing exposure to alternative ideas. Surveillance through phones, facial recognition and store cards has normalised a fascist-like biopower in a plethora of corporate and state pan-optics. As Foucault (1991; 2010) describes, constant surveillance closes freedom of expression and initiates higher degrees of self-monitoring and fear. The treadmill of low-wage capitalism is exhausting.

Heidegger's (1962) argument is that failure shows us the mode of worlding. A system failure – like the 1929 stock market collapse, or World War II or the high inflation following the COVID-19 pandemic – all show us more clearly the vulnerability inherent in modern consumerism and distribution. A pandemic shows us more clearly how intricately interconnected we are, despite the reverence of individualism as a cypher of modernity. The zone of nihilism is a zone of precarity that brings us closer to existential annihilation. The COVID-19 pandemic has brought home, to many, how close they are to homelessness. Higher numbers of homeless and under-employed people *before* the pandemic can also be understood as the iteration of systemic precarity. The zone of nihilism is not a narrow one, although moments of crisis bring it greater clarity than ever before.

That precarity always exists within modernity, but its tension is released by projecting onto the 'vulnerable' in an individualisation of blame. Blaming a company for having to file for bankruptcy is not seen as a possible outcome of a systemic vulnerability, but rather a belief in the necessary failure of the weakest. Bankrupt companies are understood as those who fail to optimise, innovate and thrive in the competitive market. Failure is co-opted by market discourse. Risk is monetarised in the double-entry bookkeeping at national level (Buch and Dages 2018: 5). By incorporating risk and vulnerability into market universalism, modern capitalism maintains a wilful ignorance about the systemic existential threats that permeate the society of fear (Beck 1992).

Heidegger (1999a) understands the threshold of nihilism as a regenerative force and new conceptualisation of our current relationship with other beings. Thus, his approach to the zone of nihilism that threatens to draw modern culture to an untimely close, is not pessimistic. He regards the line, or boundary of planetary conditions for human life, as a new beginning. The beginning is generated by the contemplation of the finite existential ending. It creates the conditions for rethinking what makes it meaningful to be human. The finitude of the second beginning opens up humanity more fully to the need for care and connectedness with our planet and each other, in the present.

> As meridian, the zero-line has its zone. The realm of consummate nihilism constitutes the border between two world eras. The line that designates this realm is the critical line. By this line will be decided whether the movement of nihilism comes to an end in a nihilistic nothing, or whether it is the transition to the realm of a 'new turning of being'. (Heidegger 1999a: 292)

Ending a civilization, with nuclear devastation or with climate change, puts our lives into a new perspective. Jünger shows us how the fear of death can be transformed. It allows us clarity and insight, and brings into focus what makes us feel refreshed, alive, meaningful and energised. The simple peacetime activities of a community in tune with its local ecology are a vast retort to zombie capitalism. Far from the overloaded, mesmerized and exhausted mood of capitalist consumerism, the fear of death and finitude is transformed into a feeling of invigoration that is on entirely different ground. It generates a new vista for behaviour, our horizon of disclosure, our cultures, our relationship to technology and, most importantly, our integrated awareness with the earth itself. As Heidegger advocates, the revaluation of values comes because we consider the zone of nihilism, not because we go *beyond* it. The zero-line is nihilistic nothingness, and it shows up as the 'turning point of being'. It generates a new mode of viability, but that has to be populated by real actions and direction now, not just vacuous future intentions.

ARMAGEDDON

Large-scale climate change could reduce the planet to full melt; a blue earth, with ocean levels between six to nine metres above preindustrial levels (Hansen et al. 2016: 376) – the thermohaline current that circumnavigates the globe having stuttered to a halt (Hansen et al. 2016). The minerals and heavy metals that the thermohaline current brings from the poles would no longer contribute nutrients to the hot equatorial region, and the heat from the equator would no longer warm the far northern landmass in Europe and North America, making them icebound. Already, climbing global temperatures mean more turbulence (Coumou and Rahmstorf 2012), with more water content in clouds, wild storms, an increase in wind and wave action, significant flooding, soil runoff and, at the other extreme, faster evaporation, an extension of long periods of drought, forest fires and desertification. Swift changes in normative temperature are already making it extremely difficult for the pace of adaption to keep up. Longer summers and variable winters are changing the patterns of insects and plants, making the seasonal cycles of birds out of kilter with their food sources. Right now, the rapid metabolism of earth's resources in the modern economy, especially the continued burning of fossil fuel, is resulting in this level of calamity. For Jünger (1993), 'over the line' is a transformation of values – a fundamentally new regime; but for Heidegger (1999a), there is no 'over' the line. To go beyond the finitude of climate change, there is no longer recognisable modern life. The planet would be a barren wilderness with surviving species, perhaps, but no modern human beings to register them. For Heidegger, the line is resolute. Any paradigm shift has to happen beforehand for reflective humanity cognizant of life to care and to change (Heidegger and Jünger 2016).

While Heidegger (1973a) criticizes Descartes for placing too great an emphasis on *rationality*, he still believes that solipsism is a uniquely human affair. Other species do not have the wit to contemplate – or to care – about life in the big picture but are completely absorbed in the day-to-day conundrums of their species-being. The enframing of the horizon of disclosure (Young 2002) generated by modern technology upscales alienated solipsism into an idealist, universalist, anthropocentric bubble. Solipsistic consciousness, or its absence, is considered irreducibly human by Heidegger and other modernists. They cannot conceive of other species having consciousness or caring about life – even Neanderthals or other early humans.

Considering life as-a-whole on the scale of 'civilisation' rather than 'individual' remains within the rubric of 'care' but attends to the care of the civilised epoch and not just immediate surroundings. For Heidegger, this upscaled solipsism is a very useful method for reflecting back upon our current-day decisions – our politics – and deciding whether they are life-affirming or nihilistic. Yet it retains an irreducible element of Kantian a priori consciousness. Human thought, for Heidegger, is a crucial and unique element of comprehending the world (1973a). The collapse of modernity brings an existential closure to a way of knowing, a way of worlding.

> The human being not only stands within the critical zone of the line. He himself – but not taken independently, and especially not through himself alone – is this zone and thus the line. In no case does the line, thought as a sign of the zone of consummate nihilism, lie before the human being in the manner of something that could be crossed. In that case, however, the possibility of a trans lineam and of such a crossing collapses. (Heidegger 1999a: 311)

Heidegger treads a tight rope, elegantly keeping solipsism intact, and with it the 'special' place for humanity, not as merely rational but as the creature that is capable of caring about Being as a whole. At the same time, he reaches out, beyond the isolated rational individual aggregated by the invisible hand and its universal market rhetoric. The universal market that transforms every element of the known into standing reserve, awaiting the pace of consumerism, is under critique. Heidegger's characterisation of the solipsist subject, *Dasein*, manages to acknowledge an older, fuller, more creative *poesis* of Beings, and of death, or nonbeing. This delicate tightrope maintains the human as a special case, differentiated and privileged with conscious thought, and at the same time, constituted by and inter-relating with all the other beings 'as-a-whole'. Humanity is in a relationship of continuous interpellation of meaningfulness as Beings offer themselves to be noticed, and authentic humans notice, care and are absorbed in the beings that surround them. The nothingness of looming death, or of the finitude of civilisation itself, is fundamental to this process, because it is the backlight that makes the vibrant luminosity of life itself stand out against the empty vacuum of eternal empty space.

Coronavirus pandemics that have breached the species barrier, ecological devastation, forest fires over entire states, reef bleaching and large numbers of species deaths, ocean rise, Arctic and Antarctic ice melt and a plethora of other related problems is backing modernity into a corner. We are the zone of nihilism. Up until very recently, the political class was practising a 'see no evil, hear no evil, speak no evil' policy. But as the impacts of climate change escalate, communities are less and less in the mood for tolerating such obvious gaslighting. People want a cohesive, sustainable direction. The clamour for economic growth and business as usual is meeting a level of resistance and awareness of system failure not seen before. Most people want a mode of distribution, of work and of community-building that is in alignment with planetary ecology, not alienated from it. Heidegger makes it clear *Über, die Linie* – concerning, the line –that crossing the boundary of 'consummate nihilism' is actually impossible. On the other side of that line there is no longer any life as we have known it. The consciousness of modernity will be a moment in the past.

Consciousness itself may persist beyond the boundary of the Anthropocene. The last Palaeolithic event that is reminiscent of this period – the last time that CO_{2e} in the atmosphere was this high – was the Eocene epoch: the extinction of the dinosaurs (Hansen 2016; WWF 2022; IPBES 2019). While most species died out in a very rapid extinction event, eerily similar to what is happening right now, a small number of species survived, and have gone on to thrive and multiply, especially after several hundred thousand years, once the planet cooled to more equable and stable temperatures. Heidegger's zone of nihilism is immanently tied to the modern anthropocentricism of Homo sapiens. Heidegger's understanding of birth and the origin of modern civilisation is wholly unrelated to maternity or seasonal fecundity, and his understanding of death is entirely unrelated to the ongoing unfolding cycles of species regeneration. He sees a lifetime from within the subjective horizon of that life, not from an evolutionary, species, or seasonal perspective. Heidegger's methodology was to expand the concept of *Dasein* from the individual solipsist subject to the greater historical epoch of a civilisation. But because he has insisted on a methodology of 'self'-reflection, he has denied the input of 'other' cultures, 'other' species' and even the differentiated otherness of the geologic earth itself from the overweening determinism of the technological horizon of disclosure. As Koko the Gorilla signed when she was told about climate change – 'humans are very, very clever, and very, very dumb' (Noé 2015; Koko 2023).

In my view, Jünger's open-mindedness is prescient, and the powers of transformation involve better recognition of shared expertise, shared tool use and a shared planet with others. Those 'others' are proving more personable than modernity knew. What needs to be done, in philosophical terms is to unpack solipsism as a field of relativity, with a better, entangled, conceptualisation of onto-epistemology. Once we have established a taxonomy of interrelated fields, similar to the indigenous understanding of the world, we are in a better position to come back to the metabolism of economics and think about production and redistribution in the context of an enormous global human population, high demand and expectation for 'civilised' lifestyles and the precariousness of supply chains and infrastructure in a post-oil and post-growth economy.

To open our eyes to these others, and to set economics in a more appropriate historical context, it is important to examine the first beginning, in terms of philosophy, mathematics, technology and economics with a wider range of inputs than selectively from the Ancient Greeks.

CIVILISATIONS AND THE DECLINE OF THE WEST

Spengler, Jünger, Heidegger and others emerged from the First and Second World Wars with a strong critique of the philosophy of technology and the ordering of order, which had justified the industrialisation of war and the Nazi death camps (Zimmerman 1990). Under national socialism and, subsequently, neoliberalism, the call of efficiency and consumerism dominate all other reasons for organising and understanding things.

Oswald Spengler wrote the first volume of *The Decline of the West* in 1918. This two-volume work refutes the concept of teleological progress. The typical discourse of progress begins with the caveman days when humanity learned to control fire and gradually humanity keeps improving, through the iron and bronze ages up through middle kingdom civilisations, until approaching the apex of humanism with modern technology – industrialism, fossil fuel, and then cybernetics, the genome and nanotechnology. Spengler (2002/1918) argues that instead of continuous progress there have been about nine great distinct cultures or civilisations. Each one has followed the same pattern of flourishing and then decline. They have developed their own unique signature style, or essence, that has informed their religion, mathematics, science, technology and culture. As this unique way of understanding the world develops, culture is creative and generative, and civilisation is formed. As they meet their peak, the civilisation begins to focus more and more on rational individual decision-making with less emphasis on the role of fate and history, and this hubris marks the beginning of the 'descent' of the civilisation. They may stay in this degenerative phase for a very long time, but essentially nothing new emerges from the fundamental horizon of disclosure, and gradually corruption, staleness and rigidity make it impossible for the civilisation to adapt to changes as they occur. The implication is that the

West is in its 'evening', or winter phase, a degenerative, rigid mode wholly focussed on individualism and inflexible rationality, which fails to creatively engage and adapt to nature, the needs of society or fate.

Spengler's (2002/1918) insight is to conceive of a civilisation as a lifespan, from its inception in an 'essential' cultural motif to its development into an enduring civilisation, through its decline into rigid rationality, ending by tailing off or in natural calamity that the rigid thinking is ill equipped to creatively deal with. The Anthropocene (Crutzen 2002) is signalled by a number of geological markers in the strata of the earth; the substantial deforestation of the Bronze Age, the soot and pollution from early industrial trains and coal engines, to radioactive fallout from the Trinity bomb in 1945, the ubiquitous plastic particles found as far away as the pristine ice caps and deep ocean trenches, the damage to the ozone layer and the greenhouse effect. Humanism has impacted the planet in profound ways, leaving a lasting marker in the isotopes, the strata and the atmosphere.

By the 1950s, Heidegger and Jünger had entered an extensive correspondence (2016), including a *Festschrift* that Jünger (1993) published for Heidegger's sixtieth birthday, and Heidegger's *Festschrift* for Jünger, 'Concerning, the Line' (1996). Heidegger (1977) wrote another important paper in 1954, 'A Question Concerning Technology', which also emerges from this discussion. Heidegger (1977) argues that the technological horizon of disclosure is not limited to the tool itself. In modernity, *technē* has become more essential, of deeper moment, than the prosthetics of tool use. Everything, whether in its raw state or as a complete product, as a 'knowing subject' or an 'object that is known', are *all* part of the machinery of calculation in the system of consumerism. Spengler (2002/1918) goes even further, arguing that we are not in control of technology but, rather, are the vector of its proliferation.

Failure is a significant trope in Heidegger's approach to truth, meaningfulness and the emergence of Being. Neoliberalism manages to subsume the failures of individuals and companies through a 'winners and losers' discourse of market competition. Losing is incorporated into the good of the larger machinery of consumerism. Instead of failure disrupting the smooth continuity of normal assumptions and illustrating the protocols and values underlying them, failure in neoliberalism actually supports diversifying the avenues of consumerism. Built-in obsolescence is the epitome of late capitalism. If a tool breaks, it generates fresh need for more consumerism, as we need to buy a new one. Built-in obsolescence has now become normalised in the consumer ethos. Tools are *made* to fail quickly, so their replacement continues to feed economic turnover. On a larger scale, catastrophes like the *Baier* shipwreck off Shetland, the *Exxon Valdez* shipwreck off the United States, two very large oil tankers that spilled hundreds of thousands of barrels of oil onto the pristine coastline and ocean are *benefits* for economic growth. The *Exxon Valdez* cost millions of dollars to 'clean up' (if that was possible), bumping up GDP that year (Waring 1990). Failure and risk are seen as opportunity for new potential profits and a stimulus for technological innovation. In this way, neoliberalism has defanged the danger of failure as a lens to refract and reflect, to generate clarity and critique about the system as a whole. But nevertheless, despite the universal claims of the invisible hand to permeate all aspects of things into the market, the neoliberal gameplan is not in control of someone's experiences when things *do fail*. Particularly when whole communities or, especially in the case of COVID-19, when all countries seriously shut down due to a virus, the neoliberal rationale feels inadequate and vulnerable, and it is possible that other, deeper ways of being in the world come to the fore.

Climate change is challenging the scales of failure and risk that have been incorporated by neoliberal market discourse. The scale of the forest fires in Australia (Van Eedon et al. 2020), and then the scale of subsequent flooding, has compromised housing and infrastructure over large swathes of the country. The flooding in Bangladesh and Pakistan suffered huge soil and infrastructure loss. In New Zealand in 2023, landslides caused by immense rainfall have destroyed the wine region of the Esk Valley and have threatened well-established housing in mountain and coastal regions. Coastlines around the world are rapidly retreating, and unfortunately it is no exaggeration to project from the handful of houses that are already falling into the sea, to wide bands of coastal cities underwater. These cannot be regarded as one-off events but, rather, as part of the frequent volatility of the new climate epoch of the Anthropocene (Crutzen 2002).

There are other technological and environmental shifts that are pressing upon 'business as usual'. Computer informatics is in the process of a significant transformation as quantum mechanics is better understood and will replace the present sixty-four-bit binary system that has underpinned computer code since Leipzig. AI and robotics are impinging on middle-class jobs, standardising and replacing the creative thinking and contribution of researchers, lawyers, judges, linguists and more. At the same time, global population is slowing and is already rapidly shrinking in wealthier countries. Japan is the best example of the effects of an aging and shrinking population on economy because it is not muddied with a lot of immigration. From a peak in 2009 of 128,500,000, Japan has 125,400,000 people in 2023, with a median age of forty-eight and a birth rate of 1.38. The population is projected to be two-thirds from its peak by 2060. The working-age population is rapidly shrinking. BAU is built on a financial sector that promotes growth through endogenous fractional banking – in a circuit where population growth underpins continuously expanding consumer and housing demand. That will stall and reverse once global populations peaks and begin to fall (Irwin 2015a). 'Degrowth' is looming, by decision and planning, or through inadvertent developments in technology, resources, population and climate change. The finitude of the zone of nihilism gives people plenty to reflect upon.

Economic growth is the core problem for modern nations to adapt to the contracting population, planetary resources and increasing turbulence of the climate. The next chapter explores the 'first beginning' in Axial cultures and how technology intersects with awareness of planetary cycles to generate sophisticated cultures who are better aligned with ecology.

Chapter 7

Ancient Technology, Time and Debt

Climate change is bringing about a series of events that are rapidly challenging the possibilities of life as we know it. Any attempt that we make to overcome the emissions causing climate change are reabsorbed in the constant pressure to increase economic growth. In this chapter, I go back to the first machine, the initial example of clockwork, to investigate how the modern horizon of thought is enframed by technology (Heidegger 1977). The Antikythera mechanism was made before 206 BC and demonstrates how the 'essence' of the technological *Gestell* is, indeed, 'nothing technological' (Heidegger 1977). Early Greek thinking relies on even earlier Mesopotamian mathematics (Freeth et al. 2021), and earlier technologies of writing, accounting and economics. These Axial philosophies lend an important counternarrative to the dominant modern discourse of exponential 'growth' by setting up complex cosmologies of cyclic temporality, that show up in religion, philosophy, technology, governance and economics. Unlike the modern cypher of aggregated utility maximising in the free market, both Mesopotamian and, later, ancient Greek communities were highly wary of unbridled greed. Mesopotamian leadership, legislation and religion sought to control the inbuilt tendencies to growth inherent in models of debt and accumulating interest. Axial philosophies already tend towards anthropocentricism, placing humanity over and above the rest of the ecological system, Yet at the same time, being more closely tied to the physical constraints of ecology, Axial philosophy actively pursues antidotes to exponential growth and aims for a type of steady state economics. These early cultural values offer ways of understanding long-term civilisation that does not produce the catastrophe of climate change.

TECHNOLOGY AND BEGINNINGS

Heidegger offers a way to think about modernity as an epoch, from its early roots in ancient Greek ideas and democracy to its recent iteration in surveillance society, pollution and resource exhaustion in the Anthropocene. In one of his most famous texts, 'The Question Concerning Technology', Heidegger (1977/1954) argues that a profound shift in worldview took place with the transition from 'craft' technology to 'modern' technology. The technological worldview both frees us and alienates us from our ties to local ecological realities and our sense of belonging to place.

Heidegger (1977) finds the roots of modernity in the ancient Greek approach to the *telos* of technology, best outlined by Aristotle (384–322 BC/1972). Aristotle carefully teases out the elements essential to creating technological items. The 'efficient' causes include the skill of the craftworker and the materials that the craft is made from. The intention, or 'final cause', of the item dictates that it becomes a 'bowl' rather than a 'chalice', or a 'spoon' rather than a 'chair'.

Something shifted with the transition to mass, industrial technology. It is not the mass production itself but, rather, that scale of production and consumption is an outcome of this new 'essence' of technology. Heidegger (1977) writes that the essence of modern technology is 'nothing technological'. Instead, the essence of technology is found in the concept of storage. Before modernity, things showed up as 'intrinsically' themselves: as the river, the soil, the plant and the lion. But once modernity was shaped by the new enframing of modern technology, all things became subject to the demands of consumerism. In other words, the lion is on display as a tourist delight, the soil is productive of various crops and the plants are offering 'ecological resource services' by absorbing carbon dioxide. The river is not itself, either, but rather potential energy, potential clean water, potential irrigation. Every aspect of the planet gets co-opted into the rhetoric of consumerism, and when it is in its simple or natural state, it is waiting as potential 'reserve' in storage for later consumption in the capitalist machine. Technology is not limited to the pace and telos of the seasons, or the skills and speed of the craft worker, but to the overall throes of consumerism.

In earlier civilisations, the seasons governed the tempo of production. Stored seed and preserved food helped a community survive the least fertile periods, but the overall constraints of the local ecology set the pace of productivity and consumption. Predicting the optimum period for communities to plant and reap was helped by calendars that were based on solar equinox and lunar periods, along with knowledge of the stars (Budiselic et al. 2020; Leather and Hall 2004).

Modern consumerism is in stark contrast to the ecological constraints of yesteryear. We pride ourselves on being free of the vagaries and dangers of local drought, pestilence, flood or famine. Instead, local communities rely heavily on trade, and when times are tough, they import food and supplies. What is more important than local ecology is storage capacity and transport infrastructure, which allow not only large-scale warehousing of materials but also 'just-in-time' (JIT) production, geared towards the pace of consumerism rather than the pace of seasonal fecundity. The philosophy of time is profoundly altered in the era of modernity.

The 'freedom' of modernity has obvious benefits. Unless the calamity is a 'national disaster' (something that is occurring more and more frequently with climate change) then communities can expunge the impact of local constraints and rely on globalised trade instead of local subsistence. The dangers associated with this freedom are less obvious. In fact, the danger is obscured and hard for local communities to heed. People are no longer aware of the impact of drought on the local trees or higher sea temperatures on fish or algae blooms. People fail to notice that long years of deforestation are also compromising soils and soil runoff. We fail to notice atmospheric pollution and its corrosion of art works, or impact on wildlife, insects, savannahs or ecological niches. We fail to notice the slow reduction in population of other species. Or the lack of clean ground water, or higher tides and the increase in heavy rainfall. According to Heidegger's theory of the technological *Gestell* (1977), we only notice natural resources, including ourselves, as potential reserves to be commodified and consumed when needed. Everything is 'standing reserve'. We may notice the soil depletion, or the lack of forest or the compromised fresh water, but only in their capacity (or failure) as commodified supplies.

EARLY GREEK TECHNOLOGY

This chapter will concentrate on the first beginning, as Heidegger puts it (1996), in Axial cultures. Two civilisations in particular bear huge relevance to our ongoing problems in contemporary modernity. Heidegger (1975) himself often cites 'early Greek thinking' as seminal to modern thought. In this chapter, I want to think about the inception of the technological horizon of disclosure in early Greek technology. Unfurling the technological horizon of disclosure in Axial philosophy opens a new vista on complex economics. The earliest inception of debt (and writing, and a lot of mathematics) emerges in Mesopotamia. Greece is the lynchpin, one could argue, between ideas, artefacts and lifestyle in the early Bronze Age civilisations, and with the threads of continuity and divergence to the modern technological horizon of disclosure. Alexander the Great reputedly sent a great deal of translated Cuneiform scripts back to his teacher, Aristotle, when he invaded Babylon in 331 BC (Lendering 2020/2000). I will argue that there are two interrelated concepts that formulate significant rupture and the root of our despair over exponential consumerism: the philosophy of time and associated techniques of debt regulation.

The pace of consumerism governs the pace of resource extraction in contemporary modern society. Despite the newly emergent discourse of 'sustainability' (Brundtland 1987; *Agenda 21* UNCED 1992), Sustainability Goals 2015), rarely are consumables limited to the pace of 'renewable' or seasonal time (Irwin 2008b). Instead, larger and larger amounts of land are cleared of forest and more acreage is devoted to industrial crops such as palm oil. However, the stress caused by consumerism on local ecosystems is not new. Early Jewish, Christian and Muslim writings advocate simplicity and piety over conspicuous consumption. At the city-state of Ephesus in Turkey, which is Ionian or ancient Greek culturally, there are stone tablets in the middle of the city boulevard admonishing citizens to live a simple life rather than one of profligate consumption. Even earlier, in ancient Sumer, the anxiety over consumerism is felt in the first written epic, where King Gilgameš and his friend Inkadu go up into the mountains and battle the ancient one-eyed giant, protector of the Teak forest (Gilgameš 2000). Indigenous Inkadu tries to stop King Gilgameš from attacking the protector of forests. By killing him, they are able to exhaustively exploit (very problematically) the timber. While staged as a heroic epic, the tale is not unequivocally admiring but, rather, casts doubt on the maturity and wisdom of the king. These earlier cultures, while aware of the dangers of over-consumerism, are not held in thrall by the capitalist market ethos. They have a world view that enables other conceptual apparatus to stabilise the relationships between consumerism, earth and world.

THE ANTIKYTHERA MECHANISM

Storage and machines characterise the technological enframing of modernity. The very earliest cog machinery is not industrial – it is found in ancient Greece. In the Mediterranean, an incredibly sophisticated clockwork machine was discovered in an ancient shipwreck. The Antikythera mechanism was discovered by pearl divers in 1901, as a badly corroded lump of ossified metal with a large cog showing at the front of it. It was found amongst many items of sculpture, coins and amphora in the wreck of an enormous merchant ship. The Antikythera mechanism has been called the first analog computer. It is an amazingly complex clock, with interlocking bronze cogs beautifully precise, enclosed in a wooden box with clock-like arms on the front face. It was made almost 2,200 years ago, more than 1,400 years before clocks were 'invented' in Europe (Edmunds and Philip 2000).

As a time piece, the Antikythera mechanism is the architype indicator of the technological world in which it existed. Along with sand clocks, sun dials and water screws, the Antikythera mechanism indicates that time management was important in Greek civilisation. Time management has also been crucial to the emergence of the industrial revolution. Middle Age clockwork changed the shape of labour and other activities, because people could make more accurate arrangements with church bells and town clocks. From timely attendance at school, factory and workforce, to production timelines and, most recently, to just-in-time production, clocks have enabled the commodification of time (Adam 2006). Time is in a crucially different relation to people in the era of modernity than in earlier epochs, when time was agrarian and dictated by the seasonal urgency to sow, harvest and store. The Antikythera mechanism is the first example of accurate machinery for time management. No doubt, the technology was shaping the horizon of thought in Axial cultures. Yet the sophisticated culture of ancient Greece was not as susceptible to the extreme ravages of consumerism as we are.

The Antikythera mechanism is the 'first beginning' of modern technology. It exhibits many of the characteristics of the industrialised world, with cogs and gears, sophisticated mathematical accuracy and the machinic reproduction of time spans. It is not a clock for minutes and hours. It is a clockwork rendering of the time span of the known solar system, down to the daily rotation linked to the zodiac and lunar cycles (Budiselic et al. 2020; Jones 2020). Yet, its fundamental worldview is, I argue, at odds with the progress driven, teleological, alienated, consumer-based world that we live in today. Modern technology is assumed to unfurl progressively. The Antikythera mechanism contrasts with progress, because it replicates a cyclic view of the universe. With remarkable mathematical complexity, it imitates the cycles of the moon, the sun and, possibly, in its lost components: the five known planets, Mercury, Venus, Mars, Jupiter and Saturn (Freeth and Jones 2012). On the front face is the Metonic and Saros cycles that trace the rate at which the sun and moon line up, over an eighteen-year cycle, and the eclipses that happen during that period.

> The Antikythera Mechanism was a computational instrument for mathematical astronomy, incorporating cycles from Babylonian astronomy and the Greek flair for geometry. It calculated the ecliptic longitudes of the Moon, Sun and planets; the phase of the Moon; the Age of the Moon; the synodic phases of the planets; the excluded days of the Metonic Calendar; eclipses – possibilities, times, characteristics, years and seasons; the heliacal risings and settings of prominent stars and constellations; and the Olympiad cycle – an ancient Greek astronomical compendium of staggering ambition. (Freeth et al. 2021: 12–13)

To further illuminate the cosmology of the machine, the mechanism is covered in Greek writing. The front plate has a description of the cycles of the five planets. On its back plate are month names of the Metonic calendar, eclipse characteristics and Saros eclipse prediction dials. On the back cover is a manual to use the mechanism in three sections: a description of the *Kosmos*, a calendar and the moon–sun cycles.

παρὰ δὲ τοῦ κοςμοῦ κεῖταį c
Alongside the Cosmos lies
(Freeth and Jones 2012, quoting line 25)

The Antikythera mechanism must have been one of the most sophisticated machines of its time period, and in extraordinary self-awareness (so to speak), it refers itself to its own cosmology, or, as Heidegger would put it, its 'world'. It clearly tracks the Olympic games which were major religious and political events. But the Saros and Metonic cycles are also, I speculate, important for timing major political and economic cycles. Just as annual seasons are understood as essential to the agricultural and fishing economies, the eighteen-year cycle gives a good 'season' for larger-term political economy. It is a time span found in many ancient world calendars, including Stonehenge and indigenous Māori *maramataka* (Leather and Hall 2004). The cycles of the Antikythera mechanism, and its earth-centric periodisation of the major celestial bodies, castes exponential growth in a completely outlandish light.

The Antikythera mechanism confounds modern notions of progress, teleology and philosophy of time in multiple ways. I want to argue that it completely reworks the limits of modern technological enframing. It also casts exponential growth into perspective – as a moment in history rather than a universal legislative norm that must be maintained to stave off the class revolution. Teleological, progressive, exponential growth seems utterly unrealistic in the narrative of a cosmology of large, long-term cycles and shorter-term seasons, months, days, hours and minutes. How our world managed to lose sight of the cyclic periodisation of time and arc off in a foreign trajectory from our planetary roots is exposed in a new way.

PHILOSOPHY OF TIME

The philosophy of time indicated by the Antikythera mechanism is quite different from the teleological or exponential model used in modern times. It is following the grand cycles of the solar system, not the hours and minutes of industrial routine. It is not engaged with consumerist instant gratification but, rather, politics and economics that are regulated by the whole of the solar system.

There were plenty of mechanisms to count hours and even minutes in ancient Greece and earlier. Many of these were hydro clocks, sand clocks, or hydro corkscrews. The word 'minute' and the base sixty used to divide hours and days is from the Sumerian language. The Antikythera mechanism was involved with time on a great scale. It was a cosmological world clock.

There is some debate about where and when the Antikythera mechanism was made, probably in Rhodes (Jones 2020) or possibly Syracuse, both part of ancient Greek, or Ionian, culture. The calendar of events included in the mechanism starts with the Olympic games in Nemea and sophisticated astronomical research has shown that the star positions correspond to 206 BC (Freeth 2002a; Freeth 2002b; Freeth et al. 2006).

The mechanism is definitely set in Greece, yet Greek maths had already embarked on trigonometry, and the Antikythera mechanism is firmly rooted in much older, Babylonian geometric mathematics (Freeth and Jones 2012; Carman and Evans 2014). What's more, Greek astronomers were already arguing that the sun, rather than the earth, was the centre of the solar system. Aristarchus of Samos (276 to 194–95 BC) considered the sun as the 'central fire' of the cosmos and had already changed the placing of the five known planets into orbit around it. But rather than exhibiting the most cutting-edge science on its front face, the Antikythera mechanism remains geocentric. The intimate knowledge of the cycles of the planets and the detailed descriptions (written in Koine Greek script) (Freeth et al. 2006) of the eclipses derives from well-established, older Mesopotamian mathematics that emerged thousands of years earlier (Steele 2000a; 2000b; Lendering 2020/2003; Freeth et al. 2021). The Antikythera mechanism can be described as the 'first' analog computer, a distinct marker of the sophisticated metal machinery that shapes modernity. But it is, itself, firmly entrenched in a much older form of technological enframing – in the cyclic world of ancient Axial cultures.

The technological enframing of the ancient world was based on the cycles of the planetary system and, particularly, the Metonic and Saros cycles. The Metonic cycle is 235 months, which aligns the cycles of the sun with the lunar cycles. The Saros cycle is 223-month period (or 18 years and 11.3 days) and shows the eclipses (Lendering 2020/2003). The Antikythera mechanism also traces the nine- and eighteen-year cycles of elliptical lunar and solar orbits (Jones 2020).

> The germ of the Mechanism was the idea of coordinating the two kinds of time reckoning that were central concerns of Greek astronomy already in the fifth century BCE, if later testimony can be relied on: the lunisolar calendrical structures that pertained to the religious, administrative, and civil life of the Greek cities, and the 'parapegmatic' years defined and subdivided by solstices,

> equinoxes, and the rising and setting dates of stars and constellations and associated with natural phenomena such as seasonal weather patterns. (Jones 2020: 138)

Freeth et al. (2021) suggest that the Antikythera mechanism fudged the debate between solar- or earth-centric astronomy. The makers calculated the erratic orbits in relation to a heliocentric system but retained the ordering of the planets in the earth centric normative cosmology of the ancients. The front cover inscription of the mechanism,

> lists the *synodic cycles* of the planets (cycles *relative* to the Sun). This is a systematic list, itemizing the synodic events and the intervals in days between them. The planets are written in the same geocentric order as the [back cover inscription]. Adding Moon and Sun gives the *customary cosmological order: Moon, Mercury, Venus, Sun, Mars, Jupiter, Saturn.* (2021: 2, italics in the original)

Significantly, the Antikythera mechanism makes use of the best astrological and mathematical knowledge possible *and retains the normative geocentric cosmology for the politics of the time*.

Important to my argument is that the cyclic nature of the cosmos governed the politics, ethics, social relationships and economics of Axial civilisations. It throws into question why early modernity – in the beginnings of democracy and capitalism – developed the idea of infinity and faith in exponential economic growth.

Greece was not a democratic culture in the same universal way that we now attempt to practice democracy (bearing in mind that women's equality is still problematic and the limited franchise of prisoners and 'foreigners', and even slavery, still permeate nearly all modern democracies). Yet despite the franchise being limited to male property owners, Greece was more fundamentally egalitarian. The disparity between the very rich and the very poor was significantly less than it is today. At the same time, it was a culture that took slaves in combat, that had debt slaves and that treated women as less-than citizens. Living conditions were sophisticated. They had the wheel and cart, traffic rules, running hot water, sewerage systems, metal instruments, knives and forks, artificial light, large libraries full of learned scrolls and most of the *accoutrements* that we recognise as the modern way of life. Already in ancient Greek culture, needless consumerism was seen as a problem. Indicated by the writing of the Stoics and public announcements such as the large stone tablet in Ephesus, imploring local citizens to refrain from over-consumption. The Stoics made an entire ethics out of living simply. Some of the most famous philosophers, like Epicurus, were applauded for their low consumer, rural lifestyle. Urban centres tended to have the same problem of forgetting to be ecologically aware, as industrial centres now. The physical ecological limits, whether seasonal or as a single extracted resource, are little known in urban centres, and people tend to consume without regard for the wider environment. I want to argue that geocentric cycles, such as the Saros and Metonic cycles, were designed to regulate consumption practices to avoid the dangers of alienated, exponential consumerism and to keep citizens bound by the annual seasons and the long-term cycles of a planetary, solar and lunar cosmology.

Scholarship on the Antikythera mechanism has concentrated on how it was made and the sophisticated mathematics involved. Prime numbers and multiplier gears are used to reproduce the planetary cycles and include the variation of ecliptic orbit with a slot and gear system. There are at least thirty cogs, and there, originally, were probably twice as many (Price 1959; Edmunds 2014).

> The researchers realized that the ratios of the gear-wheels involved produce a motion that closely mimics the varying motion of the Moon around Earth, as described by Hipparchus. When the Moon is close to us it seems to move faster. And the closest part of the Moon's orbit itself makes a full rotation around the Earth about every nine years. Hipparchus was the first to describe this motion mathematically, working on the idea that the Moon's orbit, although circular, was centred on a point offset from the centre of Earth that described a nine-year circle. In the Antikythera Mechanism, this theory is beautifully translated into mechanical form. 'It's an unbelievably sophisticated idea', says Tony Freeth, a mathematician who worked out most of the mechanics for Edmunds' team. 'I don't know how they thought of it'. (Marchant 2006)

The link to Mesopotamia is through mathematics and astronomy:

> The designer chose to make the scale primarily a spatial rather than a temporal one by inscribing thirty degree-divisions in each sector, anticipating that the scale would serve to track the revolutions of other heavenly bodies, not just the Sun. This use of degrees of longitude can be seen as an intrusion into the parapegma tradition from the later Hellenistic engagement with Babylonian mathematical astronomy. (Jones 2020: 139)

The Antikythera mechanism was not merely serving as surplus luxury goods to place ostentatiously on the mantelpiece, as some people have argued. To the contrary, my view is that the Antikythera mechanism was *worlding*. It was relying upon, and supporting, a specific approach to interpreting the ancient world; one that invoked a cyclic approach to time in its economic, regulatory and legislative governance (Marchant 2016). A cyclic approach to governance is vastly different from the exponential growth model that characterises the consumerist underpinning of teleological progress.

The cosmology involved in the Antikythera mechanism is mentioned but has not been developed with any depth (Freeth et al. 2021). To understand the worldview of the Antikythera mechanism better, I decided to investigate the culture from where the mathematics derived: Mesopotamia.

Tony Freeth and Alexander Jones, academics who have devoted their careers to understanding the mathematics of the Antikythera mechanism, make the direct link back to Mesopotamian mathematics, astronomy and cosmology, or worldview.

> The discovery that many astronomical phenomena are periodic was one of the foundations of the Babylonian astronomy of the first millennium BC. Records of dated observations beginning in the seventh century BC if not earlier, now called the Astronomical Diaries (Steele 2000b). Astronomers in Babylonia identified time intervals, generally shorter than a century, which separate very similar occurrences of a single kind of phenomenon. For example, the Saros comprising 223 lunar months, which separates lunar eclipses of almost identical magnitude and duration, and the Metonic cycle comprising 19 solar tropical years and 235 lunar months, which separates full or new Moons at which the Moon is at almost identical longitudes. . . . Besides the Goal-Year periods, many other approximate periods are attested in Babylonian texts, including longer and more accurate periods (of the order of centuries) that were built up out of the shorter ones to serve as the basis for advanced methods of predicting planetary motion. (Freeth and Jones 2012)

Freeth and Jones (2012) make similar links to the mathematics and cosmology emerging from Mesopotamian sources, probably around the time that Alexander the Great conquered Babylonia.

> Greek astronomers' records of dated planetary observations began only about 300 BC, and were never as systematic as the Babylonian records, so that their knowledge of period relations, beyond the crudest periods, was derived from Babylonian sources; thus we are informed by Ptolemy that Hipparchus, in the mid second century BC, had a set of planetary period relations that we recognize as the Babylonian Goal-Year periods. The Greeks also applied observational evidence and mathematical algorithms to obtain other period relations as modifications of the Babylonian ones. (Freeth and Jones 2012)

This is confirmed by Lendering (2020/2003) who attributes more accurate measurements of the Saros cycle and other orbits to the Babylonian Chaldean scholar Kidinnu. Alexander the Great sent translations of centuries of astronomical observations in the *Astronomical Diaries* back to Greece, and just eight months later, in June 330 BC, the Greeks had recalibrated their calendar according to the Mesopotamian data.

MESOPOTAMIA AS THE FIRST BEGINNING

Math and astronomy in Mesopotamia were incredibly sophisticated. In 2021, a translation of one of the thousands of clay tablets found by archaeologists called Si.427, from approximately 1900 to 1600 BC, is the most accurate example of the triangular triples trigonometry 'invented' a millennia later by Pythagoras. Mesopotamians used base sixty instead of base ten, which allows for significantly more accurate fractions. Base sixty is still seen in our daily clocks and in the Sumerian names for hours, minutes and seconds. They watched the night sky and recorded the cyclic repetition of planets, comets, sun, moon and eclipses, which were recorded in the *Astronomical Diaries* (Steele 2000a; 2000b). The Mesopotamian assumption was an earth-centric universe, with 'fixed stars' on the outer firmament and elliptical cycles of planets and other bodies within it.

Writing, mathematics and accounting underwent huge development in Mesopotamia. These technologies introduced a sophisticated surveillance system with normalised approximations of behaviour and productivity. The city-states occupied a fragile ecology of river delta, between the Euphrates and the Tigris. Slotsky (1997) writes, 'water levels of the Euphrates, which are recorded monthly in the diaries from –383 (Artaxerxes II. 20), and daily from –338 (Artaxerxes III. 20)'. The clay soil was irrigated carefully, because too much would draw salt to the surface and make the soil infertile. So, the numbers of grazing stock – goats, sheep and cattle – had to be carefully monitored and limited.

Five-thousand-year-old clay tablets with early cuneiform script are usually about the ownership and debt associated with stock.

92 ewes
20 rams
22 breeding lambs
24 [spring(?)] lambs
33 she-goats
4 male goats
27 kids
Total: 158 sheep;
Total: 64 goats,

Which Sinsămuh has entrusted to Dadā the shepherd. He (i.e. Dadā) assumes liability (therefore) and will replace any lost (animals). Should Nidnatum, his

> (i.e. Dadā's) shepherd boy, absent himself, he (i.e. Nidnatum) will bear responsibility for any (consequent) loss, (and) Dadā will measure out 5 kōr of barley. Three witnesses; date; Samsuiluna year 1 (?), fourth month, 16th day. (Carmona and Ezzemel 2007: 185)

Each debt tablet was written by an official scribe and stored. When animals were loaned, the returns were expected based on a normalised accounting of offspring and no deaths in the original herd. Early accounting lacks any idea of disaster, placing families at risk of becoming debt slaves if they could not meet their obligations within the specified timeframe (Oppenheim 1967). Inevitably, accounting for offspring also produces its own economic growth and would rapidly compromise the fragile ecosystem.

The debts were part of a free trade type of market. Slotsky notices that detailed records of trade in major markets were kept monthly (Slotsky 1997: 356). The value of the shekel in relation to the biggest markets – barley, dates, mustard, cress, sesame and wool – were recorded in the *Astrological Diaries*. Price was not regulated. They were, 'observed market prices not predictions (nor Temple prices). The quotations appeared too irregular to have been computed according to some abstract principle. The pace of reporting commodity prices quickened and became erratic when they were volatile, and there were reports of interrupted or suspended commodity sales at these times' (Temin 2002: 48).

By and large, the market muddled along with periods of intensity and languor, much as it does today. The records were keeping an eye on price inflation. The Mesopotamians were also closely monitoring the water levels of canals and the river, and the complex patterns of the heavenly bodies. The market was a part of the greater destiny of the cyclic cosmos.

THE FATES AND POLITICS

Polanyi (1968) argued that it was the large institutions of the temple and palace that originally developed methods of accountability and debt. The temples and palace did not tax the people but instead had their own lands and workforce who produced ale, bread and garments, along with agricultural products for sale. Polanyi's view is that the palace and temples needed to accumulate surplus value for international trade, acquiring metals, wood, dyes and other resources unavailable in the local terrain (Hudson 2002a).

> Babylonian scholars speculated relentlessly on meaningful interconnections in the observed universe, for example between constellations, cities, plants and minerals. In elaborating thousands of examples of hidden interrelations between

realities and ideas, the manifold lists of omens are the outcome of cumulative attempts to embrace the entire universe in a system of reciprocal inferences. (George 2013: xix)

The temples were centres of learning and also advised the kings of the destiny written in the stars. In Babylon, the Temple of Marduk observers wrote 1,200 tablets with daily observations. Every half year, these tablets were transcribed into larger tablets. Each tablet was organised by lunar month sections (Sachs and Hunger 1988; Steele 2000a; 2000b). These are the *Astronomical Diaries* that Alexander sent back to Greece. The month began with 'first visibility of the new moon at sunset',

> and continued with the monthly progress of the moon among the stars and planets. Nightly and daily weather conditions were written down meticulously because they had an impact on visibility. Eclipses, equinoxes and solstices, Sirius phenomena, and the appearance of comets (including Halley's comets of 234, 164 and 87 BCE) were recorded. At the end of the month, there was a final statement about the moon's last appearance and then a recapitulation of planetary positions at month's end, a list of the market values of six commodities that month, measurements of the changes in the water levels of the Euphrates River, and anecdotal historical information. (Temin 2002: 47–48)

Mesopotamia had a political system that held temple and palace separately but included rituals and obligations between them. When the king died, the new king would issue a *mišarum* edict (in Sumerian language), or in *andurārum* (in Akkadian) (Lemche 1979). These were a 'restoration' (Charpin 2010: 83; Hudson 2002a; 2002b) to 'wipe the slate clean' and revoke all debt obligations, free debt slaves and rescind any penalties or interest associated with debt. Debt jubilees persisted for a long period. 'Akkadian andurarum used for Clean Slates from the Old Babylonian through Neo-Assyrian epochs. Like its antecedents, the biblical variation annulled rural debts, liberated bondservants to return to their own families, and returned the land to families that had forfeited or sold it under economic duress' (Baruch Levine 1989 and 1996 in Hudson 2002a).

There are records of many *andurārum* or *mišarum*, and some kings, like Hammarabi, who lived long lives, instituted three or four *andurārum* during their reign. There is a list of *mišarum* from Hammurabi's reign and his son's (Toussaint 2012), but the earliest complete edict is that of Ammisaduqa (1646–1626 BC) which occurred in 1646–45 BC (Lemche 1979: 11). 'The arrears of the farmers, shepherds, knackers, people working in the summer pastures, and [other] tributaries to the palace, in order to strengthen them and treat them equitably, are cancelled: the collector must not take any legal action against a tributary's house' (in Charpin 2010: 84).

> If an obligation weighed upon an inhabitant [of the regions] of Numhia, Emutbalum, Idamaras [or of the cities] of Uruk, Kisurra, or Malgum, and he had to place his own person, his wife, or his children in servitude for debts in exchange for money, or as a security, as a result of the king's instituting redress [mišarum] for the country, he is liberated: his andurārum will be carried out. (Barmash 2020)

According to Charpin (2010: 86), a *mišarum* involves three modes of debt cancellation: cancelling non-commercial debt, returning debt slaves to freedom and returning 'alienated goods to their former owners'. The idea was to restitute the situation to its origin, or 'return to the mother'.

DEBT JUBILEE

Following Polanyi (1968), Hudson argues that debt tended to be owed to the palace or the temple rather than privatised between individuals. The debt jubilee, or *andurārum*, 'returned' conditions to 'normal' and thereby reduced the continuous growth of goods to meet debt obligations.

> Focusing on the worldly economic consequences of royal edicts so as to demystify them, Igor Diakonoff (1991: 234) has emphasized that 'the word andurarum does not mean 'political liberation'. It is a translation of Sumerian ama-r-qi 'returning to mother', that is, 'to the original situation'. It does not mean liberation from some supreme authority but the cancelling of debts, duties, and the like. Also, 'cleaning' is a terminus technicus for 'release from payments'. Dominique Charpin (1987: 39) concurs that the word for 'mother', ama, connotes 'origin' and thus should be translated as 'point of origin', so that amargi signifies a return to (or restoration of) the 'mother situation'. The implication is that economic order existed prior to the imbalances caused by debt. What was needed to restore such order was to wipe away the residue of debt bondage, land forfeitures and other symptoms of dependency brought about by the overgrowth of debt. (Hudson 2002a)

Hudson argues that the system in Mesopotamia tended to impact the major institutions rather than very many individuals, because land was not privatised as much in the early Babylonian period. However, there are tablets that demonstrate that the *andurārum* applied to private transactions too.

> Warad-Sibitti: thus [speaks] Sin-eresh. Relating to 1 mina of silver, part of the silver I delivered to Ili-wekedu: yesterday, I you, and Urisheme sat down and we closed the accounts before the god Shamash. After everything I had received was deducted, I still had a debt of 10 sheckels of silver for him, about which I seized on him in these terms: 'Until you return my money to me, you and I

> will not let go of each other. The king will cancel the debts, a great deal of time will elapse- therefore bring me my money [immediately!]' He replied to me: 'I swore by the king! I will return your money to you in five days'. I received from Paridum the guarantee that the money would be paid in five days. Pursue Paridum's pledge therefore, so that I obtain the money. Please don't be negligent. (In Charpin 2010: 93)

Clearly, the prospect of a debt jubilee or *andurārum* threatens the legitimacy of calling in the private loan. People would only lend if they were reasonably sure they could get their money back, and if an *andurārum* was scheduled, then lending would be tailored to avoid it. My speculation is that a debt jubilee would need to be unpredictable, or the donors – temple, palace or wealthy individuals – would bring debts to a close before the jubilee fell.

The sophisticated knowledge of the stars, and especially of the Metonic and Saros cycles, along with the eclipses and other astronomical events like comets, were regularly reported to the kings by the astronomers. Celestial events helped to define portents and luck for political events. In the book, *Babylonian Divinatory Texts*, George quotes from a clay tablet, 'A lunar eclipse on the fifteenth day of the third month, for example, was paired with a statement predicting the death of a king in a palace revolt (text no. 13: 21)' (George 2013: xvi).

The death of the king was associated with an *andurārum*. Hammurabi lived a long life, and so he held several debt jubilees within the span of his reign. It is very likely that an eclipse or similar dramatic astronomical moment (without the death) would be tied to the timing of an *andurārum* such as clay tablets of lunar eclipses, Late Babylonian MS3118 and MS3117 (in George 2013), or they could have been timed through divination (George 2013: xvi).

These *andurārum* helped offset the economic and ecological impact of runaway growth. They were built into the earliest economic system through a cosmology of cyclic time that helped periodisation – from minutes, hours and days, to lunar months, solar years up to and beyond the nineteen-year Saros and Metonic cycles. This cyclic approach to time was embedded in their maths, their productivity, their religion, their institutional framework and their understanding of the self, governance and power. An earth-centred, cyclic concept of time imbued their technologies, from writing, accounting and resource use, to birth, marriage and death. It was the lens through which all things were understood.

While Mesopotamia did not create a vast oversupply of abstract currency, as in modern times, exponential growth was a problem, simply because of interest on loans. The delicate ecosystem of the Euphrates and Tigris river delta cannot accommodate long term proliferation of extra goats or other stock, so debt jubilees were essential to regulate exponential growth. Debt jubilees were vital tools of governing the health of the ecosystem and redressing growing inequalities. A debt jubilee returns society to its 'original state' or *mišarum*. Debt slaves are freed and the clay tablets with recorded debt are literally wiped clean. Property that had been lost because of debt is returned to its original owners. The government debt collectors forfeit their tax collection.

Obviously, for wealthy people and organisations, there is a large loss of revenue, as wealth is redistributed back to its origin. To effectively counteract economic growth and uneven wealth disparity, it is important that richer people cannot prepare for a debt jubilee by calling in debts before the jubilee occurs. It is important that the timing is sudden and surprising. As I have argued in a couple of other articles (Irwin 2017; 2018), predictable but *irregular* omens, such as an eclipse were already vital portents for the outcome of public events. As one clay tablet explains,

> The signs on earth just as those in the sky give us signals. Sky and earth both produce portents; though appearing separately, they are not separate (because) sky and earth are related. A sign that portends evil in the sky is (also) evil on earth, one that portends evil on earth is evil in the sky'. (K.2847 Obv. 48–42; trans. Oppenheim 1974, in Steele 2000a: 22)

The stars can also indicate when to hold necessary but unpopular events, such as the *andurārum*, or debt jubilee. The shift from more than five thousand years of very successful and stable cyclic cosmology towards a more volatile and self-interested economic system without jubilees was gradual. There are religious scholars who trace jubilees in Abrahamic, Christian and Muslim traditions. A good example is Deuteronomy in the Old Testament, where a debt jubilee is set every seven years. But by the time the monotheistic religions had started to replace the older plethora of gods and goddesses, the jubilee began to lose a lot of its economic structural character, and become a moral prerogative about charity.

The modern analysis of debt jubilees in Mesopotamia tend to make too much of the public versus private debate as reason for this technique to be inappropriate in a modern context. Hudson (2002a) writes, 'An analysis of Mesopotamia's economic takeoff shows a line of development centred in the public sector's large institutions'. He goes on to argue that institutions and political control over large scale resources has become increasingly privatised.

> What narrowed in scope was the system of checks and balances that kept the early dynamics of debt within bounds. Greece and Rome did not adopt the Mesopotamian practice of debt cancellations, nor did they provide the financial escape valves that relieved rural debtors of their obligations when natural disasters (treated literally as 'acts of god') interfered with their ability to pay. When economists pick up the thread of interest-bearing debt in classical antiquity, they

> start with a system whose scale had been shrinking and becoming more privatized and less 'total' in its social scope as it had evolved and shifted its context already for thousands of years. (Hudson 2002a)

My view is somewhat different. The Christian admonishment against usury reduced the scope to charge interest on loans. Whether public or private, when loans are genuinely delivered in real silver or real goods, and require no interest, then there is no exponential growth. Since Italian bankers invented double-entry bookkeeping and loans ex nihilo in the 1460s (De Roover 1963; Buch and Dages 2018), the ravages of exponential growth on ecology have become increasingly extreme. The Mesopotamian approach to debt jubilees, and the cyclic geocentric approach to time and cosmology, are important provocations and counternarratives to modern business as usual (BAU).

IMPLICATIONS FOR MODERNITY

For our purposes, as a thought experiment about how to visualise solutions to the exorbitant disjuncture between the 'decoupled' finance and stock markets and planetary limits, debt jubilees offer a powerful, if not frightening, image. In 2022, the total nominal global (private and public) debt is 250 per cent greater than the collective global gross domestic product (GDP) trade (IMF 2022). This indicates a finance industry that is extraordinarily alienated from the pace of earth's ecosystem. We know from the 2008 crash, that despite the rhetoric of decoupling the knowledge economy from broad consumerism, trading does have deep, real-world consequences for the distribution of wealth and access to resources for ordinary people: farmers, miners, factory workers, hospitals, education systems and so on. A crash will not just affect the taxpayers (who this time have no hope of bailing the banks out) but will directly impact in relation to the type of futures betting that has gone on in the London, New York and Tokyo exchanges. With the severe damage created by hurricanes Harvey and Irma in 2017, the cyclones in the Philippines, Fiji and Indonesia, drought in the Sudan and Middle East, forest fires in California, Central Europe and Australia, and the enormous storms, floodings and landslides in Bangladesh, Pakistan, Florida and New Zealand, it could be the insurance industry that tips over the edge.

A debt jubilee is vastly different from a crash. It returns property to its original owners, redressing hardship. Whereas a crash alienates more people from their property and causes enormous disruption, *exacerbating* hardship. The most crucial thing about an *andurārum* or debt jubilee, for modern economics, is to begin to reimagine a sense of justice and philosophy of ecologically grounded time that is cyclic and earth-centred, instead of exponential.

Instead, in the information age of fast planes, fast energy and even faster computing, time and space have become both enlarged and compressed at the same moment. In a prescient analysis, Marshall McLuhan (1964) attends to the profound physiological and psychological changes taking place as computing opens up instantaneous communication across the entire planet. 'Today, after more than a century of electric technology, we have extended our central nervous system itself in a global embrace, abolishing both space and time as far as our planet is concerned' (McLuhan 1964).

McLuhan (1964) and Heidegger (1977/1954) attribute modern technology with the shift in the concept of time to the modern focus on instantaneous gratification, and the telos of progress, into the infinite. The modern enframing of thought is influencing the influencers. It enables resource extraction as a key focus of capital, without regard for social or ecological consequences. Advocates of fast capitalism, like Davidson and Rees-Mogg (1997), notice the ways technology is changing politics and work in everyday modernity.

> In the Information Age, a 'job' will be a task to do, not a position you 'have'. Microprocessing has created entirely new horizons of economic activity that transcend territorial boundaries. . . . As technology revolutionizes the tools we use, it also antiquates our laws, reshapes our morals, and alters our perceptions. (Davidson and Rees-Mogg 1997: 23)

Davidson and Rees-Mogg (1997) were advising the rich to avoid national taxes and work globally and remotely. For many people, the world of work is rapidly changing in scope, skillset and availability. Technological innovation is unprecedented in the last hundred years. Bostrom (2003a) and Stiegler (2018) believe computing power is already 'faster' than the speed of synapses in the human brain. Stiegler worries that artificial intelligence (AI) is shaping the type of information that searches produce, creating bubbles of passive consumers of information instead of active critical thinking. Bostrom (2003a) takes it further, imagining that the entire planet is merely a historical simulacrum, and that live humans are passively connected to the giant computer simulation in which they 'play' rather than 'live'. Some forecasts predict that robotic automatons will take over new aspects of the workforce, not just in factories but also in health and education.

> New pattern-recognition technologies are rendering both routine and non-routine tasks subject to automation: complex communication technologies are making computers better than humans at certain skilled-knowledge tasks, and advances in robotics are rapidly making technology better at a wide variety of manual-labour tasks. For instance, self-driving cars involve the automation of non-routine manual tasks, and non-routine cognitive tasks such as writing news stories or researching legal precedents are now being accomplished by robots.

> The scope of these developments means that everyone from stock analysts to construction workers to chefs to journalists is vulnerable to being replaced by machines. Workers who move symbols on a screen are as at risk as those moving goods around a warehouse. One report forecasts a 'depopulation of trading floors' as robots continue infiltrating the financial world; retail jobs – long a bastion of post-industrial employment – are set to be taken over by machines; and over 140 million cognitive jobs worldwide are forecast to be eliminated. (Srnicek and Williams 2015: 65)

Looming developments in quantum computing will increase the speed and scope of cybernetics in scale.

The speed and trajectory of the instant and the infinite is snowballing. Exponential growth rested on the Enlightenment-view of the elevation (and alienation) of the subject from the object, of people from nature – in close association was a monotheistic belief in absolute truth and a teleological utopian view of progress and development. This was compounded by Kant's concept of the transcendental individual, where the noumenon has no connection to empirical forces acting upon other forces but is a reified truth. Alienation from nature and a faith in progress became the underpinnings for an exponential approach to time, economics and life aspirations. The increasing speed and quantities of consumerism and technology in late modernity is on a trajectory to the infinite. Elon Musk is aiming to jump off to the rest of the Virgo Supercluster (Bostrom 2003b) after colonising Mars (Junod 2012; Carroll 2013). Alienation from the earth is complete.

All this while, remnants of older cyclic subsistence lifestyles have survived, especially in poorer nation-states. Many communities still rely heavily on a subsistence lifestyle that is far from alienated from nature and has nothing to do with GDP or growth. But as modernity continues to escalate, subsistence is under threat. In 2017, India introduced general sales tax (GST) and a commitment to enforce minimum wage requirements that had to be paid through the banking system. The aim is to bring millions of people into the modern system rather than remain in the subsistence economy. The universal approach to modern economics undermines the plethora of other successful economic systems, and locks more and more people into the exponential growth model. Hurricanes, forest fires, flooding and invasive species are proliferating, along with record extinctions of natives because of climate change (Hansen 2016; WWF 2022; IPBES 2019; Goulson 2019; Tong et al. 2019). These climate stressors are placing subsistence communities under further jeopardy as the level of natural disasters escalates.

The escalating accumulation of CO_{2e} emissions in the atmosphere is partly caused by a committed reliance on the fossil fuel industry and, just as importantly, by the drivers of finance. Fractional banking and future derivatives both dilute the value of the total pool of money in relation to the broad economy and demand economic growth to offset that dilution. This dynamic stability (Rosa 2013) is why many governments have a commitment to 2 to 3 per cent GDP growth and regard higher inflation as a problem. The discrepancy between the finance industry and the broad economy are exponentially widening in late modernity. As the global economy keeps growing, emissions keep accumulating. Weather events are escalating because the mean global temperature of 1.3°C enables the atmosphere to carry 10 per cent more water (Coumou and Rahmstorf 2012). The entire hydrological system is accelerating: evaporation, ice melt, ocean rise, ocean turbulence, perspiration, wind, storms and heavy rain. Rapid climatic changes, deforestation and pollution are already threatening many species that have not adapted to the speed of change (Hansen 2016; WWF 2022; IPBES 2019; Goulson 2019; Tong et al. 2019). Irreversible tipping points are looming. Fresh water polar ice melt will possibly halt the thermohaline current, thus collapsing the planets thermostat (Hansen et al. 2016). This rapid ice melt will expand the ocean surface by thousands of square kilometres, ironically reversing the greenhouse effect, as more ocean is highly refractive. There is high chance these changes will trigger a further rapid change in climate, making the equator unbearably hot, at around 50°C, and the polar regions extend to a substantial region of deep ice. Humanity is not immune from this ecosystem collapse. It is far easier to change the historical norms of economic growth and the institutions of endogneous fractional banking and the stock market than it is to roll back escalated climate change.

Heidegger (1977) wrote that the enframing of the modern technological horizon of thought is 'nothing technological'. Instead, it is storage. It is the storage of resources – either in warehouses or *in situ* – that are all waiting to be demanded by the call of consumerism. Modernity has had to understand everything as resources waiting for the infinite demands of increasing consumer growth because ex nihilo banking has systemically poured far too much nominal money into the economy, forcing the entire system to counteract inflationary collapse, by growing. In this situation, nothing can be understood outside of the call for infinitely faster exponential economic growth. The Antikythera mechanism and the technologies of writing and debt in early Mesopotamia all show us that the conception of time in modernity, which coalesces around the instant and the infinite are an anomaly in the history of human civilisations. It is possible to have a geocentric, cyclic, ecologically aware and highly sophisticated technological society without exponential growth. Mesopotamia and ancient Greece both show us that another worldview can produce stable, sophisticated civilisations that are based on a cyclic concept of time and an ecological responsiveness that we need to redevelop and put at the centre of our cosmology.

Chapter 8

Thermodynamic Economics

Flabby economics does not recognise inbuilt satiation: it continues to grow at all costs. Monetary volume and economic growth are driven by fractional banking, which over-runs renewable efficiencies in a gluttonous excess (chapter 2). Climate change is correlated with economic growth (Le Quéré et al. 2020; Caporale et al. 2021; Mitić et al. 2022; Onofrei et al. 2022; Noll 2023). At the same time as climate disruption is accelerating in the Capitalocene (Haraway 2016), global population is beginning to peak, and robotics and artificial intelligence (AI) are transforming the world of work, making jobs in all classes increasingly proletarianised. Energy, production, redistribution and emissions accounting, housing, infrastructure and capital projects all need to be rethought. In the shorter term, there is huge outlay necessary in creating a renewable energy network, sustainable transport and denser, better-designed and insulated, low-heat housing.

Thermodynamic or metabolic economics offers an alternative way of understanding and accounting for the energy revolution. At a moment, when resource exhaustion, pollution and slowing population growth are bringing the wild west of economic growth to a collapse (Steffen et al. 2015; Irwin 2015a), entropy helps to explain the exhaustion of these systems and changes the rhythm and value of economic production. On its own though, economic metabolism obsesses on accounting for emissions. The danger remains of embedding the 'ordering of order' (Heidegger 1977: 17) in knowledge and economic systems, with mankind as the Master of the Universe.

The planetary metabolism *includes* economic transactions. Economics does not *universalise* the planetary ecology. A universal market guided by the invisible hand is a dangerous mythology that overstates the transcendental alienation of human constructivism at the expense of ecological systems. Multiple disciplines lend concepts about the entanglement or intraconnection of humanity with ecology that help to re-orient the normative morality of economics. An indigenous lens of full ecological kinship with other species – with mountains, rivers, the rocks themselves – can transform our value system towards deep ecological well-being, of humanity, and the planet we inhabit. Feminist philosophy of embodiment, and the entangled intra-action of quantum physics also give vital concepts to update and conceptualise more integrated modes of understanding culture and nature. The contemporary science of thermodynamics and quantum physics have a more complex approach to the intraconnection of things, than the linear causality of Newtonian physics (Prigogine and Stengers 1984). This chapter examines the laws of thermodynamics and how that enables a materialist approach to economic metabolism. The last part of this book delves deeper into intra-action as a guiding ethos for economic principles.

EXISTING MODERNITY

Despite common-sense narratives that the spectrum of politics is widely divergent, I contend that at a macro level, both 'left' and 'right', capitalist and socialist, adhere to the concept of economic growth. The Keynesian justification for the multiplier effect (1933) is that workers will also benefit from access to the advantages of globalised modern consumerism. The safety net of education, health and welfare, along with consumer goods, delivers a share of the profits of capitalism to the working classes. Keynes (1933) explains that state provision generates consumer demands as the marginal disposable income of the poor does not get stored in bank deposits but goes immediately into greater consumerism. Consumerism stimulates the economy and new jobs. Milton Friedman (1970) leaves out state provision of education, health and welfare, but talks of 'growing the pie'. The argument is that economic growth benefits all parties, even those with a tiny fraction of a share (see Klein 2007; Edmans 2020). Communism, too, is committed to macroeconomic market distribution, even if the state micromanages local distribution. Modern communism, too, conceives of everything as a resource waiting for the demand to enter the machinery of redistribution. According to modern modes of political economy, the technological enframing of the modern horizon of disclosure (Heidegger 1977) and its commitment to consumer growth, has universal extension.

In this context, degrowth has been treated with a great deal of suspicion, as an anathema to the common-sense commitment to economic growth that permeates modern society. For policymakers, the neoliberal narrative of rational utility maximising consumer decisions and the aggregate of predictable behaviour through the invisible hand, is difficult to first learn. But once it is understood, neoliberalism provides a compelling narrative that shapes policy in all realms – making it difficult to unlearn. The invisible hand serves to even out a 'perfect' market, that ensures the fairest distribution of scarce resources amongst all stakeholders. Competition and rational individual utility maximising are essential components to the discourse of the neoliberal market. A slow economy is seen as a problem that disturbs the energizing innovation of dynamic stability. Steady state or degrowth cannot avoid the pitfalls of *reducing* opportunities and wealth across the most vulnerable as well as the wealthiest echelons of society. Reducing consumer demand would also reduce employment opportunities and exacerbate the vulnerability of the poorest. Small wonder that despite the obvious correlation between consumerism and emissions there is reluctance to engage with degrowth. But the narrative of growing the pie (Friedman 1970) and the multiplier effect (Keynes 1933) obscure deep failures in equitable distribution (Piketty 2022), political voice and pollution. These inequities need to be uncovered and replaced with better policy directives. By growing the pie without improving equitable distribution globally, some regions are extremely vulnerable to the impacts of climate stressors – both incremental and extreme events (Steffen et al. 2015; Irwin 2022b; 2023b).

Despite widespread absence of discussion, the correlation between economic growth and emissions growth cannot be ignored. Finally, the Sixth Assessment Report of the IPCC has acknowledged the discourse of degrowth, referring to it in the chapters on 'Impacts, Adaptation and Vulnerability' (2022) and the 'Mitigation of Climate Change' (2022). It comments that 'prosperity and the Good Life are not immutably tied to economic growth'. The IPCC has been challenging the business as usual (BAU) model for decades, but its solutions have tended to be neoliberal ones up until now. Overtly embracing 'degrowth' and 'enoughness' are a major new signal that is already making an impact. In 2022, the European Union (EU) delivered a multimillion Euro grant to a degrowth research initiative at the Universitat Autonoma de Barcelona.

There is an extensive literature in 'steady state' or 'degrowth' economics. Frederick Soddy (1933) wrote a critique of exponential fractional banking as early as 1926. After the student revolution in 1968, there was a proliferation of books with alternatives to classical economics, including Georgescu-Roegen's book *The Entropy Law and the Economic Process* (1971); Meadows et al.'s *Limits to Growth* (1972) and Schumacher's *Small Is Beautiful* (1973), which all take a neo-Malthusian view of limited resources. Martinez-Alier and Schlupmann kicked the ball rolling again in 1987 with *Ecological Economics*. Daly's *Beyond Growth* (1996), Gorz's (2008) *Ecologica* (2008) and Leff's *Degrowth* (2009) all engage with Georgescu-Roegen's work on thermodynamics, applying the concept of entropy directly to energy use and productivity. These ideas are gaining much more traction, as can be seen in

Kallis's *Degrowth* (2018) and Hickel's *Less Is More* (2020). Also recently, Tim Jackson's *Prosperity Without Growth* (2017) and Kate Raworth's *Doughnut Economics* (2018) make the case for slowing down consumerism and for higher awareness of fixed resources and the impact of humanity on the ecological capacity of the planet. In related ecological economic work, Rachel Carsen (1962) and, later, Vandana Shiva from 1989 until now, have written numerous books on the toxic impact of fertilizer and pesticides on industrial, profit-driven agriculture poisoning soil and ecosystems.

Fundamental concepts in the literature are often framed as limited resources, or the pace of regeneration (Stockholm Resilience Centre nd). The most fundamental concept, which was phrased as sustainability, before the term was absorbed into the neoliberal framework (N. Gough 1987; A. Gough 1997; Irwin 2008b; Hickel 2019; Higgs 2023), is of metabolic throughput. The conceptual basis of metabolic economics emerges from ecology, from Marx (1887/1867), from energy analysis and from the science of thermodynamics.

Planetary ecology is intricately interlinked. Starting from the capture of solar energy through photosynthesis, there are cascades of material and energy throughout the variety of symbiotic relations between species, geology, water and atmosphere. Waste for one species is often nutrients for another. The ecosystem has evolved unique methods to take advantage of entropy in all its forms.

The concept of entropy is present in Marx's analysis of 'metabolic rift'. Marx wrote about the impact of the land clearances and the beginning of urbanisation and industrialism. Wark explains the cascade effect of entropy in throughput economics:

> Marx's example of metabolic rift was the way nineteenth-century English farming extracted nutrients such as nitrates from the soil, which growing plants absorbed, which farmers harvested as crops, which workers in the cities ate to fuel their industrious labors, and who would then shit and piss the waste products out of their private metabolisms. Those waste products, including the nitrates, flow through run-off and sewers and pour out to sea. Whole industries for making artificial fertilizer would arise to address this rift – in turn causing further metabolic rifts elsewhere. (Wark 2015: xiv, in Marks 2019)

The industrial metabolic rift creates gaps all through the ecological trophic cascade of symbiotic waste and nutrients in the ecosystem (see below) (Dekker et al. 2016). The rifts are 'solved' through global production practices. Global industrial production involves high levels of entropy, accelerating disorder and new forms of order that impact flows throughout planetary ecology.

The laws of thermodynamics which, not coincidentally, emerged during the early phase of industrialisation and combustion engines, are important for understanding economic flows. The first law of thermodynamics is that nothing is ever lost. Matter and energy change form but nothing is extinguished. The second law of thermodynamics is that every time matter/energy goes up a scale in order, disorder also increases. Entropy occurs as a by-product of the increase in order (Prigogine and Lefever 1973).

My thesis is that the ecological scale of entropy and interrelated closed systems can be considered to reduce environmental impacts. They constitute material production, alongside reduction in throughput, lowering emissions from energy sources and the circular economy. Trophic metabolic cascades can be understood in the highly industrialised world as *entrophic cascades*. Waste material from one source of productivity is raw materials for another. One line of production intersects symbiotically with others. Entrophic cascades revise efficiency and waste in terms of environmental impact and marginal returns rather than simplistically drive the unending expansion of profit.

To fully grasp the potential of the 'degrowth' literature, it is necessary to revisit some of the erstwhile externalities of classical economics (Meade 1973). These include finance, energy throughput, pollution, enoughness and behavioural diversity.

BANKING

To fully engage with thermodynamic economics, it is important to include externalities of pollution into the analysis. But equally important is the critique of modern fractional banking, which, as I showed in chapter 3, forces economic growth into the system.

Ex nihilo finance builds in a reciprocal, spiralling multiplier effect that oversupplies money. Prolific money supply raises prices, increasing profit, which stimulates more consumerism. More consumerism grows the economy, providing more jobs to a swelling population who needs more housing. More people require new ex nihilo loans. Modern finance systemically cogenerates exponential economic growth. It utilises Enlightenment beliefs about individual free will and rational choice, and justifies capitalist distribution through the concept of the perfect market. However, Adam Smith had no theory of financial instruments (as outlined). Neoliberalism, following Von Mises (1953/1912), prefers to see money as yet another commodity that should 'balance' according to the invisible hand, as is claimed for other items on the market.

The value of money is usually understood as a proxy for bartering. Marx (1887/1867) regards money as a fetish of capitalism. He writes, 'A commodity is a product intended principally for exchange, and that such products emerge by definition, in the institutional, psychological, and economic conditions of capitalism' (1887: 6). The value of money approximates different commodities, making it easier to exchange across multiple goods and services. Von Mises's book *The Theory of Money and Credit* (1953/1912) makes a similar point but adds money itself as another commodity to be traded on the market. This view is why neoliberals rarely mention money at all, seeing it covered by the general theory of the market.

In contrast, Simmel (1978/1907) argues that monetary value is irreducibly constructed by society. Value is, 'never an inherent property of objects but is a judgement made about them by subjects. Yet the key to the comprehension of value, lies in a region where 'that subjectivity is only provisional and actually not very essential' (Simmel, 1978: 73). This provisional and cultural valuation of money means it always has a very fluid relationship to value. This fluidity is taken advantage of by the banks, when they create new money and introduce it into the existing system.

The London Interbank offered rate (LIBOR), or the overnight cash rate (OCR), is generally understood as the 'price' of money, as banks have to borrow to cover their daily requirements from customer withdrawals, along with normal business overheads. But the price of money does not reflect the unique ability of a bank to generate a new credit line when a suitably employed customer walks in, and guarantees it. The initial loan-created ex nihilo then gets redistributed 'endogenously' throughout the banking system. This is almost a separate system of money from fiat exchange. However, gradually, ex nihilo credit makes its way to service real goods and transactions (rather than inert nonproductive existing housing) and eventually starts to saturate the real economy.

As we saw in chapter 3, fractional or endogenous banking is a very unique commodity. It is literally written into the ledger account when a suitable customer walks through the door. Soddy (1933) argues that the sudden creation of new money dilutes the value of the entire system – known as inflation. Interest rates also force value dilution into the system. Interest is also 'fractional' or re-applied to the loan as the money multiplies across the endogenous banking system. As the capital is re-leveraged across multiple institutions, each time it is reloaned, it attracts a new layer of interest payment. Double-entry bookkeeping generates the new money in the form of a debit to the borrowers bank and a credit in the seller's bank, making the new deposit available for reloaning on further goods and services. A new house mortgage is eventually multiplied over many banks and subsequent fractional higher risk loans. This 'fractional' multiplier, along with all the interest attached to each cascading loan, all needs to be accommodated by the real economy. Economic growth needs to cover the dilution in value and the new demands to repay interest and loan amounts.

Thus, the multiplier effect advocated by Keynes (1933), as a plausible growth in consumer demand that trickles down throughout the workforce, is also needed to generate real work that can offset the dilution of value generated by new money from mortgage loans, plus the value of the interest. Growth is forced into the system. It offsets the new increase in overall monetary circulation and the new demand from fractional interest that goes with it. This is what Rosa (2013) despairingly calls 'dynamic stability'.

Irving Fisher (1936) points out that there is always a 'credit overhang'. There is more credit than transaction value in the fiat circulation of trade. Even if the trading economy grows, it cannot keep up with the extra demand to fulfil the new interest requirements of new credit. The circularity of the system stimulates new loan creation faster than the old ones can be fulfilled.

The correlation between economic growth and emissions growth is firmly established (Le Quéré et al. 2020; Caporale et al. 2021; Mitić et al. 2022; Onofrei et al. 2022; Noll 2023). Jevons paradox demonstrates that the compulsion for further growth outstrips efficiency gains from renewable energy. It creates continuous new demand that absorbs new streams of energy without sufficiently replacing sources from coal and other fossil fuels. Economic growth is not merely a narrative that serves political good will. It is built into the system via the current finance paradigm. To critique economic exponential growth, understanding banking is vital. It is necessary to interrogate the structure of banking as it now functions. It is also important to consider what is necessary about raising capital.

What might be a viable alternative to fractional banking and built-in exponential growth? The literature on thermodynamic economics makes a start on a new and compelling narrative that adequately understands the flows of finance and money. It includes pollution accounting. It comprehends more of the complexity of human behaviour. It includes the impacts on other species and ecosystems. It introduces a discourse that might inform future economic and policy paradigms. In its 1970s versions, thermodynamic economics grappled with broad and complex ideas. Georgescu-Roegen (1971), Gorz (2009/1971), Leff (2009), Martinez-Alier and Schlupmann (1987) and Altvater (2002) all make a strong case for the complex interactions of a variety of levels of economics. I want to focus on entropy, negative entropy and the development of new theory based on both entropy and the ecology of trophic cascades (Dekker et al. 2016) to think about granular levels of micro- and macroeconomics.

FLABBY ECONOMICS

Fossil fuels are the prime example of entropy, upon which the theory of thermodynamics emerged. The work achieved by the combustion engine releases immediate heat and large amounts of CO^2, which have slowly accumulated in the atmosphere (as a reservoir) and created global warming. Fossil fuels have largely replaced labour with energy. This finite resource has leveraged available productivity, in the short term, a thousand fold.

> The shock of our present moment is that this steady and linear increase in productivity, conceptualised as the natural progress toward a tomorrow greater than today, was only ever the product of a temporary influx of energy from a diminishing resource. Rob Nixon writes, 'in this interregnum between energy regimes, we are living on borrowed time – borrowed from the past and from the future', with the continuation of the status quo only accelerating us 'towards an abbreviated collective future as fossils in the making'. (Nixon 2011: 69, in Marks 2019)

Dense, rich and portable fossil fuel energy comes from a finite reserve of extinct and buried ancient forests. It appears normal within the lifetime of all people alive today, but the availability of this finite resource is a small glitch in the larger pattern of history. Usually ignored as a factor in modern economic productivity, the extraordinary impact of dense and portable fossil fuel energy has underpinned economic theory since the turn of the twentieth century.

The progressive utopia of Keynes (1933) and liberal economic growth is built on the multiplier effect, which promises social stability via expanding consumer capacity and wealth. However, Keynes's multiplier relies on ignoring the exponential nature of growth, from slow, minute additions, early at the outset, to later larger and faster expansion in resource use and consumption. After the 1970s, exponential growth has become rapid, so now, the disparity between monetary value and the real productivity associated with finite resources is dangerously vast. Modern industrialism and global trade emerged from the alienation of the land clearances, which, as Marx (1887/1867) points out, created a metabolic rift in the ecosystem. That metabolic rift was the result of alienation. The increasingly abstract quantity of nominal money is that rift writ large. It is especially vast in the futures derivative stock market and fractional banking. Abstract expansion has more and more to do with 'value' attributed to intellectualised financial mechanisms, such as credit default swaps, and less and less to do with the 'value' of finite planetary mineral deposits and solar-tempered seasonal growth.

Neoliberal economics reified growth almost to a theology, subsuming all elements of life to potential resource awaiting consumer demand. Universalism dictated there is no 'outside' of the market. Every item, idea and innovation can be marketised. The neoliberal market encompasses nation-states, natural resources, human capital and even the space race. The market sets a price on the value of commodities, and this is supposed to moderate scarcity. Scarce resources are expensive and thus unavailable to the majority of consumers. Pricing scarcity is supposed to generate a fair system across borders and class divide. Importantly for the theory, consumers are individuals, universal in their characterisation of rational utility maximising. Group identity, such as cultural/gender/class expression could warp the fair operation of the invisible hand. Social and political (and state) interests are treated with great suspicion (Benson 2020). Universal rational individualism is foundational. Fukuyama (1992) strongly advocated this utopian view of market growth and argued that capitalism 'won' the dialectic with communism when the Berlin Wall fell in 1989. Since then, the only political questions have been issues of consumer fashion.

Certain factors are, however, externalised and ignored from market calculations. Pollution has been relegated to the natural commons. The needs of future generations have been 'discounted' so that they are almost entirely unvalued in present-day cost–benefit calculations (Von Mises 1953/1912; Nordhaus 2018; Hickel 2018; Masini 2021). Martinez-Alier (2012) calls the land clearances, discounting future generations, along with the needs and rights of other species and ecologies, 'ecological distribution conflicts' based around 'accumulation by dispossession' and 'accumulation through contamination' (2012: 349). Globalisation has been stuck in this universal market orientation despite growing awareness of the impacts of pollution on the planetary commons in United Nations (UN) policy, from the first UN Environmental Education conference at Tbilisi in 1977 to the latest COP on climate change. Two of the three major externalities in the neoliberal market theory – future discounting and pollution – have become compelling forces in current climate-dominated politics. The needs of other species and the ecosystem are still largely ignored.

STEADY STATE ECONOMICS

In the early 1970s, there was a strong push to find solutions to the oil, pollution, resource and overpopulation crisis. Barry Commoner (1971), Meadows et al. (1972), Daly (1973) and others in the Club of Rome developed the concept of steady state economics. Meadows et al.'s *Limits to Growth* (1972) is a seminal work in this literature, examining the limits of raw materials and energy to consider how global society can remain within planetary limits. The idea was that freely available contraception (or in the controversial case of Ehrlich 1971; famine) would reduce population growth and consumption, and thus, economics could attune to the limited resources of land, minerals and renewable energy.

> A steady state economy must obey five rules. First, renewable resource extraction cannot exceed regeneration rates without eventually driving resource stocks to zero. Second, waste emissions cannot exceed waste absorption capacity, or else waste stocks and the harm they cause will continuously increase. Third, with current technologies it would likely be impossible to meet the basic needs of current populations without certain non-renewable resources, such as fossil fuels. The rate at which society consumes these resources therefore cannot exceed the rate at which it develops renewable substitutes. Fourth, neither resource extraction nor waste emissions can threaten the ecosystem functions essential to human survival. Finally, human populations must be stable. The most obvious approach to achieving the first four of these goals is to mandate limits on throughput. How to achieve a stable human population is more controversial. (See neo-Malthusianism; Joshua Farley in D'Alisa et al. 2015)

The aim of steady state economics is for 'balance' between consumption and resource and energy availability, and thus production. Steady state theories have been criticised for tending towards forms of centralised control, remaining within the modern 'ordering of order'. Steady state economics engage with the issue of exponential growth as function of resource exhaustion. Mostly the literature does not deal with fractional banking and how that drives inflation and growth in resource use and pollution. The exception is Herman Daly (1996) who looks into Frederick Soddy's seminal book *Wealth, Virtual Wealth and Debt: The Solution of the Economic Paradox* (1933/1926), which examines banking ex nihilo, in great detail.

Soddy (1933), the Noble Prize–winning chemist, describes how a closed system cannot remain in homeostasis as some kind of 'perpetual motion machine'. If we are thinking about production and economics, then continuous input is required. Soddy argues there is no 'steady state'.

> We thus arrive at the conclusion that any sort of perpetual motion is impossible. A continuous stream of fresh energy is necessary for the continuous working of any working system, whether animate or inanimate. Life is cyclic as regards the material substances consumed, and the same materials are used over and over again in metabolism. But as regards energy, it is unidirectional, and no continuous cyclic use of energy is even conceivable. If we have available energy, we may maintain life and produce every material requisite necessary. That is why the flow of energy should be the primary concern of economics. (Soddy 1933: 56)

Fossil fuel is trapped sunlight through photosynthesis into ancient forests and then condensed through thousands of years of geological pressure. It is the most dense and portable energy available. The shift towards renewables improves the energy throughput but introduces new issues of storage and mobility.

Soddy (1933), Broadman (1966), Georgescu-Roegen (1971), Gorz (2009/1971) and Leff (2009) apply the concept of thermodynamic chemistry to economics. They argue that entropy is endemic in the economic system. Slowing the rate of entropy is important for preserving finite natural resources and the health of the ecological system upon which humanity survives. Thermodynamic economics analyse how production practices rearrange basic resources. In the process of greater production (or order) they also create enormous pollution and end use waste (or disorder). The higher the mixture of different raw ingredients in productivity, the more problematic. Georgescu-Roegen (1971), Gorz (2009/1971), Martinez-Alier and Schlupman (1987), Leff (2009), Higgs (2014), Kallis (2018), Hickel (2020) and others argue we are well over planetary carrying capacity, and advocate reducing throughput, or degrowth, rather than steady state economics.

Valdivieso (2008) regards Gorz as the founder of political ecology. Gorz argues that entropy can be seen on a large scale, in the acceleration of technology and its impact on work and consumerism. Like Heidegger, Gorz sees consumerism as a vital element in the flows of capitalism but he inverts the pace of productivity from consumer demand to the pressures of profit. The 1950s signals an entropic shift, where most households (in 'developed' nations) reached market saturation, and subsequently,

> The centers of capitalism were faced with the necessity to *produce consumers for their commodities,* needs to match the most profitable products. Following its spontaneous, capitalist dynamic, production had ceased to correspond to preexisting needs: in as much as such needs persisted (notably in housing, sanitation and public health) their satisfaction was not profitable, or not sufficiently so, for capital. And conversely, the most profitable products did not match unsatisfied needs*: these needs had to be created.* (Gorz 1985, in Evans 2010, italics in the original)

This is a vital argument and especially important in the light of the vast quantities of extra money pumping through the economic system.

Climate change can also be seen as entropy. CO_{2e} emissions are entropic heat from production, transport, industrial agriculture and energy use. But in a significant feedback loop, CO_{2e} traps the planet's solar load within the atmosphere as an *internal reservoir*, by capturing solar heat with greenhouse gas molecules rather than letting it reflect out of the atmosphere. The ongoing entropic cascades of global warming include higher mean ocean levels, with greater ocean turbulence, that is accelerating the entropic effect of eroding the coastline all around the planet. Warmer atmosphere means faster evaporation and a higher water load in the atmosphere (presently 10 per cent higher than the Holocene), making heavier downpours of rain, greater flooding and more soil run-off. These impacts are another type of metabolic entropic cascade.

Exponential growth shows up in the extraordinarily rapid increase in global human population over the last two hundred years. But according to the UN *World Population Prospects 2022*, the average world birth rates will drop to a replacement level of 2.1 births per woman in 2050 (2022: 15). The era of exponential population growth is already slowing towards a peak and subsequent decline (Irwin 2015a; Vollset et al. 2020). Most countries offset their aging population with young migrants. This will become harder as the transition to low- or below-replacement birth rates is also taking place in middle-income and increasingly low-income nations. Along with faster robotics and AI, population degrowth is part of the equation when considering economic futures.

Aging and contracting population means lower consumer demand overall. Some parts of the market (aged care, for example) are likely to expand, short term. In a similar fashion, 'green growth'(Stern 2018) will dominate markets during the infrastructure transition to more efficient cities, production, and distribution methods.

ENTROPY AND TIME

Thermodynamics emerges from molecular chemistry and physics. The First law of thermodynamics is that energy can neither be created nor destroyed, but it can be transferred from one form to another. The Second law of thermodynamics is that the entropy of any closed system always increases until all the molecules reach heat (or energy) equilibrium (Clausius 1856). Often thermodynamics is characterised as linear – the physics of time, which ultimately will reduce all order to uniformity, or the heat death of the Universe. But the linearity is not as straightforward as that large picture describes. In more immediate terms, the second law of thermodynamics is a spontaneous event which produces order in the singularity and at the same time generates an increase in entropy in the universe (Alastair Fleming personal communication). The relation between order and disorder is usually understood as oppositional, like a spectrum of inverse proportion, so that an event is either more ordered or more disordered. But I find it much more useful to understand it as a paradoxical increase of both at the same time. An increase in order in the singularity also produces an increase in disorder (aka heat) in the general universe. The third law of thermodynamics is an extension of the second; multiple closed systems, singularities, or bodies, will keep exuding heat amongst each other and to the wider universe (or reservoir) eventually reaching heat equilibrium.

A good illustration of entropy is the generation of the first elements at the very beginning of the universe in the Big Bang. In the origin, there was only Hydrogen. The huge energy explosion of the Big Bang created conditions where hyper-accelerated Hydrogen atoms crashed together. Whenever four protons of Hydrogen crashed together, they lost a little fraction of energy and at the same time forged a new element: Helium. The Helium atom has slightly less mass than four Hydrogen atoms. That loss is entropy. So, while the system has increased in order, it has also lost some mass/energy to low level heat that is now not readily utilised by the overall system. The system has simultaneously increased in organisation, and also increased in disorder.

> If (as is usual) the energy transfer is not reversible, the system's entropy will increase with time. This is so although the total energy of the system remains constant: the available energy falls. This concept follows from the second law of thermodynamics, and we can expand it to equate entropy with disorder: only a system with increased order shows a rise in entropy. (Deeson 2007)

Occasionally entropy is described as a two way process. But entropy follows the 'line of time'. Once two separate materials are mixed – cold milk and hot coffee, for example, it is not possible to unmix warm milk-coffee to the original state of singular materials. Furthermore, to 'increase' the ordering of a system is not possible without increasing the disordered heat loss required to make that order possible (this is the Bolzman distribution). Thus, in the literature, an increase in order is usually associated with a 'rise in entropy'.

The prevailing norms of the mechanical universe is for cause and effect, where forces act upon other forces in a linear and logical relation. Theoretically, cause and effect can be reversed. But thermodynamics is importantly different from the cause and effect of forces acting in a linear fashion. When order is increased (through matter and forces) there is *also,* irreducibly, an increase in disorder. Newton's mechanics had no idea about entropy. The idea of forces acting upon other forces is reversible; it makes no calculation of the loss of a fraction of force to the wider reservoir with each interaction and development of new order.

Thermodynamics replaces reversible Newtonian *mechanical* clock time with the thermodynamic arrow of time. Georgescu-Roegen understood the implications of the arrow of time in physics on the metabolics of open economic systems. 'Irreversibility means that a system cannot follow the same path in a reverse order. Irrevocability means that a system cannot pass through the same state more than once' (Bobulescu 2017: 1). Prigogine takes the statistical analysis of Bolzman and explains how important nonreversability is.

> Classical physics was 'considered *deterministic* and *reversible*. Processes involving randomness or irreversibility were considered to be exceptions, near artefacts. Today *we see everywhere the role of irreversible processes, of fluctuations*. The models considered by classical physics appear to us now to correspond only to limited situations which we can create artificially, for example by putting matter into a box and waiting for it to reach equilibrium. The *artificial* may be deterministic and reversible. The *natural* contains essential elements of randomness and irreducibility. This leads to a new view of matter in which matter is no longer passive, as described in the mechanical review, that is associated with spontaneous activity. This change is so deep that I believe we can really speak about *a new dialogue of man with nature*. (Prigogine 1982: 46, italics in the original)

Most importantly, the Newtonian 'idea of an unchanging, permanent substrate for matter has been shattered' (Prigogine 1982: 46). Instead, when 'far from equilibrium', nature has to be understood as a 'self-organising system,' which has its own agency, and informatics.

It is possible then, to begin to think about degrowth and 'negentropy' as a combination of the laws of entropy with ecological metabolism (Riedl 1992). Taking the core concept of metabolic economics in Marx (1887/1867), itself firmly ecological, I combine it with the ecological concept of Trophic Cascades and the second law of entropy, that with every increase in order, there is also an increase in disorder. Instead of remaining confined with the second law of thermodynamics – that multiple closed systems will lose heat to a large reservoir – I want to think about the disorder in a new way. Trophic cascades occur when a keystone species gets reintroduced into a complex ecosystem (Dekker et al. 2016).

The prime example is when wolves were reintroduced to the Yellowstone National Park. People were frightened of the wolves, and assumed they would kill a lot of the animals in the park. Instead, the wolves tended to kill old, sick, or maimed individuals from a herd, improving the overall condition of the species. The Elk changed their behaviour, which meant that the rivers were not grazed as hard along the river banks. Trees grew back, and the soil runoff and silt into the water decreased. The cleaner water improved conditions for fresh water fish, molluscs and other species. With more salmon, were more bears, and a plethora of other species. The trophic cascade associated with the reintroduction of the keystone species improves conditions for soil, water, plants and animals at all levels of the ecosystem (Dekker et al. 2016; Farquhar 2021).

When one system increases in order, and at the same time increases disorder, or waste heat, the entropy could be a useful biproduct to another system. A change in scale establishes a context or milieu that enables diverse species to take up the excess disorder from one system and incorporate it as fuel or food for a different species. Photosynthesis for example, takes the entropy from the sun, and converts sunlight into new order, new vegetation, new materiality. I call these phenomena of intersecting closed systems that 'feed' one another with waste entropy, 'entrophic' cascades. It is an interlocking ecosystem of diversity. Permaculture is a good example of this intersection of diverse closed systems.

Diversity is usually considered an ethical position, a democratic and inclusive attitude to those that differ from the norm. For modernity, the umbrella of universalism still extends over all deviations, collating outliers into universal categorization. But I want to conceptualise diversity in another role altogether; as a fundamental element of the universe, where massive variations in scale and characteristics enables negentropy to capture the entropic disorder, or 'heat death' of other systems, and convert them into usable energy and matter. Diversity is not merely an ethics associated with equality and sameness. It is criteria for a healthy ecosystem and a healthy economics.

To fully understand the repositioning of diversity and thermodynamics as core elements of political economy, it is useful to unpack three macroeconomic trends; the J curve of exponential growth, the steady state straight line, and the diminishing curve of degrowth.

ECONOMIC METABOLISM

Human societies have always been thermodynamic. Georgescu-Roegen (1979) s that at the outset of culture, the Titan, Prometheus dedicated humanity with fire. Culture emerged because fire enabled people to cook, forge tools, keep warm, and lengthen hours of useful light. As hunting was supplemented with agriculture, Georgescu-Roegen argues, animal husbandry is another recipe for thermodynamics. Animals eating grass and plants is a form of burning fuel, just as is fire. The next cultural recipe has been coal, which utilises concentrated captured solar energy in the form of fossilised forests. The dense energy input of fossil fuels has accelerated productivity and leveraged labour power as never before. The fourth industrial revolution is information, which might also be seen as an accelerant.

Four key concepts are commonly associated with evolution of economics from the perspective of thermodynamics. (1) The association of the technology of fire and energy with the development of civilisation, (2) acceleration of productivity, (3) the inevitable fall in energy return on investment (EROI), and (4) a shift in the *raison d'état* of efficiency and innovation from increasing throughput to decreasing throughput, waste and the circular economy. I want to add a further concept that combines entropy with ecology, to conceive of (5) 'entrophic cascades'. This is an ecology of diverse productivity, where a closed system utilises waste as fuel or resource, from another closed system. Circular design facilitates decreasing the rate of entropy by participating in entrophic cascades.

Şorman analyses modern productivity in terms of the kilojoules of energy that has leveraged labour and produced high levels of technical capital to make the current cultural conditions possible.

> Current societal functions (service and government, production of food etc.) and their associated metabolic patterns (joules of fossil fuels used for the maintenance of the health system, the hours of human activity used for producing a certain quantity of food) are based on the exploitation of fossil fuels as a principal energy source. Fossil fuels are a source with a high output yield and quality. The advent of fossil fuels has dramatically reduced the amount of energy, labour, and technical capital going into the actual production of useful energy itself. Thus, modern societies have been able to achieve their current level of complexity with the surplus of time that cheap sources of energy yield. (Şorman 2015: 3)

Kilojoules of work can be applied to an analysis of labour too.

> One barrel of crude oil can perform about 1700 kW h of work. A human laborer can perform about 0.6 kW h in one workday (IIER 2011). Simple arithmetic reveals it takes over 11 years of human labor to do the same work potential in a barrel of oil. Even if humans are 2.5x more efficient at converting energy to work, the energy in one barrel of oil substitutes approximately 4.5 years of physical human labor. (Hagans 2020: 4)

The shifts from hunter/gathering to agriculture, and then industrialisation has vastly increased kilojoules of work, transforming expectations and capacity. Each era of 'progress' has accelerated energy throughput and entropy.

Energy return on investment (EROI) has become a crucial element of accounting for thermodynamic productivity. Mostly it has applied directly to energy generation. Increasingly, as the impact of greenhouse gases affects the climate, the concept of EROI is being put in place to account for emissions. As early as 1880, Podolinsky analysed farming in terms of energy flows, and this has since been developed by Charles Hall in 1986, into the concept of energy return on investment (Hall et al. 1986). Oil has been evaluated in terms of EROI since then.

Multiple oil fields have been in decline since 2005, requiring higher levels of energy investment to extract one barrel of oil. The EROI has gone from roughly one barrel of oil to extract 100 barrels, to one barrel to extract and process just six barrels (Hall et al. 2014; Brockway et al. 2019). The rate of entropy is increasing, and the order is decreasing in relative proportion.

The danger of poor EROI returns is both in climate change and resource exhaustion. Since 1896, Arrhenius argued that the combustion of coal and other fossil fuels will increase the levels of CO_2 in the atmosphere (a relatively closed system). The gas triggers a change in the input and output of solar energy through the earth's atmosphere. The impact of both climate heating and EROI results in an Industrial overshoot and collapse trajectory *unless* the remains of fossil energy is used to transform infrastructure and technology towards a more renewable energy source (Hagans 2020). Georgescu-Roegen's recipe for low energy economics is solar panels. But unless they produce enough energy to reproduce themselves, they are 'parasites' and will collapse when the oil economy collapses.

Of course, it is not only energy sources that needs to transform, because entire cities have been built on the premise of versatile, cheap, and dense fossil fuel transport. Low entropy cities need to become walkable (Vergunst and Ingold 2008), which means a proliferation of services like schools, healthcare, parks, and small food shops throughout the suburbs and changes in the ways we work, commune, and travel. This is the basis of Carlos Moreno's fifteen-minute city.

Waste is another form of entropy, that is not very amenable to easy solutions. An increase in order is also an increase in disorder – or waste. Every time a product is assembled from raw material sources from vastly different parts of the planet, and arranged into complex technologies, entropy also grows. The inevitable waste is causing enormous problems in the ocean and with overcapacity of landfill sites. Robert Ayers and Benjamin Warr (2009) argued that 'spaceship earth' could put waste into huge 'baskets' that breakdown and can be mined, or recycled, over time. The circular economy is where products are designed to be easily disassembled at the end of their life cycle and reassembled into something new. This could substantially reduce some of the entropic disorder by enabling materials to be reassigned to new typographies. As a fundamental goal of design circularity, entrophic cascades could help shift the ethos of material productivity. However, I would suggest that circular design cannot be understood in the same category as the Invisible Hand, as an overarching solution to waste and disorder, that allows consumerism to remain the driving force of economics. Rather, the circular economy is a facet of thermodynamic economics. It aims to reduce throughput, and optimise efficient innovation in product design, longevity and recycling. The slogan 'reduce, reuse, recycle' is in a specific order of effectiveness. Reduction of consumerism is the best way to reduce waste, reuse objects is better than producing endless new ones, and recycling, even at its best, still only takes out a percentage of waste from the entropic throughput of industrial production.

Martinez-Alier and Schlupmann (1987) argue that the economy is not circular, but entropic. They argues that neoclassical and Marxist economics need to consider social metabolic flows of energy and materials. Circular design assumes a closed system. However, energy is not recycled, and materials are limited to partial recycling. Entrophic cascades of interlacing production processes, which utilise each other's waste byproducts will also reduce the rate of entropy but is likewise an element of the metabolics of the economic system, not a universal solution to entropy. Resource extraction and waste disposal inevitably show up in ecological systems. Scarce physical resources and pollution, and I would add, climate stress such as drought, are disclosed in 'socio-ecological distribution conflicts'. Rather than conflict being driven by economic surplus, as Marx (1887/1867) surmised, Martinez-Alier and Schlupmann (1987) argue that conflict is over the increasing use of biophysical and mineral resources as social metabolism speeds up. Clean water shortages are likely to cause conflict.

The law of thermodynamics suggests that entropy proliferates with more productivity throughput. To slow down entropy the obvious solution is to reduce throughput. In other words, more than circular design: to slow down consumerism. Broadman (1966) applies thermodynamics most eloquently,

> By contrast, in the spaceman economy, throughput is . . . to be regarded as something to be minimized rather than maximized. The essential measure of the success of the economy is not production and consumption at all, but the nature, extent, quality, and complexity of the total capital stock, including in this the state of the human bodies and minds included in the system. In the spaceman economy, what we are primarily concerned with is stock maintenance, and any technological change which results in the maintenance of a given total stock with a lessened throughput (that is, less production and consumption) is clearly a gain. This idea that both production and consumption are bad things rather than good things is very strange to economists, who have been obsessed with the income-flow concepts to the exclusion, almost, of capital-stock concepts. (Broadman 1966: 8)

Kate Raworth (2018) has articulated the parameters of finite resources, arguing for a doughnut economy, that recognises the limits on resources, waste disposal, and pollution, and like Broadman, resituates society within the planetary ecological sphere, rather than the universal market sphere.

Ecological interconnection can inform economic systems through a recognition of the need for diversity in circular design and production as entrophic cascades. The seasonal cycles of crops are one mode of recognising ecological regeneration. Larger metabolic cycles of debt jubilees, for example, can also be factored into economic theory. The recognition of large scale ecological metabolic cascades in both resources and economic processes keep the interconnection with planetary systems, despite the proliferation of new technologies, new materials, new chemistry, or new financial mechanisms.

To summarise, there are several important points to make about economics from a thermodynamic perspective: metabolics, productivity acceleration, infrastructure, energy throughput, EROI, circular design, and entrophic cascades all reprioritise economic systems. There is a shift in the concept of efficiency from innovation, proliferation, and profit to reducing consumption of capital stock. Biomimicry in technological innovation, and in the interconnection of entrophic cascades, enable economic theory to embrace diversity as a core concept. Dense energy input of fossil fuels, and accelerated productivity, are an increase in the second law of thermodynamics. An increase in order (productivity) *and* an increase in disorder (heat, and pollution). 'Circular economy' importantly reduces but does not eliminate disorder, or waste. It is not a universal theory that replaces the invisible hand, but it significantly changes the role of efficiency, from the profit motive to the ecological health motive. Energy inputs are constantly required in any given system. The recycling facility of circular design may reduce rubbish, but it cannot halt the irreducible disorder of entropy, as heat, emissions, rubbish, and pollution. Entropy can be seen in EROI or Energy Return On Investment, which gets poorer over time. The costs of extraction escalate both in terms of money and entropy. Contrary to theory of the neoliberal market, the aim of efficiency and innovation needs to aim to maximise outcomes for *minimal* throughput, not *greater* throughput.

FUTURE DISCOUNTING

Neoliberal economists believe that the Invisible Hand sets prices in the universal market through the aggregation of individual, rational, consumer demand. Thus, the market is the widest circle in a Venn diagram, with every element of the planet at the disposal of marketisation. The market envelopes the earth. Abundance is utilised by global demand, or remains *in situ,* as a resource available for consumerism when required (Heidegger 1977). Scarcity is dealt with in two ways; firstly, through setting a high price, which gives scarcity high value, and thus decreases demand; secondly, through innovation, which creates new materials to replace antiquated or unavailable ones.

Paraphrasing the neoliberal assumptions of Solow, Hayek (1960) and other neoliberals, Georgescu-Roegen drips with sarcasm, 'The world can, in effect, get along without natural resources, so exhaustion is just an event, not a catastrophe' (1986: 12). Neoliberal faith in economic innovation believes that alternative sources will be invented. Broadman (1966) calls this type of thinking, 'Cowboy economics'. 'The emphasis on the growth of GDP is reminiscent of the wide open frontier, promotes more and more throughput and ignores effluviant' (Broadman 1966: 7).

A lot of the debate about climate change rests on these ideas about technological innovation and future discounting. Technological fixes are built into the discourse of 'net zero' where there is a thirty-year-old expectation that technologies capable of large scale carbon capture will become available. Hopeful rumours occasionally pop up, but other than the photosynthesis of plants, these technologies are inefficient and too costly for large scale implementation. Another argument that crops up is that because of inflation, future generations will be able to deal with pollution problems more cheaply (Nordhaus 2018).

Discounting future generations is an old debate and goes back to the earliest industrialisation. In the late 1700's William Wordsworth mounted a campaign against coal smog (Yoshikawa 2020). Then after World War I, in 1919, Otto Neurath (1973) forged the Socialist Calculation Debate against Von Mises (1953/1912), and Hayek (1960) and the neoliberal theology of rational pricing as fair distribution being promoted by the Mount Pelerin Society.

> Neurath's point was that it was impossible to base decisions on prices when the issue was, for instance, whether to save labour by using more coal now, and therefore have less coal in the future. A factory owner would just look at the present prices of coal and labour. But there were other considerations to be taken into account. We were not sure about the amount of coal deposits remaining, we were uncertain about future technologies, and there were no prices for future options of coal deliveries in twenty years. Important decisions had to be taken outside markets, by what we now call deliberative, democratic planning. (Martinez-Alier 2012: 346)

Neurath's bioeconomics is based on an analysis that capitalism has been so ragingly successful because it utilised cheap and energy rich fossil fuels. High quality energy replaces or leverages traditional labour and productivity. Fossil fuels are a limited resource which are available at most for a couple of centuries. Using them up now fails to accommodate the needs of future generations.

Georgescu-Roegen acknowledges that technology and innovation are effective levers of economic growth but could not be as operative without the 'mineral bonanza' (1986: 13). Furthermore, following Broadman (1966), Georgescu-Roegen argues that the market mechanisms of supply and demand are limited to the immediate pressures of consumerism. Supply and demand fails to accommodate the needs for an intergenerational allocation of limited mineral deposits, and other nonrenewables like fertile soil. Georgescu-Roegen (1986) says that energy is either a mineral deposit 'stock factor' or triggers transformations within the economy 'flow factor'. He rejects man-made capital in the economy 'fund factor' as physically impossible. Modern market consumerism is 'a dictatorship of the present over the future'.

Cowboy economics discount the needs of future generations in what has become known as the *intergenerational* allocation problem (Pearce and Turner 1990). This is becoming an acute debate in climate discourse, as current emissions keep accumulating. Current consumption is making the climate increasingly dangerous for thousands of years. Current consumerism makes it next to impossible for future generations to use even meagre amounts of fossil fuels in future. Broadman (1966) shows how exponential growth facilitates and exacerbates that discounting in economic terms. It is a lack of imagination and care for our grandchildren.

> Even if we concede that posterity is relevant to our present problems, we still face the question of time-discounting and the closely related question of uncertainty-discounting. It is a wellknown phenomenon that individuals discount the future, even in their own lives. The very existence of a positive rate of interest may be taken as at least strong supporting evidence of this hypothesis. If we discount our own future, it is certainly not unreasonable to discount posterity's future even more, even if we do give posterity a vote. If we discount this at 5 per cent per annum, posterity's vote or dollar halves every fourteen years as we look into the future, and after even a mere hundred years it is pretty small – only about 1 and 1/2 cents on the dollar. If we add another 5 per cent for uncertainty, even the vote of our grandchildren reduces almost to insignificance. We can argue, of course, that the ethical thing to do is not to discount the future at all, that time-discounting is mainly the result of myopia and perspective, and hence is an illusion which the moral man should not tolerate. (Broadman 1966: 10–11)

The argument for discounting that Broadman explains so well here, is clearly a result of an expectation for continuous exponential economic growth. If the banks stopped forcing too much money into the system, and inflation died almost away. If there was no possibility of compound interest; then the neoliberal rational argument for discounting would dry up.

Instead, the future solves its own problems according to William Nordhaus (2018), the controversial noble laureate in climate economics (Masini 2021). Reminiscent of Solow, Nordhaus (2018) believes that technological innovation can compensate for issues of resource availability. The problems of pollution, and the devastating damage of climate change will be easier to solve in future, he thinks, because inflation and growth will have devalued money by such a huge degree by then, that it will be 'cheaper' to remedy any problems caused by the present day generation. For Nordhaus, discounting future generations is justified.

Not just future generations are at stake. Other species suffer deeply because their needs are 'discounted' or more accurately are *not* counted, because 'other' species are assumed to have no rationality, no emotions, nor any capacity to feel pain, communicate, or retaliate for the deprivation of their ecosystems and species health. The savage deforestation of the Amazon Rainforest during the Brazilian Presidency of Bolsanaro is a good example.

Furthermore, while GDP growth is the ethos of government, large institutions, and corporations, then Jevon's paradox remains at work. Any improvement in efficiency is absorbed and exceeded by greater consumerism. Thus, the shift to renewable energy has thus far created efficiencies which enable more productivity, rather than a reduction in throughput. This mis-step has annihilated effective emissions reduction. 'Efficiency' is understood as driving technological innovation to *expand* consumerism. We need to resuscitate the older concept of efficiency as innovations that reduce energy and resource extraction and reduce the need for more consumption.

DEVELOPMENT IS COLONIALISM

Indigenous subsistence economies have not been regarded with modern respect for the incredible systems, technologies, and sophisticated ecological knowledge that enable ancient cultures to live within the ecological parameters of their territories (see Pascoe 2014; Yunkaporta 2019). Instead, subsistence economies have been described as poverty stricken, ignorant, backward and desperate. They are often disregarded as 'terra nullus' so that corporations can mine or log with impunity on indigenous territory.

> Post-development scholars have deconstructed the assumptions, colonial relations and violence embedded in supposedly neutral, 'scientific' metrics such as the World Bank's poverty indicators, GDP league tables and the like (e.g. Escobar, 2011, McAfee, 1999). These metrics frame alternative, non-Western ways of living as 'backward', posing market-driven growth as the only alternative, serving the interests for Northern economies to expand their markets and access to cheap labour and raw materials (Galeano 2004). International development programs, thus, sustain colonial relations in the post-colonial era (Escobar 2011; Esteva 2010). (Hanaček et al. 2019)

Neoliberalism frames the capitalist norm as business as usual (BAU), as though there is 'no alternative'. However, steady state and degrowth literature regards the last two hundred years as an energy rich anomaly. That energy is rapidly becoming exhausted, making it important to reconsider economics with lower energy systems. Indigenous knowledge is proving increasingly valuable as contemporary culture struggles to cope with the impacts of climate change, and the devastating floods, fires, drought, and storms that are impacting local conditions.

THE SOCIETY OF FEAR

Under the rhetoric of progress and growth, a synonymous dystopian rivulet has turned into a flood. Market saturation, offshore jobs, resource exhaustion and pollution have combined with environmental impacts. Poverty and precarity make it harder to cope with the predicaments of vulnerable industrial agricultural monocultures, soil runoff, deforestation, escalating global warming, sea level rise, ice melt, drought, desertification, forest fires, storms and flooding. Exponential economic growth has also been the victim of its own success with financial crises ripping through the system at unforeseen moments.

> From the 1970s onwards, with the oil crisis and the transition from the Keynesian-Fordist system of accumulation to that of flexible accumulation, founded on the tertiary sector, the rates of growth and productivity in advanced capitalist societies have gradually decreased. In contrast, social and ecological costs, connected to the hyper-complexification of the bureaucratic and economic-financial 'mégamachine' increased. The crisis of the 1970s marked the passage to a second phase of an S shaped cycle of accumulation: the phase of 'declining marginal returns'. This phase is accompanied by a reduction in social well-being, which has become even more acute with the 2007 financial crisis. (Bonaiuti 2015: 3)

Stiegler's (2018) concerns about the impact of accelerating AI and robotics, combines with alarm about the looming drop in working aged people, as populations peak, age and shrink (Irwin 2015a; Vollset et al. 2020). Climate catastrophes are destroying infrastructure, agriculture and housing in entire regions with flood, fire and drought. All these factors change the contours of work and distribution. Finance and share markets double the double bind, which is both loosely and strongly connected to these disparate factors, and exacerbate the heightened levels of risk. The belief in mastery over nature through technological analysis, innovation and solutions (Weber 1948/1919) are giving way to fear and recognition that, after all, nature is not within human control (Beck 1992/1986; 2009).

From the perspective of thermodynamics and the law of entropy, these catastrophes are understandable, if not predictable. Government dispenses of money during COVID-19, and the high turnover of housing and large increase in ex nihilo credit shows how the market of money equilibrium is a myth. If the proliferation of money during COVID-19 were utilised as market rhetoric dictates, then prices would have managed shortages and absorbed wages in a theological 'equilibrium'. Instead, money behaved as flow. The huge slosh of new money moved like a swollen river and came to rest in the stagnant marsh of property, stocks and shares, Bitcoin and fine art prices, each a 'reservoir of value' which flooded accordingly.

ENTROPHIC CASCADES

Marx's analysis of 'metabolic rift', that ties household and smallholding economics to the ecological cascades of nutrients, and ecological growth and decay, is an early explanation of economics as throughput. We might also call it entropy.

In the second law of thermodynamics, as matter is organised into higher order systems, entropy translates some of the matter/energy into 'unusable' heat that gets expelled into the greater universe (ultimately generating the heat death of the universe). In other words, as order is created, disorder is also created. 'Negative' entropy would aim to reassemble disorder into new forms of order, in *entrophic* cascades.

Negative entropy is not outlined as a 'law' of thermodynamics. It cannot be understood as a reversal of entropy, which is associated with the irreversible arrow of time. Broadman (1966), had an excellent understanding of thermodynamics and put forward a good argument for reducing the throughput of economics to reduce entropy (as pollution and waste). Stiegler (2018) has taken this ethos of slowing down consumerism and economics and avoiding the stagnating connotations of 'degrowth', named it 'negentropy'.

Stiegler (2018) argues that the sophistication of robotics and computing is accelerating. AI is so fast that the human brain cannot keep up. Humanity has the creative capacity to ask existential questions and critically think through the modes of social normativity (including critique of racism, sexism and ecocide). Stiegler warns that people are becoming more passive, and our thought patterns are shaped by the black box of computer algorithms which channel our search results more than our thinking does. The algorithm reinforces like-minded thinkers and limits the availability of contrasting viewpoints. People become gradually absorbed in a conservative 'bubble'. Robotics and AI are proletarianizing wider ranges of work. AI makes creative engagement less necessary and undervalued.

Negentropy aims to shake up the passivity of allowing AI to shape thought and reinvigorate the creative dynamic of noetic thinking. Stiegler (2018) argues that negentropy is like the pharmakon – a tiny amount of poison is medicine, but toxic overload will kill the entire system. In my view, suggesting that noetic thought is overloaded by the algorithms of information technology is an important critique. However, the suggestion of negentropy as a pharmakon that might solve the issue of passive receptivity is optimistic. Just like the circular economy, it slows entropy, but is not really a reversal of the technicisation and proletarianisation of thought.

Entrophic cascades are interrelated with the ecology of discrete systems. It is not merely the reduction in throughput which reduces entropy. Negative entropy actually takes the entropic tailings, the emissions and waste products, the heat death, the disorder, and turns it into something productive, creating higher order materials out of by-product. This often takes place in an asymmetrical shift in scale between one system's increase in order and its commensurate disorder and another system's uptake of the 'heat' or 'unusable' energy from the first system. The most obvious example I find of actual negative entropy is photosynthesis. Plant life is the most incredible and prolific counter example to the second law of thermodynamics. In other words, photosynthesis takes that excessive 'heat' or 'disorder' that is gradually heating the universe (a.k.a. sunlight) and utilises it to increase order and absorb carbon, to create plant matter. Plant respiration produces oxygen, as disorder, which shifted the makeup of earth's early atmosphere some 2.5 million years ago (Hodgskiss & Sperling 2022). As a result of early plant life, the earth suffered the oxygen explosion that required an evolution in animal physiology to cope with the rapid change in circumstances. Evolution is fitting in with the environmental circumstances. Survival of the fittest is a misapprehension of Darwin's theory of evolution. It is not a case of competition but rather adaptation to make use of entropy as a form of nutrient. Once animals had evolved to make use of oxygen, photosynthesis produced an 'entrophic cascade' that shaped the conditions for other species in planetary ecosystems.

The term 'trophic cascade' derives from ecological literature, made most famous by the example of wolves in Yellowstone National Park. The keystone species was reintroduced to the park in 1995, triggering a cascade of ecological change, including more aspen, willow and cottonwoods and an increase in beaver colonies from one to nine. One ranger said, 'It is like kicking a pebble down a mountain slope where conditions were just right that a falling pebble could trigger an avalanche of change' (Farquhar 2021). The earlier lack of wolves meant elk proliferated. Elk did not have to forage far from the river, destroying trees and beaver habitat. Disturbing the sanguine elk population to become more vigilant and roam further from the river for foraging allowed the willow stands to thrive. 'As the beavers spread and built new dams and ponds, the cascade effect continued, said Smith. Beaver dams have multiple effects on stream hydrology. Beavers even out the seasonal pulses of runoff; store water for recharging the water table; and provide cold, shaded water for fish, while the now robust willow stands provide habitat for songbirds' (Farquhar 2021). The wolves are now the primary killer of elk, now that winters are less fierce. The wolves even out carrion-bounty across the seasons for ravens, eagles, magpies, coyotes, bears, wolverine, lynx and beetles.

Trophic cascades can occur in political ecosystems too. Combined with concepts of thermodynamics, entrophic cascades are modes of social organisation or production that intersect with other practices to reduce overall entropy in the cultural system.

Indigenous modes of governance and land management give us further concepts that could overlay our existing ownership model. The Māori method of *whenua tuku* is the multiple layers of community attachment to the same piece of land (T. Smith 2000).

Whenua tuku is a useful Māori concept of land 'ownership' which is based on a long-held and intimate relationship with a place. Specific knowledge and rights to make use of certain resources available belong to some families but not others. Thus, one piece of land is 'owned' by multiple groups at once, but each group has a different relationship with that land. All have an obligation to take care of it. That means that each utilitarian relationship with that land has to meet the needs of the other people's rights to it – as well as the needs of the ecosystem itself. This is a complex diversity of rights and utility, that works optimally when it benefits all, through layers of entrophic relations.

This already takes place in some ways in Scotland, where farmers own the top several feet of soil, the mineral rights are retained by the Crown and the population has the 'right to roam' over the same fields. *Whenua tuku* enables multiple land uses to occupy the same space. Everyone's primary concern is ecological land management. Each user of the place is required to take care of it in a manner appropriate to their knowledge and material impact. Instead of 'freedom' and 'private property', the overarching motif of 'rights' is combined with ecological care. This entwines the layers of community with the environment.

Permaculture is another example where symbiosis is actively utilised to limit the metabolic rift. Permaculture aims to lower rates of entropy within an ecosystem. There are existing technological examples of negentropy too. Solar paint and solar panels mimic photosynthesis to a degree. They use chemical reactions to convert the sun's entropy, or sunlight, into viable electricity. Circular design is based on entrophic cascades, by designing for better recovery of raw materials and enabling the mining of waste materials in old consumer goods for new products.

It is important to consider what entrophic cascades are *not*. A bioeconomic approach to fuel replacement, for example, has converted hundreds of thousands of hectares of arable land towards crops to replace fossil fuels. This assumes that bioeconomic technologies are another innovative solution for continuous economic growth (Asveld et al. 2011). A metabolic rift is 'solved' with bioeconomic crops. But this approach misses the point entirely. Any monoculture does not take the overall ecosystem into account. An entrophic cascades approach requires that all aspects of the ecosystem are considered when new utility is required from a place. Without the ethos of intraconnection, bioeconomics can easily be co-opted back into the economic growth model.

Profitability is not necessarily the primary motivator for productivity governed by thermodynamic economics. Efficiency has been interpreted in classical and neoliberal economics as improving technology or processes in order to make more profit. The concept is far broader. Efficiency is about improving processes or reducing resource use to lower the need for throughput, not to lower costs and keep production high.

Better emissions and EROI accounting, combined with a more overt recognition of the limits of marginal returns as opposed to an ethos of continuous growth, would shift the economic landscape in subtle and achievable ways. As EU national border tariffs are introduced to account for embedded emissions in steel, cement, fertilisers, aluminium and electricity in 2026, these emissions-accounting initiatives are likely to gain more ground. Preparation for the Carbon Border Adjustment Mechanism (CBAM) and domestic net-zero targets are beginning to improve emissions accounting, via the scopes 1, 2 and 3 categories. While there is still a lot to be done, especially on embedded emissions, large businesses are onboard and it looks very promising.

In modern society, the normative ethos of ecological integration – suffusing through religion, institutions, language and technology – has been marginalised and forgotten. So, we need to take ideas from disparate disciplines and adapt them to the current situation. Climate change is demonstrating very clearly that planetary ecology cannot continue to bear the consequences of exponential economic growth. In this context, revisiting the origins of our contemporary modes of governance and redistribution are valuable. Both Mesopotamian debt cycles and *whenua tuku* cater to raising capital for large-scale projects and, at the same time, avoid exponential economic growth. Permaculture, slow food and co-operatives all offer modes of organising productivity and redistribution that are more amenable to entrophic cascades and circular design. These modes of social organisation already exist. They need better visibility and respect, in an era where global corporate lobbying has captured the political and social imagination. Of business, 97 per cent are small family-owned firms. Global consumerism need not dominate the economic imaginary.

ABUNDANCE VERSUS SCARCITY

Market economics is based on price to set the distribution of scarcity. There is regulation or limitations on scarcity, not limitations on abundance.

Abundant natural resources in a global market are understood as wealth to be sold elsewhere – for example, a plentiful forest is a 'standing reserve' of timber (Heidegger 1977) rather than a healthy regenerating ecosystem. Ecological economics has a different imperative – only take what you need. Leave ecological abundance intact. This is the ultimate outcome of a metabolic approach to slowing economic throughput, resituating 'innovative efficiency' to mean maximising outcomes from minimal inputs. This contrasts with increasing sales through techniques such as built-in obsolescence. The diversity of intersecting entrophic cascades in design and system structures facilitates more abundance rather than more exploitation of existing reserves.

Bataille (1991) makes the case for distinguishing between the 'restricted economy' and the 'general economy'. The restricted economy is the utilisation of (scarce) resources as they circulate throughout economic transactions. The general economy is the overall flows of energy that percolate the natural system, from photosynthesis to all other forms of material energy. The scarcity of the first is a subsection of the abundance of the second (Zwier et al. 2015).

Abundance and diversity are a form of negative entropy. The large variety and scale of species enables some to benefit from the 'waste' or disorder of others. Slaters and insects benefit from the leaf litter of trees, microbes and mushrooms benefit from soil and rotting wood, trees benefit from entropy of the sun and soils, and plants benefit from carnivores which keeps grazing animals to lower numbers, in an ecological trophic cascade (Dekker et al. 2016; Farquhar 2021). Humans are adaptable, and if we situate ourselves in the diversity of ecological concerns, instead of regarding abundance as a standing reserve of resource to plunder, then we can return to a healthier mode of planetary ecological and technological integration. This does not mark the end of global trade (which predates industrialism by millennia), or a return to romantic pastoralism. Instead, it is a new ethos that regards abundance as an ecological necessity, not an economic opportunity. In practical terms, fish quotas are a good example of the shift in thinking I am advocating. At present, fish can be caught if they are sufficiently plentiful and larger than juveniles. But this is a 'false economy'. The larger the adult fish (who have a long lifespan in most species) the more fertile, so killing *large* fish is counterproductive to regenerating fish 'stocks'. It would be better to take small numbers of juveniles and leave the older adults behind. Furthermore, fishing until a population is so small that it is endangered is short-term thinking driven by an immediate profit motive and with no regard for the 'rights' of either the fish or future generations. It is better to take a bare minimum, leaving ample fish populations, that are thriving so that plentiful fishing remains across time. An indigenous economics of abundance makes much more sense than an economics of scarcity. There is an important story about the Aboriginal Eora fisherwoman Barangaroo, who was fiercely angry when early settlers in Australia caught four thousand salmon in Sydney. She regarded this huge catch as devastating for the salmon population and arrogant. The settlers regarded the huge catch as a victory and as resources to trade and make profit.

A thermodynamic metabolic economics is based on efficiently slowing throughput and utilising entropic by-products to enable wider cascades of order. It needs a diversity of scales, and a preservation of abundance, to capitalise on the by-products of other kinds of ordering and disordering. The circular economy gives new aims for technologies to be negentropic by enabling easier disassembly and extraction of raw materials for future consumer goods. Metabolic economics pays attention to the mode of energy, the sufficiency of raw materials, the impact of emissions and waste and the health of diverse ecology and future generations.

The short-term availability of intensely rich fossil fuel energy is coming to an end, as EROI diminishes and the impact of CO_{2e} increases. BAU has been understood as a universal category that swallows the resources of the earth as standing reserve in the unfolding ethos of consumerism. But the end of fossil fuel–driven economics shows, more clearly than ever, that the pace of consumerism needs to accommodate planetary parameters, and the energetic rates of sunlight (which drive the hydrological and atmospheric systems). Diversity is a fundamental aspect of the economies of the future. At the moment, classical or neoliberal economics are supposed to be universal – to cover every contingency. The externalities, such as women's free domestic labour – are just not taken seriously as an economic contribution (Meade 1973). Social, perhaps, but not rationally or financially significant. My suggestion is to recognise an 'ecology of economics'. That is an ecosystem of intersecting strata of economic interests. It does not rule out raising large-scale capital, although loaned from collective existing wealth, not ex nihilo value usurped from the public. The ecological health of the entire system is the *raison d'être* of economics rather than the narrative of economic growth. Ecologies of knowledge trace epistemological ideas back to ancient Axial cultures, to subsistence economies, to gift-giving status economies, to feminism, to photosynthesis and Mycozia. Complex symbiotic distribution systems across species are a lovely example (Leff 2010; 2015). The strata of economics are boundaried by ecological health rather than a myth of the universal market where the invisible hand plays politics over all species, genders, ethnicities and classes.

BIOECONOMICS

Climate change policy literature sets out a strong need to transition towards a postcarbon economy. What that looks like has not been the focus of very sustained public discussion. Partly, the lack of public debate on 'degrowth' is because politicians are determined to avoid engaging with a 'retreat' to sustainable sufficiency instead of a continuous commitment to 'progress' and economic growth. Bankers have a policy of silence. The discourse of 'retreat' relies upon the Enlightenment distinction between nature and culture. Bioeconomics is at the boundary between these domains and illustrates the tensions in normative assumptions that plague the global postcarbon transition to a poorly defined alternative, to 'business as usual'. Bioeconomics is part of the materialist economic alternatives that have a focus on the metabolism of resource maintenance, extraction and productivity. It is a form of thermodynamic economics. However, unless there is an onto-epistemology of intraconnection that replaces the Enlightenment trope of the rational individual, bioeconomics is likely to continue to be co-opted into the paradigm of economic growth. Like nearly every other form of productivity, bioeconomics cannot be decoupled from emissions. The orientation of intraconnection can be found in a variety of fields, including indigenous cosmology, feminist embodiment and quantum physics (Part III).

Bioeconomics is in the same genre as steady state, thermodynamic and degrowth economics, surveyed above. Neurath (1919/1973) and Georgescu-Roegen (1971) are important developers of bioeconomics. Economic theories that engage with the metabolism of environmental and production cycles have a strong critique of the classical externalities that ignore the impact of pollution on the environmental commons. Bioeconomics emerges from the agricultural sector and specifically deals with the realm traditionally understood as the intersection between culture and nature: the human intervention in healthy soils, clean streams, horticulture and stock management.

Discussion of barriers to the transition to a postcarbon economy are emerging at the global policy level (IPCC 2021) rather than within national democratic communities. These barriers include a lack of clear vision for a safe future, a lack of finance to implement green infrastructure and new modes of production, and the continued large-scale oil subsidies, and protection of lobbying from the existing petrochemical industry. The contested definition and implementation of 'bioeconomics' shows these debates most clearly.

To complicate the emergence of clear direction in the transition to a postcarbon economy, barriers from existing multinational corporate lobbyists are proving significant. The IPCC outlined these dimensions in the 2013 report.

> Systemic barriers constrain the implementation of adaptation options in vulnerable sectors, regions and social groups (high confidence). . . . Prioritisation of options and transitions from incremental to transformational adaptation are limited due to vested interests, economic lock-ins, institutional path dependencies

> and prevalent practices, cultures, norms and belief systems (high confidence). (IPCC AP6: 28)

The economic metabolism of steady state, thermodynamic, circular, and entrophic cascades entails new methods to permeate normative economics. When the orientation changes to intraconnection and a cyclic pace, then bioeconomic approaches to economics contrast with classical and neoliberal economics.

Classical economics in Adam Smith and later, rely on the concept of the Enlightenment individual, who leverages rational decision-making to his or her own advantage. This is a crucial trope in the justification of aggregated consumer decisions, which are balanced by price, to create the fairest form of redistribution of scarce resources. Without the 'individual rational utility maximiser' the entire edifice of the 'market' guided by the 'invisible hand', disintegrates (Irwin 2014). As examined in chapter 2, classical and neoliberal economics are unable to address the problems of pollution and climate change. As an example, the Emissions Trading Scheme has proven inadequate to the task of reducing emissions (Atkinson and Tietenberg 1991; Skoerseth and Wettestad 2008), yet it continues to be promoted as a market solution. The alienation intrinsic to the Enlightenment concept of individual subjectivity results in an increasingly tenuous connection to reality, as the truth gets tied to the simulacra (chapter 5). These are the same forces that enable economic monetary growth to expand at a startlingly disparate rate from the real, or broad, economy (chapter 3). Instability is produced by this alarming disparity between the nominal value of the finance industries (including banking and futures derivatives) (BIS 2023; Gaspar et al. 2023). The finite limits of planetary ecology, show up acutely in the sphere of agriculture and bioeconomics.

Bioeconomics is a big sector in New Zealand, where large-scale farming co-operatives have encouraged the global industrialisation of agriculture through powerful import/export blocs. Dairy in New Zealand consumes the bulk of the fibrous by-product of palm oil – contributing to enormous acreage of deforestation in Borneo and the Amazon. These co-ops also encourage monoculture grasses with high milk yield that rely heavily on superphosphate fertilizer. The fertilizer has been mined-out in the Pacific Island, Nauru, and is now produced from poor-quality lignite coal. Both the direct fossil fuel use through fertilizer and monoculture grass management exacerbate low soil friability and high methane in cattle digestion. New Zealand has very poor water quality in streams, rivers and lakes primarily because of poor stock management (Joy 2022). Bioeconomics can address these problems by including energy throughput, pollution, soil ecology, plant and species diversity, greenhouse gas emissions and new green technology at the centre of agricultural discussion rather than a focus on pure profit.

AGRICULTURE AND BIOECONOMICS

As a primary sector in many nations, agrifood is a good example of the politicisation of agribusiness. Agriculture is at the sharp edge of the competing discourses of extractive economic growth and organic, renewable farming. The challenge is to transition from fossil fuel–led industrialism to ecologically sound, postcarbon agriculture. This is achievable through the sophistication of both high-tech and low-tech bioeconomics as long as ecological intra-action is at the heart of decision-making. 'Bioeconomics' (Bonaiuti 2015; Mayumi 2001) is in the same conceptual family as thermodynamics (Georgescu-Rougen 1971) and steady state economics (Daly 1973; 1996). If care is not taken, all of these metabolic economic theories can be co-opted into extractive productivism (Safian 2022; Zwier et al. 2015; Bugge et al. 2016).

Bioeconomics is not the same as the green economy (Stern 2006). The green economy aims to create a new infrastructure boom that enhances postcarbon economic growth. Nor is it limited to concepts of the circular economy, which assumes a design function of products that can be broken down into constituent parts and efficiently recycled into new products. Bioeconomics encompasses a plethora of ideas from traditional subsistence farming, permaculture, through to high-tech bioengineering, all within a postcarbon framework.

Tensions arise about whether bioeconomics has a necessary focus on ecological health, or merely postcarbon, knowledge-based, extractive industrial production. Much of the debate remains stuck in the 'growth paradigm' that is anxious to avoid agricultural sustainability as a 'retreat' from economic productivity growth (Bugge et al. 2016; Safian 2022; Zwier et al. 2015). Although the debate is framed around technological sophistication, it is shaped by the economic growth paradigm (Bayne et al. 2018).

Bioeconomy seeks to heal the metabolic rift of extractive agriculture (Marx 1887/1867). The land clearances removed human consumption and expulsion from the ecological cycle. Produce is consumed in urban settings and sewerage becomes a problem to be washed away rather than a crucial source of organic nitrates to fertilise the land. This metabolic rift is an example of the gaps that require artificial and industrial scale compensatory measures, such as lignite coal fertiliser.

Bioeconomics covers the spectrum from food security to global revenue. Bugge et al. (2016) examine the competing aims, values and drivers of three different visions that have not been sufficiently teased out in the discourse: bioecology, biotechnology and bioresources. The biotechnology vision has a focus on economic growth and job creation. It commercialises research and technology to create new modes of value, through science funding, patents and enterprise. The bioresource vision also assumes economic growth as a fundamental paradigm but tries to connect it to sustainability. Bioresources are converted and upgraded with technology and labour to add value. New technologies focus on replacing nonrenewables such as fossil fuels with cellulose, biofuels and other circular economic models. Both biotechnology and bioresources remain firmly entrenched in the technological horizon of disclosure outlined by Heidegger (1977). In contrast, the vision for bioecology is oriented towards biodiversity, avoiding soil and water degradation and conserving ecosystems. Development of trade occurs through high-value products and what Bugge et al. (2016) calls 'territorial identity'. It is characterised by organic agri-ecological practice, transdisciplinary, localised, sustainable and circular production modes. Bioecology gives space and respect to the more than human. As such, while it might include both low tech and high tech, it is a fundamentally different onto-epistemology than the alienation and resource extraction of the other two modes of practice.

A variety of localised strategies that aim for organic healthy soils, lower erosion with better mixed planting, cleaner waterways with riparian strips, lower stock numbers, diversity of plants for feed and lower emissions generate better-quality bioeconomics. It is impossible to be prescriptive about the thermodynamic throughput of a given locale – the issues are site specific. A general approach to high-nutrient food production through principles such as permaculture can enhance agricultural production and skip corporate imperatives for nitrate- and pesticide-dependent crops that destroy soil and water health, insect populations and biodiversity (Shiva 1989; 2015).

The core concept of bioeconomy refers to the pacing of production according to solar input and photosynthesis. This contrasts with the industrial intensification of fossil fuel–driven agribusiness. The postcarbon transition conceives of bioeconomy as a more organic but technologically savvy approach to contemporary agriculture. Bioeconomics aims for sophisticated enhancement of agricultural productivity while avoiding fossil fuel–based fertilizers, nonbiodegradable polymers and other toxic pollutants that disturb rather than enhance overall ecological health and sustainability. Bioeconomics is seen as the global solution to environmental and societal challenges. It 'refers to an economic system that reduces the dependence on non-renewable resources and shifts activities towards the utilisation of biological resources to provide services, products and processes' (Safian 2022: 6).

Permaculture and bioeconomic agriculture do not need to be regarded as a 'retreat' from high-output crop production. They have high-yield productivity. Overall, biodiversity in the ecosystem can be included in production practices without needing to reduce farms to monocultural intensification.

The general approaches to 'bioeconomy' have been characterised by Hausnost, as a pair of axes – from industrial (bio)technology on one axis and agro-ecology on the other (Hausnost et al. 2017). This characterisation remains mired in the 'economic growth' mindset that permaculture is a 'retreat' from progress and growth, whereas biotechnology – which might include genetically modified insect predators, for example – 'advances' continuous economic growth. Healthy agriculture aiming for clean friable and nutritious soils and water as a basis for vigorous, biodiverse ecological communities, can embrace sophisticated technology within eco-ontological boundaries. Awareness of ecological intra-action affects how the entire socioecological system flourishes. It is not limited to a reductive, isolated dimension that is the focus of specific technological innovation. Bioeconomics without an eco-ontological stance is simply an extension of the corporate leverage of the metabolic rifts created by the land clearances and large-scale industrial extractive agriculture of the eighteenth to twenty-first centuries (Marx 1887/1867). In one way, postcarbon bioeconomic agriculture will necessarily return to the 'organic economics' of preindustrial communities, insofar as preindustrial agriculture was 'constrained by the annual amount of solar output and the photosynthesis process in plants' (Safian 2022: 4). Extractive agribusiness is not, many would argue, the shape of bioeconomics in the future (Bugge et al. 2016; Campbell 2020; Campbell et al. 2009; Zwier et al. 2015). Bioeconomics fully engages with the metabolic thermodynamic throughput conceptualised by Broadman (1966), Georgescu-Rougen (1971) and others. Without analysis of entropy and intraconnection, bioeconomics is simply old-fashioned agribusiness. It is open to sophisticated (old and new) ideas that enhance ecosystem health to produce nutrient-rich and diverse crops. The old division of economic contraction versus growth is then redundant.

The need for continuous growth issues from global population growth – which is aging and peaking – and the structural imperatives of the finance industry. If finance is achieved without creating new money ex nihilo, nor charging more interest than justified by the costs of an intermediary for loans (Jackson and Dyson 2013), the demand for economic growth disappears. Agriculture, like the rest of society, is not liable to constantly extract increasing revenue from fixed resources. There will be no need to offset the exponential expansion of the money pool and the structural demand for growth. People can relax and settle into the seasonal cycle without continuous pressure to expand production.

DECOUPLING THE KNOWLEDGE ECONOMY

The initial postcarbon transition promoted by the neoliberal-dominated global policy regime in the 2000s was framed on decoupling new knowledge and technology from material consumption and emissions. Bio*technology* was tied to the knowledge economy, in an effort to include it in the decoupling of the economic growth in manufacturing from emissions growth. The 'New Perspectives on the Knowledge-Based Bio-Economy (KBBE) conference organised by the European Commission (EC) in 2005 promoted bioeconomics as part of the knowledge economy. 'As opposed to traditional bioeconomy, The KBBE interpretation of bioeconomy anchored it firmly in biotechnology driven innovation to support economic growth, energy supply, and living standards. It pointed to the use of biomass as a resource substitute for the production of biofuels, biopolymers and chemicals' (Birner 2018 and GBC 2015, in Safian 2022: 6).

Barriers to transition pathways from fossil fuel–based industrial agriculture to bioeconomic agriculture is seen in competing terms. For different nations, bioeconomy serves a variety of purposes that often overlap. These include food security, domestic responses to global challenges, new biomass value chains, including raw materials for downstream value-added productivity and enhancement of global revenue (Safian 2022: 9–10). High-tech industrial biotechnology is the biomass stripped to chemical and molecular constituents and reassembled – for example, pharmaceuticals, nutraceuticals, biofuels and cellulose plastics.

In this discourse of high-tech industrial bioeconomics, sophisticated technology is counterpoised against the seasonal pacing of permaculture, as though high technology is in opposition to biodiverse farming. It is underpinned by the Enlightenment dualism of culture from nature – where industrial bioeconomics makes 'artificial' products that fit human needs in contrast to natural ones. Intraconnection cuts through the dualism of nature versus culture (see, next, chapter 9). Ecosystems thrive with sensitive human intervention. This confusion is exacerbated by the economic growth imperative, that co-opts sophisticated production and especially the replacement of fossil fuel derived chemical products with biomass, such as plastics. The economic ethos of agribusiness belongs to the pace and requirements of corporate six-month profit returns rather than the long-term integrity of intraconnected planetary ecology. If the economic growth imperative were reduced in importance because finance institutions were no longer stimulating inflation with new ex nihilo endogenous money, then the schism between competing discourses within bioeconomy would largely disappear.

The tension between economic growth and environmental health needs to be more overtly explored in public discourse. At present, as Safian laments, 'policy remains dominated by productivist thinking, which makes little creative effort to reconcile growth and environmental imperatives' (2022: 1). The dominant policy norms remain bogged down in the expectation of continuous growth which is pitched against (high-tech) nutrient-rich sufficiency.

INTRACONNECTION AND THE DISCOURSES OF BIOECONOMICS

To open up the future of a globalised postcarbon economic infrastructure, thermodynamics, degrowth or bioeconomics is not enough. A clear eco-ontological orientation is required to shape the ethics of policy and economics, at both micro and macro levels (Victor and Jackson 2015). Much of the debate on bioeconomics pitches sophisticated technologies against the pastoral traditions of yesteryear. This dualism is a misnomer that does no justice to the sophistication of indigenous ecological husbandry, nor to the values that can inform future technological development. It merely perpetuates the dualism of culture versus nature and assumes a Promethean optimism about technological innovation as a solution to all issues.

Extractive economics has been built upon the doubled alienations of the Enlightenment (Stiegler 2018) – that is, a physical and attentive alienation from ecological place and an internal alienation from emotions, the body and the community. Contested debates in bioeconomics have been part of industrialisation from the outset. The proliferation of large volumes of sheep encouraged the original land clearances in the United Kingdom. At that time, the Enlightenment pitched the isolated rational individual against natural objects. Indigenous concepts like the Celtic communication with local species, or the Māori cosmology of kinship in *whakapapa*, have a more integrated approach to knowledge and locality. *Whakapapa* is not merely anthropocentric either, but, rather, it is the genealogical connection of all species, water, earth, atmosphere and stars in a cosmology of intraconnectivity (T. Smith 2000; Mika 2017). This orientation of intraconnection is shared by nearly all indigenous communities, albeit with different narratives and emphases. The assumption that indigenous communities are backwards and low tech is inaccurate. At no point does an orientation of intraconnection deny or exclude sophisticated technology. However, the ethos of respect for ecology suffuses technological innovation so that it works *with* the constraints and pace of natural cycles instead of overpowering them (chapter 7).

The modern alienation between nature and culture continues to be a problem in conventional modern BAU and in bioeconomics. The following chapter examines the rupture between nature and culture from the perspective of technology and its enframing of epistemology (Heidegger 1977). It shows how technology does not need to be pitched in a dualism against nature. Technology can be ecological, collapsing the gap between nature and culture that has underpinned modern alienation. Technology is not necessarily implicated in the widening gap of the simulacra-society from integrated ecological concerns. Indigenous people regard ecology as technology. The tree is medicine, it is the ancestor of the community, it is intimately intraconnected with the language and ways of knowing that permeates society. The technological enframing of knowledge, in that case, is an ecological horizon of thought.

Part III

FOUNDATIONAL ENTANGLEMENT

Chapter 9

Technology, Learning and Dissolving Alienation

In the philosophy of technology literature, technology is a phenomenological intervention that shapes human awareness, thought and language. Technology enhances knowledge and the accuracy of perception, but it also divorces people from immediate apprehension of the ecological surroundings. The telescope, for example, can bring distant stars into view, but reduces awareness of the sounds, smells, and embodied and emotional comportment in the present surroundings. Technology enframes the way people think (Heidegger 1977). The anthropologist, Lotka (1925), argues that key technologies from Paleolithic rock art, through to the printing press are all externalised repositories for knowledge, which reduce the need for memorisation. The representations are a 'screen' of culture that signifies things and ideas and constructs interpretive values. Technologies change the way we experience, remember and think. This approach to technology as an exosomatic repository of knowledge, consolidates the schism between nature and culture. Stiegler (2018) argues that the technologies of exosomatic knowledge, such as computers and artificial intelligence (AI) are increasing the entropy of thinking. People are stagnating in front of the screen. However, indigenous technologies *are ecological*. There is no representational lacuna between the thing and its interpretation, whether as a repository of knowledge or a phenomenological intervention in physical apprehension. The Gordian knot separating culture from nature is sliced through.

For thousands of years, civilisations have externalised repositories for knowledge retention from rote learning and memorisation to art, clay tablets or written documents. Nowadays computers play this role too. The exosomatics of memory in technologies such as rock art, writing, the printing press, or the computer create a lens – or screen – through which people understand their world (Lotka 1925; 1945; Leroi-Gourhan

1993/1943 and 1945; Stiegler 2018). This view of exosomatic techniques such as a book or a film as a 'screen' that 'represents' natural objects contributes, I argue, to the schism between subject and object, so eloquently explored by Descartes (1980/1641) and then Kant (1992/1781). It resides within the modern technological enframing of the horizon of thought (Heidegger 1977).

As computing accelerates the speed of information and money flows, and as more aspects of work and writing are automated by AI, such as ChatGPT, Stiegler (2016; 2018) argues that the human brain cannot keep up. AI is taking over the active decision-making of researching and *noesis*, or thinking. Exosomatic devices mean the brain does not often practice mnemetic techniques and noetic activities, impoverishing the ability to think creatively and originally. Ross explains, 'knowledge is always subject to decay (into dogma and automaticity)' (Ross in Stiegler 2018: 27). As Heidegger describes in *Being and Time* (1962/1927), the surrounding objects and tools shape cultural and noetic expectations and enframe the horizon of thought (Heidegger 1977). In a seemingly unstoppable and determinist fashion, the alienation from the cycles and limits of nature are exacerbated. Modern exosomatic technology progresses to the zone of consummate nihilism (Heidegger 1996). There seems to be no fundamental recognition in the modern worldview of the ecological consequences of technological damage, such as CO_{2e} emissions, on the environment. Economic growth continues its exponential curve, and emissions grow exponentially with it.

Banks create new lines of credit and dilute the value of the existing pool of economic transactions, and governments collaborate by encouraging growth across the national economic system, to offset the ravages of inflation. Computing, as an exosomatic device, steadily extends its influence. Ultimately, Nick Bostrom (1998; 2003a) argues that AI enables the complete absorption of the world into a simulacrum that governs all elements of a person's experience of reality. The total alienation from ecology by the simulacra is discussed and advocated by Elon Musk, Sam Alton, Peter Theil, and other Silicon Valley billionaires.

The alienation between subject and object, culture and nature (Merchant 1980; Plumwood 2002; 2010), has enabled modern capitalism to thrive. That schism relieved economic theory and, thus, finance, business, and corporate practices of considering or accounting for environmental pollution or the impact of extractive economics on local flora and fauna. Alienation has enabled private financial institutions to generate enormous loans and devalue the broad economy, forcing economic growth to offset the threat of runaway inflation. The relation between economic growth, and more resource extraction, correlates with growth in pollution and emissions (Le Quéré et al. 2020; Caporale et al. 2021; Mitić et al. 2022; Onofrei et al. 2022). Holzmann et al.

(2023) have just released a study showing that extinctions rates are also accelerating as climate change accelerates.

The determinism of the modern technological enframing of the horizon of disclosure is built on 'forgetting' nature as an integral part of the socius. The alienation of the individual from the object is writ large in late modern extractive economics. Modern disclosure *seems* to be completely deterministic. Yet, as Heidegger himself notes, agency to create change is always irreducible.

> Always the destining of revealing holds complete sway over man. But that destining is never a fate that compels. For man becomes truly free only insofar as he belongs to the realm of destining and so becomes one who listens and hears [Horender], and not one who is simply constrained to obey [Horiger]. (Heidegger 1977: 25)

Importantly, existing indigenous modes of comprehension break down the gap between subject and object, culture and nature. Lotka investigated rock art but through a modern lens of understanding. Indigenous ways of knowing show how the surrounding ecological features *are* technological. Nature is not categorised as pure and unadulterated. Pure nature is not contrasted to polluted, technological culture. Nor is rational, ordered culture superior to the wild ravages of nature.

For indigenous cultures, the intraconnection between all things makes the medium of time, space, and subjectivity (Kant 1992/1781) intricately woven together, like a skein of fabric. A tug on one thread warps and shifts the entire fabric of the universe. Nobody can extricate themselves as an individual, outside of the material of space-time, as proposed by Kant's transcendental rational free individual. This means that the entire skein of Ta/Va (time/space) (Māhina 2010) shifts as a result of each person's change in perspective and actions. We all affect each other.

Each person is not reduced to the a priori of universal rationalisation. In fact, comprehension and thought, for Māori – and other indigenous people – is an embodied, emotional, and gut-wrenching process (Mika 2017; T. Smith 2000). Embodied, emotional thinking has awareness of the breath, or *mauri*, of the self and others, including other species, waters, rock strata, atmosphere, and stars (Ngata 2011/1944). Each person individuates themselves and has 'self' awareness – without cutting the self off from the wider material world. Thinking *relies* on layers of meaning attached to natural objects, extra to the internal mental and physical processes of the individuals' somatic or subjective bodies. Memory and learning are triggered by 'exosomatic' events (Stiegler 2018), whether they be ideas written in the technology of a book, or the technology of the trees, birds, or river.

CLIMATE AND ENTROPY

Climate change is a good example of entropy. The second law of thermodynamics indicates that any increase in order irreducibly results in an increase in disorder. As economic production grows, the rate of entropy inevitably grows also. Modern economics externalised the accounting of pollution, or entropy, from the cost–benefit equations and 'forgot' to notice the impacts of pollution on the environmental commons. The ordering of consumerism is producing the disorder of the planetary ecosystem. Twelve thousand years of peaceful Holocene has been replaced by the turbulent weather conditions of the Anthropocene. The forecast is to become much worse as the level of carbon dioxide in the atmosphere drives up radiative forcing (Crutzen 2002; IPCC 2021; Hansen et al. 2023).

The impacts of fossil fuel pollution have generated dismay since the outset. As far back as the 1770s, Wordsworth was complaining about coal pollution from steam engines (Yoshikawa 2020). In 1896, Arrenhuis figured out the greenhouse effects of chemicals in the atmosphere and made remarkably accurate predictions about climate change. Yet, after 150 years of warnings, its only in the last decade or two, as system collapse looms (Bendell 2023), that serious attempts at replacing fossil fuels with renewable energy is taking place. Even now, the correlation between economic growth, consumerism and emissions is still largely unacknowledged. In the initial quarter of 2020, during the global pandemic of COVID-19, global emissions dropped by 17 per cent (Le Quéré et al. 2020; Myllyvirta 2022a; 2022b). The pandemic is the first real achievement in global reduction of CO_{2e}. Usually, energy savings created by renewable generation gets converted into more consumerism. New energy sources lift the overall global production base and continue to increase emissions, plastic pollution, resource exhaustion and entropy. As the global economy recovered after COVID-19, total yearly emissions were only 1 per cent lower than previously. In 2023, the IEA found CO_{2e} emissions grew in 2022 by 0.9 per cent, or 321 million tonnes – accumulating to an unprecedented 36.8 billion tonnes. The introduction of renewables and other efficiencies has 'successfully decoupled' the rate of economic growth from emissions growth by 550 million tonnes (IEA 2023). However, Jevon's paradox still prevails; economic growth results in a net growth in CO_{2e} emissions. Greenhouse gases continue to accumulate faster, causing global warming to escalate (Hansen et al. 2023).

Interestingly, the governmental economic response to the pandemic was to increase the monetary flow by providing salary cover and lowering Reserve interest rates to almost zero or below zero. At first, the abundant money had almost no inflationary effect except on 'financial reservoirs' like housing, Bitcoin, the stock market, and fine art. Minor quantitative easing enabled the consumer economy to keep going, despite a large loss of jobs and the disruption of entire industries, especially aviation and tourism. According to Keynes's multiplier effect, government spending decelerates a recession and enables economic growth to continue expanding. Most importantly, in my view, was the reduction of the OCR to almost zero, making loans cheaper than ever before. This stimulated a huge housing turnover. High numbers of new ex nihilo mortgages introduced vastly more new money into the system. There is a lag while this money circulates through endogenous banking and gets sopped up by housing, art, and stock markets, but eventually it soaks the fiat money exchange and stimulates widespread inflation.

The pandemic made clear the vulnerability of the global food system. Storage and shipping were delayed and unreliable. Land alienation continues to be a significant catalyst for industrial consumerism across the colonised world. Of the 8 billion population today, 57 per cent are urban and have limited access to subsistence food production (Leff 2009). Whether people are urban or rural, they are so invested in industrial production that they are 'alienated' from the impacts on the local ecosystem. People have lost their sense of attunement, in both practical and theoretical terms. Heidegger (1977) attributes this to the technological enframing of the way we know about the world; every aspect of our selves and environment is reduced to 'standing reserve' waiting to be called forth by the demands of consumerism.

Heidegger's (1977) important concept of the technological *Gestell*, or modern enframing of the horizon of disclosure, is a deterministic evaluation of epistemology. Heidegger regards the enframing of modern thought as almost impossible to overcome from 'within' because every new knowledge is subsumed into the machinery of global consumerism. Yet Heidegger also regards it as impossible to overcome from 'without' the Western culture of modernity. He fails to engage with indigenous thought in any serious way.

> It is my conviction that a reversal can be prepared only in the same place in the world where the modern technological world originated, and that it cannot happen because of any takeover by Zen Buddhism or any other Eastern experiences of the world. There is need for a rethinking which is to be carried out with the help of the European tradition and of a new appropriation of that tradition. Thinking itself can by transformed only by a thinking which has the same origin and calling. (Heidegger 1990/1966: 63)

His work on Buddhism is interesting (Ma 2008; Nelson 2023) and adds dimensions to his work on *Gelassenheit*, or 'release', and attunement. The engagement with Daoism also informs his work on poetics as an opportunity for deeper thinking than consumerism provides. For Heidegger, there are limited means to overcome the epistemological straight jacket of technological modernity. The primary focus is to remember that which is forgotten, which he describes as the 'question of being'. The methodology is historical, philosophical, and creative. As outlined in chapter 6, he sets it out as the 'two beginnings' (Heidegger 1999a; 1999b). The origins and the finitude of modern civilisation, both offer ways to reflect on the current state of events.

Overcoming the modern technological horizon of disclosure with its focus on the pace and demands of consumerism is proving every bit as difficult as Heidegger (1977) argued. Stiegler (2017 and 2018) has engaged with the problem by tracing the emergence of technology and art back to the earliest Palaeolithic evidence. Following Leroi-Gourhan's seminal work (1993/1943 and 1945), Stiegler (2018) argues that cave paintings, and later technologies, film and artworks, enable an 'exosomaticisation' of memory. This is an important development, which Leroi-Gourhan (1993/1943) argues creates the seminal moment that is the distinguishing feature of Homo sapiens. Ideas are not merely embedded in the body and personal experience. They are exteriorised into the works of technology or art. This is a crucial development of the mind, which enables learning to take place in completely new ways. Instead of learning through immediate experience or first-person relating of knowledge, ideas are embedded into artworks and technologies. The transmission of ideas is no longer one-on-one but across the entire community, beyond immediate time and space. Exosomatic technology and artwork create an 'interposed membrane' or 'screen' of interpretation, according to Leroi-Gourhan, that informs the way people interpret the world (1945).

Stiegler (2018) traces how exosomatic memory is produced through the image on a rock wall, the written word, the printing press, the cinema, to contemporary smartphones, computing and artificial intelligence. He makes a compelling case that the acceleration of technologies is impacting the way people think.

Heidegger's (1977) critique of modern technology as the enframing of the horizon of disclosure helps us understand why knowledge of the cause, and drivers, of climate change have not resulted in an immediate change of course. Technological knowledge is framed around consumerism and economic growth, and thinking outside of that horizon of thought is difficult to imagine. So far, in late modernity, we understand that fossil fuels cause climate change. Economic growth is the driver that maintains fossil fuel use because the exponential increase in energy demand exceeds attempts to convert to renewables. In my view, it is the alienation of modernity from the environment that 'forgets' the intrarelationship of humanity with all other beings that is ultimately the problem. 'The Question Concerning Technology' (Heidegger 1977) hints at modes of exceeding the determinism of modernity. Heidegger resolutely remains within the 'two beginnings' (1982) of the modern Western tradition, to interrogate and question why modernity is 'forgetting to ask the question of Being' (1962; 1973a). Heidegger initiated an important pathway to exceed the nihilism of modernity but was unable to produce any real solutions. Modernity remains entrenched in the discourse of consumerism, where everything is understood as a resource, waiting for the demands of consumption. Stiegler (2018) opens up some productive modes of thinking about technology from a new perspective, beyond the historical limits of the modern epoch.

Perhaps it is too much to hope that Stiegler's argument would short circuit Heidegger's deterministic enframing of thought by the essence of modernity. Heidegger (1977) argues that the modern epoch reduces everything to waiting in reserve for consumer demand. Stiegler (2018) certainly exceeds the 'origin' in ancient Greece, and finds technology operates as a 'screen' which effectively shapes the way that human beings understand the world. He shows how modernity is accelerating with artificial intelligence (AI), and *noeisis*, or 'thought', is decaying into passivity and dogma. Stiegler introduces the *pharmakon* as a negentropy to reduce the impacts of accelerating technologies.

THE SCREEN OF TECHNOLOGICAL DISCLOSURE

Lotka (1925) develops the idea that cave painting is a crucial moment in the development of human epistemology. Ideas reside in artwork and can be transmitted across generations. It exhibits an important conceptual development in human cognition and at the same time enables the process of human evolution as Homo *sapiens*, the animal that *knows*, to develop. The earliest cave paintings make concrete ideas that are circulating in the community. Their conceptual articulation forges a lens or screen through which ideas about the environment are generated (Lotka 1925 and 1945; Leroi-Gourhan 1993/1943 and 1945).

Stiegler (2018) argues the screen is exacerbated yet further, by the acceleration of cyber space and AI. These authors concur that exosomatic art and technology 'represents' the bison, the deer or the concept, which is at a remove from the original empirical thing. The image constitutes a form of estrangement – membrane – and occurs in all exosomatic techniques. The acceleration of contemporary technologies amounts, Stiegler argues, to accelerating entropy. The Anthropocene (Crutzen 2002) is an *Entropocene* (Stiegler 2018: 36). The Anthropocene has increased order, or technological sophistication, and at the same time increased disorder, or pollution.

Stiegler's aim is to develop techniques of 'negentropy' to counteract the entropic tendencies of forgotten disorder – emissions, pollution, plastics, toxins, and dogmatism – into the environment. Exosomatic memory is an important concept to understand the impact of accelerating technology on learning and knowing. Likewise, his concepts of the *pharmakon* and 'neganthropy' provide an antidote to the ravages of the Anthropocene. These techniques can be added to Heidegger's modes of exceeding the enframing of modernity through asking the question of being, through the two beginnings of origin and finitude, through *Gelassenheit* or attunement, and poetical contemplation of death, or nonbeing, which throws the meaning of being into relief.

During this zone of nihilism, the insight of Heidegger's modern technological enframing of disclosure continues to dominate. Economic methods of consumer distribution continue to expand under the ethos of 'growth'. The Anthropocene continues to accelerate. Exosomatic memory is important for explaining the evolution of becoming-human, but whether the 'screen' of technology and art are essential to the generation of culture, and how that screen is a vehicle for exteriorising modern technology, evokes many new questions which need to be unpacked. Particularly if these exosomatic devices could constrain modern consumer pollution instead of exacerbating it.

THE ENTROPOCENE

Homo sapiens evolved some four hundred thousand years ago in the mid- to late Pleistocene, along with our 'sister' species, Homo Neanderthalensis and Denisovans (Stringer 2016). All three species have a large brain with 'upper parietal expansion', upright bipedalism, basicranial flexion, 'faciality' that is smaller and more mobile. Childhood is longer, with 'lengthened post-natal growth period and life history'. Homo sapiens have a specific ear canal and dental density. The Mugharet-es-Skhul fossils demonstrate evidence of intentional burial, some 100 to 130,000 years ago (Stringer 2016). These things all suggest, like the Neanderthals and Denisovans, a human ability for conversation, thought and learning that comprehends 'life' and death that predates the emergence of civilisations during the more recent Holocene. The Holocene is the last warm period of the stable Pleistocene, and when the major ice sheets had receded about twenty thousand years ago. Modern Homo sapiens rapidly evolved complex agriculture, housing, towns, cities, and other cultural technologies.

The Anthropocene epoch is distinguished from the Holocene because the anthropocentric impact of humanity has produced plastic detritus that is now a definable layer in the geomorphology of the planet. That is in addition to the significant changes in climatic conditions (Crutzen 2002; Stefan et al. 2011). Crutzen defined the change of epochal conditions for its anthropocentric, or human, causality. There are now rubbish gyres miles wide in the Pacific Ocean, microplastics in the deepest ocean trenches and plastic in the Arctic and Antarctic and at the top of mountain ranges.

While plastic is relatively recent, like most species, humanity transforms its ecosystem. Humanity seems to like tundra and have been clearing forest for thousands of years. Human intervention remained within the boundary of planetary tolerance until shortly after the industrial revolution. By now, the Anthropocene has already traversed several 'tipping points', and climatic conditions are becoming more chaotic and turbulent. This contrasts with the Holocene, where the ocean's thermohaline current operates as a thermostat that has kept a stable and moderate temperature over most of the planet's surface (Rahmstorf 2006; 2009). The Anthropocene results from human-generated entropy.

The first law of thermodynamics is that in molecular exchanges, matter is never lost. However, the second law is entropy, when matter is exchanged, heat is externalised to the wider system. For example, if two atoms are conjoined into a higher-order element or molecule, some energy as heat is dispersed to the wider universe.

Entropy is the simultaneous increase of order and disorder, as two or more discrete elements mix together in an irreversible process, creating a new composite. At a molecular level, hotter atoms whose frenetic activity bounce frequently off each other, gravitate towards cooler, sparser areas, in a gradual tendency towards thermodynamic equilibrium. In a closed system, heat is lost to the wider environmental 'reservoir' (Kondepudi and Prigogine 2014: 124). The stable consistency of the earlier simple order atoms become irreversibly mixed together, like milk dispersing through coffee for example. Returning the milk to its original state is impossible. Thermodynamic entropy introduces the arrow of time into physics (Prigogine 2014: 2).

Lotka (1925 and 1945) and Soddy (1933/1926) applied the thermodynamic statistical insights of chemistry to early emerging disciplines of ecology and economics in the 1920s. Georgescu-Roegen (1971), Gorz (1971), Daly (1973) and a few others (including Karl Popper 1974) followed up with a second wave of analysing economics in terms of thermodynamics and energy throughput in the early 1970s. Theoretical definitions of thermodynamics and entropy tend to discuss either matter or energy. Its only in recent quantum experiments on the behaviour of particles and waves that matter and energy are beginning to be thought of as different 'complimentary' presentations of the same phenomena. The muddle has made applying thermodynamic principles contested. Georgescu-Roegen developed a controversial 'fourth law' of thermodynamics, that 'proclaims for matter what the Entropy Law proclaims for energy', 'an isolated system, instead of tending toward heat death [when all energy is unavailable], tends toward chaos "when all matter-energy is unavailable"' (Guillen 2000: 91). The transition from the discipline of molecular chemistry to the complexity of economics and ecology needed to expand the concept of entropy from closed system experiments in chemistry to accommodate open and interactive, large-scale systems.

> Matter and energy are so closely linked to one another that the formulation would not be altogether complete if we limited the meaning of entropy to that associated with the degradation of energy. We must inevitably take a further step, even at the risk of blurring the concept, to take into account the degradation of matter, which like energy, is dissipated and irrevocably lost to man in every interaction between the two. (Guillen 2000: 89)

Following this line of thought, the Anthropocene is entropy writ large. The improvement of modern human mastery, or the increase in order, comes at a cost: the increase in disorder, or pollution, into the larger environmental 'reservoir'. Not all people are contributing to the modern ordering of order. Indigenous and subsistence communities continue to live with a light footprint, low entropy lives. It is not all of 'humanity' that causes the Anthropocene but modern pollution, modern entropy. Hence, Stiegler (2018) does not characterise it as part of the Anthropos (humanity) but as the Entropocene.

Negative-entropy, or negentropy, could be considered an oxymoron by the chemical experiments in Clausius (1856) type stochastic thermodynamics, where a hot body gradually dissipates heat and gets cooler, but a cold body only gets warmer if energy or work is applied to it. Negentropy should not be understood as the reverse of time-directional entropy. It only makes sense in the real-world complexities of intersecting ecological biodiversity. I view negentropy as entropic cascades, which capture the stray energy that is externalised from an entropic system and convert it into useful ordering. An excellent example is photosynthesis, which captures the sun's entropic radiation and converts it into glucose and plant material. Ecological systems are full of examples of this kind of negentropy.

Stiegler (2018) created the new concept of 'neganthropy', which attempts to reframe the eschatological collapse of the technological Anthropocene. Humanity can reduce the thermodynamic throughput in consumer systems as a negative impact of anthropocentric activity, or neganthropy. Stiegler's idea echoes Lotka's ideas about the self-regulating rhythms of energy in ecological systems (1925) to regenerate the cultural noosphere and the atmospheric commons. The noosphere is the sphere of noesis or thought. It is similar to Heidegger's *Gestell*, the enframing of the horizon of thought. Neganthropy aims to halt the degradation of thinking, that Stiegler argues is occurring. Artificial intelligence and other modern technologies reduce the sphere for original, creative, critical thinking. 'Fake news' and AI stereotyped bubbles of like-minded people atrophy the exposure of people to challenging new ideas. The golden age of the algorithm sets the pathway for the human mind to decay into rigid and dogmatic thinking.

RATIONALITY AND THE SPECIES BEING OF HOMO SAPIENS

Since Aristotle (384–322 BC/1972), the species being of Homo sapiens has been attributed to rationality, which was thought to be peculiar to humanity. Rationality is often associated with the use of tools. Hobbes defined humanity as the 'tool-wielding animal'. Kant (1992/1781) believed in the Christian hierarchy of the great chain of being, with humanity at its peak. Humanity is also often associated with higher-order speech, which enables us to share and extend rational knowledge into a universal domain. The assumption of human superiority over other species rests on a triage of exemplars: tools, language, and rationality. The way that humanity alters its environment has been attributed to these 'superior' attributes. Over time, though, it has become apparent that other species share many of these facilities. Animal, plant and even microbe communication is becoming better understood and accepted. At least eight words of meercats are understood by linguists studying meercat culture (Rauber et al. 2020). As Rauber et al. put it, 'The ability to recombine smaller units to produce infinite structures of higher order phrases is unique to human language, yet evidence of animals to combine multiple acoustic units into meaningful combinations increases constantly' (2020: 1). In the flatlands in Cambridgeshire, for example, there are large, man-made clay pits that were 'restored' into lakes. People saw herons fly over these newly created lakes and drop live fish in them to develop fish stock. There are several species of eagle, and gulls, that use rocks to crack open eggs or shellfish. Bonobos and chimpanzees use tools (Koops et al. 2015) and elephants hold funerals (Goldenberg and Wittemyer 2020). In Australia, three species of hawk spread forest fires to drive out prey (Bonta et al. 2017).

Karl Popper, the regent of rationality, was early to read Lotka (1925), and recognise that tool use is not unique to Homo sapiens. Anthropocentric superiority would not hold long under that criterion.

> Animal evolution proceeds largely, though not exclusively, by the modification of organs (or behaviour) or the emergence of new organs (or behaviour). Human evolution proceeds, largely, by developing new organs outside our bodies or persons: 'exosomatically', as biologists call it, or 'extra-personally'. These new organs are tools, or weapons, or machines, or houses. . . . The rudimentary beginnings of this exosomatic development can of course be found among animals. The making of lairs, or dens, or nests, is an early achievement. I may also remind you that beavers build very ingenious, dams. But man, instead of

> growing better eyes and ears, grows spectacles, microscopes, telescopes, telephones, and hearing aids. And instead of growing swifter and swifter legs, he grows swifter and swifter motor cars. (Popper 1974: 238)

Most writers that acknowledge the legitimacy of other species' ability with tools and language also acknowledge their legitimate occupation of planetary territory, whereas Popper reverts to rationality as the justification for human superiority and 'advancement'.

In the hands of Karl Popper (1974), Lotka's (1925) ideas about exosomatic technologies justify an optimistic view of progress, and an 'objective' epistemology. In his book, *Objective Knowledge: An Evolutionary Approach* (1974), Popper argues that the exosomaticisation of thinking into artefacts such as a book enables progressive 'correctitude'. Subsequent reading and critique reduce the 'faulty' analysis of empirical events. Exosomatic recording or storage allows ever-closer representation of objective reality (Popper 1974: 98). Popper is refuting Hume (1985/1750; 1958/1739–40), who raises an important critique of normative epistemology. Hume regards normal interpretation as more habitual and entrenched in the linguistic and social infrastructure than it is related to the best interpretation of empirical events. Popper disparages Hume's critique of normative repetition or habit as 'primitive', presumably because it is not exosomatic but rather culturally experiential.

Popper (1974) uses the development of exosomatic *technē* as a reinforcement of rational universalism; 'without the development of an exosomatic descriptive language – a language which, like a tool, develops outside the body-there can be no object for our critical discussion'. An 'external' or exosomatic language becomes a sceptical apparatus, where a 'third world' can germinate, allowing a context in which 'the problems and standards of rational criticism can develop' (Popper 1974: 120). He relates tool use to rationality, innovation and progress, arguing that 'instead of growing better memories and brains, we grow paper, pens, pencils, typewriters; dictaphones, the printing press, and libraries' (239).

The technological optimism of Popper (1974) contrasts with the pessimism of Stiegler (2018), although the methodology of both is essentially similar.

EXOSOMATIC *TECHNĒ* AND LEARNING

Memory is usually understood as an internal process of mind. The discourse of exosomatic memory and thermodynamic energy throughput emerges from Alfred Lotka (1925 and 1945), Frederick Soddy (1933/1926) and Leroi-Gourhan (1993/1943 and 1945). Stiegler (2018) proposes that memory is externalised and embedded in art and technological artefacts. Writing a text, for example, makes concrete the ideas circulating in the community, as they are creatively expressed by the individuated author. The text emerges from a particular cultural milieu. It makes sense to readers in other cultures, and in other times, enabling the transmission of ideas beyond immediate experience and communication.

Popper (1974) interprets the development of exosomatic artefacts as progressively more accurate and rational. But fiction and poetry are also written words. The internet has become an exosomatic vehicle for the proliferation of 'fake news'. Rationality is one aspect of communication via the written word, but in my view, Popper's Enlightenment assumptions about continuous improvement, progress and honing rationality are overblown.

Exosomatic memory helps the transmission of ideas across time and distance. Learning, in modernity, is usually associated with schooling. Exosomatic memory operates at a deeper and larger level of discourse. The narratives are embedded in the practical associations, the utility and the stories and philosophies that interpret artefacts, subjects and environment in a constellation of ideas. Educating each student is not merely developing a range of credentials to help enhance their value as a 'human resource' in standing reserve for the demands of the machinery of consumerism.

In the debate over 'What Is an Author?' between Roland Barthes (1977), Michel Foucault (1998/1969) and Jacques Derrida (1978; 1985/1982), the text as an exosomatic artefact is actively (re)interpreted by each reader. Literary criticism had tended to highlight what Barthes called the 'intentional fallacy', where the author intends a particular meaning and readers are responsible for uplifting the original intention. Regardless of the origin of the text, the historical milieu of the reader brings to light new aspects and emphases that change over time (and place and culture). The role of reading is also a role of individuation (Siegal 2022), as an active pursuit of thinking and differentiating the reader in continuity and contrast with the text, and its prevailing interpretations.

Foucault's (1991) disciplinary society indicates a more insidious view of exosomatic *technē*. Behavioural discipline and self-knowledge are shaped through architecture, furniture, classroom layout, and surveillance technologies that operate as a panoptical gaze, always exposing personal behaviour to public visibility. The technologies of surveillance, from architecture, exams, bank cards, and cameras, make the awareness that erstwhile private behaviour is always on display. Self-discipline is shaped by the recognition of surveillance. Technology proscribes and enframes agency and events. Formal transmission of knowledge through modern institutions like education, operate within these larger, enduring contexts of exosomatic technologies.

Heidegger (1962) maintains that we are each 'thrown' into an 'always-already' techno-social environment, and the technologies themselves shape the way we think about the world. Children pick up the widely disseminated assumptions embedded in the environmental artefacts, utensils, tools, machines, architecture, infrastructure, food and work ethos. Cohesion and individuation in the community knowledge system are crucial psychological processes of development that forge mature thinking in the cultural milieu (Siegal 2022). In continuity with this surface layer of gaining credentials, the student individuates himself or herself in cohesion with, and in distinction from, the greater communal milieu into which he or she is 'thrown'.

Universal rational individual*ism* as set out by the Enlightenment, is vastly different from the concept of individ*uation*. Individualism proposes that all humans are essentially the same, and that learning develops inherent propensity to rationality. When done objectively, all rational propositions turn out the same, and so each individual is ultimately alike with every other rational individual. Individuation has an entirely different premise. Humanity is communal and intraconnected with all other elements of the ecosystem (O'Brien 2021; Siegal 2022). But each human has different experiences, and his or her personality, knowledge and expectations are forged through those experiences. Careful reflective thought about exosomatic encounters helps to both connect and hone the specifically individuated interpretation of the life-world.

Exosomatic repositories of knowledge helps to shape the personality, or 'individuates', each person. More than the exosomatic enculturation of separate individuals, over long periods of time, exosomatic *technē* shapes the cultural personality, which Stiegler calls 'transindividuation'. Culture is an 'open system' which requires input from the wider ecological milieu. Given how long these exosomatic processes have successfully taken place (at least 130,000 years), neither of these processes of transindividuation, or individuation, necessarily result in an entropic deterioration of knowledge or the environment. But exosomatic *technē* are the evolutionary 'origin' of the modern technological enframing of disclosure described by Heidegger (1977).

During the Entropocene, exosomatic memory corrodes into dogmatic cultural conditions (Stiegler 2018: 192) alienated from ecological rhythms. Given the existential threat presented by the Entropocene, understanding whether, and how, this exosomatic process has gone awry is crucial. The alienated mode of being is accelerating, according to Stiegler (2018: 37, 46 and 90–91). Understanding the consequences of these modern technological processes is important if we are to reconnect with the possibility of an ongoing future.

Accelerating cyber technology presents humanity with a new type of exosomatic process, that is further reshaping the way people think. Stiegler argues that exosomatic memory is essential to what makes us human. The evolution of technology is creating a repository of knowledge that is global in orientation. 'Technological facts are such concretions: *always partial* (even negative) *concretions* of tendencies that constitute *territorial exorganic* functions. Today, the latter tend to become *totally deterritorialised* as they become *functions of the biosphere* on a *planetary* scale' (Stiegle 2017: 192).

Data collation, artificial intelligence and the marketisation of everything – from shopping preferences, credit ratings and political orientation, to health, fitness and menstrual cycles – all facilitate a globalised production of consumer society. It obscures the nuances of embodied environmental attunement. The acceleration of cyber technology is contributing to the enframing of modernity. It exacerbates an already alienated production of culture. Modern culture is increasingly disconnected from an embodied attunement with the complex biodiversity of the local and planetary social and ecological milieu.

Exosomatic memory is the crucial moment that generates a specifically human epistemological breakthrough in evolution. It is groundbreaking because it surpasses immediate experience and enables cross-generational learning. This helps generate the production of culture, or transindividuation.

At various points in the evolution of communication technologies and knowledge transfer, profound shifts have occurred that have changed the technical method but also the way that people think. The printing press, for example, transformed culture, class access to information and types of knowing (Derrida 1978). The way exosomatic techniques now show up through the internet is complex. The globalisation and speed of the network has accelerated and deterritorialised the type and availability of information storage, searches, and access. It has changed the nature of physical libraries, which are throwing out old books. At the same time, the physical comportment of reading has changed. Stiegler (2018) describes IT reading as passive, receptive and guided by AI rather than an active, energetic engagement with the search for knowledge. '(A)ccumulated knowledge, has become dead, stereotypical, automatized' (Stiegler 2018: 194). With examples of AI such as ChatGPT, the art of writing – and the thinking that takes place through writing – are going to be increasingly automated. The internet is an entropic feature of epistemology. It undermines the ancient negentropic traditions of culture generation through creative crafts, art, and technology across a broad sphere of practices.

Stiegler (2018: 37, 46 and 90–91) argues that the speed of the internet is overwhelming the human ability to think. We get carried along by associated pages in our internet searches, which algorithms bring up on the basis of our history and categorised identity assumptions. Big data analyses – of past searches, lengths of time, number of clicks, similar demographics, areas of IP address and more – sift searches in a way that reinforces a 'bubble' of identity-stereotypical search results. Decision-making is robotised, and people are becoming passive consumers of internet information. Our attention is commodified. People are glued to the screen, passively absorbing entertaining information, without engaging their imagination, creativity, passion, embodied vibrancy, or the immediate preverbal awareness of other bodies and emotions. The constant input of information leaves no room for dwelling, contemplating or boredom. And those crucial quiet periods are when the mind makes new associations and insights. In front of the screen, decision-making is, in vital ways, disactivated. Critical thinking is not intrinsic to the internet. The source of information, or its validity, whose interests it serves, what bias and assumptions are embedded, are often excluded or obscured by pseudonyms and anonymity. The source of information, or what information the person is exposed to, emerges from the unregulated 'black box' of machine learning and the sheer quantity and speed of traversing information. Attention span is diminishing, and a certain passivity or brain fog is taking over ever-lengthening screen time.

Stiegler (2017) makes an important analysis of the technological and aesthetic screen as a Paleolithic, as well as contemporary, lens on the world. To echo Baudrillard's (1994) analysis of the emergence of the hyperreal, the screen is the medium of truth, the medium of the representation, that is already at one remove from the real itself. The screen is part of the dissemblance that mimics and is also a parody of the real. Baudrillard's four stages from intricately bound to the ecosystem, to the hyperreal, can be understood as the entropy of the formulation of truth. It goes from mimicry, or representation, to a focus on the conduit of truth, or the screen or *technē*, to the dissemblance, and then the hyperreal. Stiegler is attending to the deterioration of knowledge into dogmatism, which he associates with the screen, the dissemblance of AI, and the zombie consumerism of information that nudges people further and further into alienation and conspiracy theories.

Stiegler (2017) echoes Heidegger's concept of the modern technological horizon of disclosure (1977) with the role of AI in the enframing of thought. The modern technological horizon of disclosure is enframed, as Heidegger argues, by 'nothing technological' but rather the consumerist ethos that everything is waiting, in 'standing reserve' until its demand from consumerism. The oeuvre of exosomatic learning, from Lotka's anthropology in the 1920s through to Stiegler in 2018, all conforms to the modern dualism separating culture from nature, via the membrane or screen of representation. Modern worlding is a cultural affair *about* nature, not as part of it.

NON-ALIENATION AND EXOSOMATIC MEMORY

Te Haumoana sweeps his hand up the hill towards the steep ridge of straggly trees that bask in the late westerly sun. 'Those are the Tainui trees'. These unremarkable bushes with their wispy pale flowers glow softly as they meld into the steep mountainside. 'They came here with the Tainui canoe'. 'What??' I was surprised. How could those obscure bushes arrive on the great ancestral voyaging canoe, the Tainui? But I had missed the subtlety of Te Haumoana's point. The Tainui trees were not resources shared and carried on the canoe. They *were* the canoe.

The origin story most accepted by early European historians is that seven voyaging canoes, known as the Great Fleet, arrived in New Zealand, about AD 1100 or 1200 from Hawai'iki. The Tainui moseyed down the West Coast of the North Island, or *Te Ika o te Maui*, stopping at various suitable anchorages, such as Kawhia, Raglan and Mokau, as it explored the coast. Mokau was an existing settlement. Today, the huge anchor stone of the Tainui canoe lies safely in a concrete plinth, in the village of Mokau, ensconced so it cannot be stolen again by marauding European museums.

After its long journey, Te Haumoana explains that the *muka*, or twine, holding the ornately carved bow to the dug-out hull was coming loose. Someone pulled it out to replace it, and stuck it in the ground, where it struck. The living trees are descendants of the canoe itself.

The landscape is saturated with such mnemetic devices. The Tainui trees are an immediate, living relic of the great ancestral canoe, tying the local community to their heritage in the Pacific, to the skill and prowess of navigating by the stars, ocean currents and knowledge of the range of birds, driftwood, phosphorous, wind, weather and clouds (Lewis 1972; G. Irwin 1992). Colonial Europeans think of a vast empty ocean, with occasional islands. Polynesians regard the Pacific as an ocean abundant with islands. The grove of Tainui trees is part of a system of signs through which the world is knowable.

Later, Te Haumoana White, the *rangitira*, or 'chief', of the Poutama iwi in Mokau, on the West Coast of the North Island, in New Zealand, contacted me to say that *Pomaderris apetala* does not originate anywhere in Polynesia. It is native to Australia, hinting that those adept sailors made recurring visits to the ancient 'grandparents' of indigenous peoples (Susan Gameraigal personal communication 2020) drawing swiftly through the water in their aerodynamic craft (Irwin and Flay 2015).

The trees themselves provide exosomatic memory that can be explained to newcomers without needing direct experience of canoe voyaging. The Tainui trees enable subsequent generations to enter the cultural semiotics of being Māori and being Polynesian. There is no alienation between technology and biology. There is no screen. No alienation is constituted by economic methods either. Te Haumoana's alertness to the Tainui trees is an important example of how integrated knowledge, attachment and caretaking the land stand in contrast to modern modes of alienated epistemology, storage, and redistribution.

The Polynesian gift-giving economy of community distribution is based on local seasonal productivity. In contrast, modern economics has generated a vast distribution network built on widespread communication, transport and energy networks that 'lift' local communities from reliance on immediate production. It inserts them into a global network of commerce. The freedom this network produces is important; any immediate ecological emergency is alleviated by access to global provision of goods and services. However, as Marx (1887/1867), and later Heidegger (1973b), pointed out, the price for a networked society is alienation from the ecological impact at a local, and now a planetary, level.

Te Haumoana's narrative of the Tainui trees predates and exceeds modern epistemology in numerous ways. Te Haumoana's indigenous perspective offers insight into a mode of knowing, learning, and exosomatic repositories that are non-alienated from local ecology. Yet they are capable of engaging with contemporary science and technology. The screen, or membrane, is no inhibition.

INDIGENOUS INTRACONNECTION

More immediate than the concretisation of discourse in art and technological artefacts, the trees themselves are the exosomatic memory that help introduce subsequent generations into the cultural semiotics of being Māori. Tainui trees are the embodiment of ecology, land, ocean, and the technology of the canoe. There is no division between *technē* and biology. The Tainui trees symbolise genealogies, voyages, an ocean of islands, connection between different nations, medicinal knowledge, and the expertise of ocean navigation. The trees have utilised Pacific people to expand throughout the Pacific, cultivating culture on new islands. There is no alienation involved in this mode of redistribution. Exosomatic memory resides in the *technē* of the Tainui trees themselves.

The Tainui trees are called *kūmarahou* in Māori. The Latin nomenclature of *kūmarahou* as *pomaderris apetala* and its biological claves add to the narrative by clarifying the forgotten Australian origins of the tree. That Australian origin confirms and extends the range of ocean-faring undergone by the ancient ancestral Tainui canoe. Exosomatic memory is extended by new layers of meaning within contemporary epistemological systems. The new layer does not succumb to the modern, alienated enframing of knowledge. This is partly because the Māori epistemological system is built on the connections of *whakapapa*, or genealogy, rather than universal systemisation. *Papatūānuku*, means equally 'earth mother' and 'conceptual foundation' (Mika 2017).

> There is a unity evident here of language and world. When, for instance, whakapapa is referred to, immediately the fact of interconnection arises also. We would not so much talk about whakapapa, then, as be claimed by its fact: We might mean 'whakapapa' as genealogy, but a persistent ground of thought (Papatuanuku) establishes itself in our speech or perceptions in various ways. As we mention 'whakapapa' we are reminded of our reliance on the world and, indeed, on the ancient origins of our thinking and our existence. (Mika 2019: 22)

New layers of meaning add other strata to the genealogical lineage (Smith T. 2000; Mika 2017), without overturning older modes of knowledge. The debate between Hume (1985/1750; 1958/1739–40) and Kant (1992/1781) and Popper (1974) about universal epistemology or 'more accurate' narrative explanations of empirical evidence are unnecessary. Genealogy is capable of adaptation and evolution. The origin of the Tainui trees in Australia, traced through the Latin nomenclature *pomaderris apetala*, do not obliterate the truth of the genealogy of the great canoe. It simply adds new genealogical connections. Māori reached Australia, they interacted with indigenous people there, they discovered new medicinal plants, and they incorporated their new knowledge into existing *technē*.

Rock art and stone tools remain more obvious in the archaeological record (Weule 2020). Their materiality is part of the image, or screen. Recent analysis of Paleolithic rock art noticed that the representations of animals use the curvature of the cave, and natural cracks in the stone, to help forge the image. The rock is the bison, the ochre is the deer (Asiain et al. 2023).

Plants like the *kūmarahou* and *nīkau* can become more thoroughly integrated into native flora and fauna over time, losing their significance as imported *technē*, as people forget their own role in introducing exotic species. But with observation, these species reveal their oddity (having no other biological relatives in the local ecosystem). The insertion of the new species is a new *technē*. Remembering, also, that the humans are themselves a new species in New Zealand. Humanity is itself a new *technē*, a successful vehicle in the proliferation of these plants, canoes, and other animals such as the *kiore* rat, and *kurī* dog. As a *technē*, new species are a lens or screen to understand the world. It is also another layer, like a cultural and ecological palimpsest. The new species is another stratum, just as the layers of geological rock illuminate the ancient origins of volcanoes or tectonic plate movement. The *nīkau* is the only palm in New Zealand prior to colonisation. Its name is clearly Polynesian, rather than uniquely Māori, because there are 'no coconuts' so far away from the equator (Māori do not need the word *nī* because coconuts do not grow so far south and are not part of the diet). Despite its forgotten introduction, the *nī-kau* plant can still reveal its genealogical and exosomatic significance (Irwin 2001) without recourse to narrative explanation if the system of semiotics is understood.

Of course, local native flora and fauna are also *technē* and shape the worlding of the culture in similar ways. Kohn discusses something akin to this in his book *How Forests Think* (2013). Kohn says, 'humans and nonhumans use signs that are not necessarily symbolic – that is, signs that are not conventional – [which] demonstrates why these signs cannot be fully circumscribed by the symbolic' (2013: 15). The representational, or the symbolic, is part of the 'open whole', Kohn argues. 'The "Open Whole" aims to show that the recognition of representational processes as something unique to, and in a sense even synonymous with, life allows us to situate distinctively human ways of being in the world as both emergent from and in continuity with a broader living semiotic realm' (Kohn 2013: 15–16).

While the concept of the 'screen' developed by Lotka (1925) and Leroi-Gourhan (1993/1943 and 1945) is interesting in the development of the concept of exosomatic *technē*, it is also immensely useful for analysing how and why modernity remains locked into its consumerist epistemology. The screen of representation misses the intrinsic integration of indigenous people with their territory. The ecology itself is brimming with meaning for indigenous people, without a membrane separating technology from the environmental surroundings. Plants, animals, rocks, and water *are technē*. Ideas are embedded in the plants, rocks, geological strata, ocean, and stars to generate further cultural dissemination and regeneration. They are co-beings, bound in kin relationship with indigenous people, who require respect, as do fellow human beings. The agency of trees, which utilise humanity as a vector of transport, are as active as the humans who navigated the seas to arrive in Aotearoa. Landscape that are understood as exosomatic indigenous artefacts, such as the Tainui trees or the *nīkau*, are found all over the world. The meanings they inhabit are still present in the ancient place names for mountains, rocks, and geological features, or the multiple descriptions of wind, ocean currents, types of rain, and the flora and fauna.

In contrast, Heidegger (1977) argues that the phenomenological experience of *modern* technology tends to enhance certain affects and obscure others, shaping the way people think as it does so. The motorcar has accelerated the mobility of communities and increased their range, at the same time as it wasted muscle and limited people's ability to walk far on variable, uneven, uncomfortable surfaces. The modern technological horizon of disclosure subsumes all aspects of knowing into a system of consumerism. Mountains are no longer ancestors but, instead, storage repositories of wood, fresh water, and precious minerals waiting in reserve for the demands of consumerism. Rivers are not living beings (Te Urewera Act 2019) but potential hydro dams, and tourism spots. Forests are 'ecological services'.

Neoliberal economics, following Enlightenment philosophy, has 'universalised' global markets, reducing the need for local populations to worry about local ecological conditions. The impacts of pollution and resource exhaustion in local places is externalised from productivity measures. Global claims of atmospheric pollution have, likewise, been largely ignored – until the recent cyclones and forest fires forged an upswell in public alarm. The failure of supply chains to negotiate COVID-19, and the turbulence of the Anthropocene, have exposed modern people to planetary ecology. Now, modern reliance on logics that ignore ecology look ridiculous and antiquated. Nevertheless, because alienated individuals remain the key unit of modern economics, there is still no route for policy, economic or finance practices that adequately account for the environment, either at planetary or local levels.

Chapter 10

Quantum Physics and Ontological Turn

Modern physics is catching up with the indigenous way of integrating all things, in what is sometimes called 'the ontological turn'. Quantum field theory introduces a new level of thinking, beyond the atomised individualism of Descartes (1980/1641) and Kant (1992/1781) and the enlarged *Dasein* of Heidegger and modern philosophy. Karen Barad (2007) explores the implications of quantum entanglement as a mode of ontology. The rules of linear cause and effect fail at this level of physics. The superposition in quantum mechanics sets out the potential – or the virtual – indeterminate positions that sit within scales of probability, rather than specific, locatable points. The entanglement at the subatomic particle level of electrons and quarks shows how existential reality is more like diffractive waves in the ocean or a 'field' than separate atomic particles. Time and objectivity work in different ways through this lens of existence. The whole notion of progress, of telos and of finitude needs to be reworked and thought about in new ways. Intimations of these changes in thinking are already taking place, with lineage to the interwoven genealogy of all existence in indigenous taxonomy (Irwin and White 2019; Irwin 2022a; T. Smith 2000; Kohn 2007; 2013; Mika 2017; Māhina 2010); feminist thought on the fluid interchange between body and environment (Kristeva 1982; Irigaray 1985; Irwin 2023c); technological prosthetics and the self (Barad 2007; Haraway 2016; Irwin 2005); open-ended ideas about difference and repetition (Heidegger 1969b); non-teleological change, the random and the necessary 'post'structuralism (Whitehead 1969; Prigogine and Stengers 1984; Deleuze 1994; Deleuze and Guattari 1999; 2004; Foucault 1979; 1990; Derrida 1978) and new materialism (Dolphijn and van der Tuin 2012; Braidotti 2013; 2018). None of these arguments are universal. They each outline the boundaries and interlocutions of an-organic life. This is a new metaphysics that indicates a shift in perspective that is yet to be fully realised in political economy.

Throughout this book, it can be seen that liberal individualism, informed by the Enlightenment, still maintains normative power in policy and political discussion. Liberal individualism continues to inform the Intergovernmental Panel on Climate Change (IPCC). Even promising materialist and metabolic forms of bioeconomics can succumb to the Enlightenment schism between nature and culture, subject and object. Solutions to the problem of rational individualism and its sceptical gap with objects in the environment come from a variety of disciplines. This chapter sketches some of the most important sources of critique and the most fruitful conceptualisations of intraconnection. Feminist embodiment, indigenous cosmology and quantum physics are three disparate sources for a similar idea – an ontology of intraconnection that informs the onto-epistemology of an ecopolitical future.

INTRACONNECTION

The focus of the Western Enlightenment has been on the rational individual. But what happens when the self-contained integrity of the 'individual' can only be understood as an illusion? The individual is always embedded in a context. Enlightenment emphasis on rationality does not exclude irreducible embodiment. The body is exchanging parts with its ecological context at subatomic, institutional, intrapersonal, and linguistic levels. There are three unrelated discipline areas that all confirm intraconnection in very different ways: feminist embodiment, indigenous kin taxonomies and quantum physics.

Feminist embodiment shows how the fallacy of the self as a solid object is an illusion (Kristeva 1982). Bodies exchange matter with the outside world all the time – through breath, food, drink, sweat, saliva and other excreta. Ingesting and excreting in a constant exchange of the inside with the outside threatens the notion of solid individual personal integrity. Kristeva (1982) argues that the abject revulsion towards bodily fluid emerges from the anxiety about the porous containment of the body-self as it fails to maintain separation from its surrounding environment. The breaches in bodily integrity amount to a breach in modern etiquette and dignity (Irwin 2023c).

Philosophy of technology asks a similar question: In the case of a dental filling, for example, where does the *technē* start and the self finish? Haraway (1991) rewrote the rules of engagement when she collapsed the schism between culture and nature, technological object and personal subject, in her text *The Cyborg Manifesto*. It is difficult to understand embodied technology as an external tool when it is an extension of the phenomenological body. The technologies of glasses, clothes and so forth constitute the self. Or as

Gregory Bateson wrote in relation to his walking cane, 'Where' exactly along its sturdy length, 'do I start?' (Kohn 2013: 6).

Most indigenous cultures emphasise how intrarelated individuals are – within their immediate family, their wider kin ties in the community and with concepts like *whakapapa* – to the other species, strata, water, atmosphere and stars of the Milky Way (Ngata 2011/1944; T. Smith 2000; Mika 2017; 2019). Indigenous philosophy, and in this case Māori and Pacifica philosophy, shows how everything is connected in a taxonomy of kinship relations, from the stars, through to the ocean, geology, trees, birds, insects, fish and so on. Humanity is in a genealogical connection with other species and strata (Irwin and White 2019; Irwin 2022a; Smith 2000; Kohn 2007; 2013; Mahina 2010; Mika 2014). These connections are visceral and embodied (Mika 2017; T. Smith 2000). Intraconnection underpins the physics of the universe.

Whakapapa is a key idea to conceiving of these carefully articulated interrelationships. In modern society, *whakapapa* is often associated with the English term 'genealogy' and applied to long lists of illustrious ancestors. It is usually a type of nuanced identity politics, 'whereas', Cherryl Wairea i te Rangi Smith writes, 'it actually maps out the nature of existence' (C. Smith 2000: 46). According to Ngata (2011/1944), *whakapapa* is an important way to conceive of the intra-actions of all beings, which includes the human, all species, rock strata, waters of different kinds – from springs to rivers, wetlands and oceans, types of clouds and rain, the winds and stars. This means that anthropocentric perspectives are embedded in the network of kin relationships across all the other species and strata. Mahuika (2019) explains, 'The Māori natural and spiritual worlds were expounded in genealogies that gave meaning to landmarks, fauna, flora and phenomena, the roles of men and women, and divine descent. These whakapapa chronicled evolutions from the beginning of time, providing explanatory frameworks for the world and everything in it'.

For Māori, the connections between an individual's perspective, experience and thinking and the rest of the world are visceral and embodied (T. Smith 2000; Mika 2017). The concept of *whenua*, for Māori, means both earth and placenta. The word *wai* for 'water' also means 'who' you are. The inseparable connection to ecology is built into ordinary language and cosmology (Mika 2019). Traditionally, the intraconnectivity occurred at every level of comprehension, from the gods through to ordinary everyday language and protocols (see Sahlins 2017).

Indigenous cultures share and venerate their locality with the local flora, fauna, and geology. The production of knowledge is stimulated by the ecological, so the schism between culture and nature is irrelevant and unknown. For example, the Tainui trees on the West Coast of New Zealand, are evidence of this argument – the trees help people learn. They are memory storage that is held by the community, not just one individual, and so they are a widely spread source of cultural narrative and truth (see previous chapter). People do not need to believe in rationality, monotheism or universalism to make claims to shared wisdom (Irwin 2022a). Te Haumoana White describes the Tainui trees as living examples of the Tainui canoe. This is a *whakapapa*, or taxonomy of relationships, to which people connect in specific ways.

Other species, and even anorganic things, like mountains, rocks, and rivers, all have personhood. Kohn explains the indigenous approach for the Amazonian *runa*. *Runa* means 'person':

> [For the *Runa*] all beings, and not just humans, engage with the world and with each other as selves – that is, as beings that have a point of view. Runa ways of knowing others, then, are predicated on what I call an 'ecology of selves'. In this regard, they share something in common with Jakob von Uexkull, an early twentieth-century pioneer in the study of animal ethnology. Von Uexkull insisted that ecological relations 'are not the product of mechanical cause-and-effect interactions among organisms as objects. Rather, they are the product of the interaction of the phenomenal worlds – what he called 'umwelt' – that are particular to the perceptual and bodily dispositions, motivations, and intentions of different kinds of beings. (Kohn 2013: 4)

Nietzsche (1909) says something similar in his last, unpublished, manuscript. Nietzsche is reflecting on Descartes dualism between the mind and the body, and considers the way different species and things experience the environment in very specific ways. Dogs have much better hearing than humans, and a different range of light receptors in their eyes. Butterflies and many insects see ultraviolet light, so blossoms are full of these spectra of colour of which humans are unaware.

> There are no such things as 'mind', reason, thought, consciousness, soul, will, or truth: they all belong to fiction, and can serve no purpose. It is not a question of 'subject and object', but of a particular species of animal which can prosper only by means of a certain *exactness,* or, better still, *regularity* in recording its perceptions (in order that experience may be capitalised). (Nietzsche 1909, book III: 480)

The triumvirate of light, plants and insects – plants needing pollinators, insects looking for sustenance and the vector of ultraviolet light spectrum – create a good fit and mutually reinforce the specifics that optimise their mutual connection and interdependence. It is a specific worlding.

In contrast, Heidegger remains caught up in Kant's a priori medium of truth. This is where the a priori function in tandem; subjectivity and time, or subjectivity and space, emerge commensurately and have no existence without each other (Kant 1992/1781: preface). In Heidegger, the common vernacular for 'existence', *Dasein*, is used to mean authentic human subjectivity. *Dasein* is the conduit of time and awareness, and brings consciousness, thinking and language to bear on the world.

> There is the pure possibility that man might not be at all. After all there was a time when man was not. But strictly speaking we cannot say: There was a time when man *was* not. At all *times* man was and is and will be, because time produces itself only insofar as man is. There is no time when man was not, not because man was from all eternity and will be for all eternity but because time is not eternity and time fashions itself into a time only as a human, historical being-there [*Dasein*]. (Heidegger 1973: 84, italics in the original)

The way 'technology worlds us' is understood by Heidegger as particularly human. Heidegger (1973a) believes that animals are 'world poor'.

But Von Uexkull's view of the animal's phenomenal disposition – that is, worlding – is not unusual in the indigenous frame of reference. The *runa* of the jaguar in the Amazon funnels time and space through the existential lens of a very large cat. Kohn heeds the warning of his friend Juanicu, 'Sleep faceup! If a jaguar comes he'll see you can look back at him and he won't bother you. If you sleep facedown he'll think you're *aicha* [prey; literally 'meat' in Quichua] and he'll attack'. As Kohn registers, 'How other kinds of beings see us matters. That other kinds of beings see us changes things' (Kohn 2013: 1). Kohn's (2017) ecology of selves hints at how the worlding of different species and strata influence and shape each other in an intricate *whakapapa*, or genealogy, of relations. Selfhood emerges as an event (Heidegger 1999) but always in relationship with historical context that is shaping and shaped by all the beings that occupy that *umwelt*.

The ecology of selves disrupts the anthropocentric superiority of the individual human 'who knows'. Other species have complex and specific worlds that humanity are largely unaware of. Feminist embodiment challenges the integrity of that individual and draws attention to the slippery, porous interaction between the environment and the self as matter gets ingested and expelled. The intraconnection indicated by quantum physics is on another scale, but the analogy of an ecology of selves reverberates.

Quantum physics reinforces this entanglement between apparently separate 'things' at a microscopic level. Subatomic particles are not confined to concrete objects, as we commonly know them. Subatomic particles can occupy space without conforming to the boundaries of objects that appear solid from a human perspective. For example, a muon – first registered by the CERN Hadron Collider in 2012 – is a subatomic particle that ripples through the entire earth without disrupting the integrity of the planet. For a short period, the muon is 'within' and part of the planet until it passes through, uninterrupted, and continues its pathway through space. In a variety of ways, quantum physics messes with the assumption of forces acting upon other forces in a neat continuum of cause and effect, that is the basis of the mechanical universe set out by Descartes (1980/1641), Newton (1687) and (differently) Leibniz in the seventeenth century.

THE MANY AND THE ONE

In the 1640s, Descartes was a key figure in developing Enlightenment philosophy and mathematics. For Descartes, the individual is the basic unit of society, and he set out to deduce the most 'pure' characteristics of the universal individual. To do so, Descartes carefully stripped away the deceptive input of circumstantial, emotional, or bodily information. His aim was to discover something akin to Aristotle's *animal rationalis*, as the 'essential' species-being of humanity. These are the elements of humanity that are peculiar to the human species and endure across all historical and personal circumstances. Descartes sought to define the universal rational individual mind as the essence of the human. His claim, 'Cogito ergo sum' – 'I think, therefore I am', is one of the most famous premises of modernity. The logical deduction of 'ergo' or 'therefore' as cause and effect between thinking and being, is crucial to later scientific method.

Descartes suggested a deep scepticism about everything the body experiences on the grounds that bodily signals can be completely wrong: a ghost, a dream, or the trick of a demon (1980/1641: 7). To make his case that deductive rationality is the most secure source of knowledge, Descartes started off the *Second Meditation* with a thought experiment.

> I suppose therefore that all things I see are illusions; I believe that nothing has ever existed of everything my lying memory tells me. I think I have no senses. I believe that body, shape, extension, motion, location are functions. What is there then that can be taken as true? Perhaps only this one thing, that nothing at all is certain. (1980/1641)

Thus, the Enlightened individual of modernity is sceptically held aloof from natural objects. Embodied, phenomenological sensations are merely functions. They do not even have the significance of a medium for truth, as does subjectivity, space, and time in Kant (1992/1781). It is a sceptical remove about embodied information that perceives things, creates the dualist schism of alienation between subject and object, culture and nature. The alienation between subject and object requires another mode of thinking about objects that aims to ascertain their universal characteristics. Every individual is plagued by similar issues of faulty interpretations of bodily phenomenological experience, so each person is sceptically alienated from others and from objects in the same fashion.

Descartes's (1980/1641) rational individual is self-reflective, elevating the mind over the body, in order to deduce the universal essence of the object under consideration. The object is assumed to be a discrete entity – the chair is not part of the floor, nor can it be confused with a table. The Enlightenment category aims to discover the universal essence of 'chair-ness'. Enlightenment thinkers lean on the explanation proffered by Aristotle (384–322 BC/1972) in the *Metaphysics*, that to completely understand the essence of a thing, one has to investigate every single example over the entire history of time and space. Chair-ness can only be completely understood if each and every chair is perceived. Only when the universal essence has been ascertained can someone deduce what is essential to a particular example, and what is a historical 'accident'.

Nobody, of course, other than a god, is alive long enough to ascertain each and every example of chair-ness. And you would be fairly busy, if at the same time you were trying to classify each and every bird, fish, reptile, mammal, rock, gas and star system. Universal essence is an ideal, nowadays approximated by large statistical samples, but never achieved. Yuk Hui makes an interesting observation. He claims, 'universalism remains a Western intellectual *product*. In reality there has been no universalism (at least not yet), only universalization (or synchronization) – a modernization process rendered possible by globalization and colonization' (Yuk Hui 2017a, italics in the original).

Not only is sceptical rational deduction about the universal essence time-consuming and unachievable but the application of rationality as the human species-being to the modern market context also requires a great deal of reflecting upon one's self. In a consumer context, how else can one know what one desires? Desire, as an aside, is not strictly rational, and almost certainly would not meet Descartes's (1641) strict criteria. But it is impossible to make rational consumer decisions, unless you have confidence in what you, yourself, want. Adam Smith's (1904/1776) invisible hand turns these individual rational decisions into a type of generalised aggregation that 'balances' the market to ensure fair distribution. The invisible hand enables the market to be seen as a universal arbiter that crosses all disciplines and includes all aspects of life in the rational processes of capitalist enlightenment.

Karen Barad (2010) describes the Newtonian method and its causal assumptions as creating a God's-eye view, that enables rational individuals to see the universal pattern to everything. With this deductive logic, leverage to change the unfolding cascades of cause and effect becomes possible, and humanity becomes the master of the universe (Weber 1948/1919). Crucially, the Newtonian mechanical universe also positions objects as isolated and exterior, reinforcing the philosophical split of subject from object with physics and mathematics. 'Matter is discrete, time is continuous. Place knows its place. Time too has its place. Nature and culture are split by this continuity, and objectivity is secured as externality' (Barad 2010: 249).

Universal claims to truth have encouraged physics to find a theory of everything (TOE). Newton's laws of physics, and especially the notion that cause and effect are simply forces acting upon other forces, may still operate at the clumsy classical level, but the theoretical basis that underpins it has become much more complicated and nuanced.

> How much of our understanding of the nature of change has been and continues to be caught up in the notion of continuity? For Newton, physicist extraordinaire, inventor of the calculus, author of biblical prophesies, uniter of heaven and earth, continuity was everything. It gave him the calculus. And the calculus gave voice to his vision of a deterministic world: placing knowledge of the future and past, in its entirety, at Man's feet. Prediction, retrodiction. Time reversal, time universal. (Barad 2010: 249)

The ability to utilise calculus about material matters in the environment enhances the belief in the determinism of cause and effect, and with it, the illusion of man's mastery and control (Weber 1948/1919). 'Determinism rules. Nature is a clockwork, a machine, a windup toy, the Omniscient One.' Barad is scathing. 'The universe is a tidy affair' (2010: 249).

THE ONE AND THE MANY

The tidy affair has become much more complicated. Quantum particles behave in 'weird' and wonderful ways. They entangle over long distances. Subatomic particles remain smeared out in virtual probabilities rather than exact locations. The particles change behaviour if the observer is involved in measuring the experimental outcomes. It muddles the premise of time-space as mooted by Enlightenment physics. Kondepudi and Prigogine outline the field theory of quantum physics, where everything is fundamentally flow: 'According to quantum field theory, all particles are excitations of quantum fields' (2014: 49).

Embodied feminism explores the intra-action of ingress and egress between the body and environment (Kristeva 1982; Irigaray 1985). Indigenous cultures speak to the imbrication of the self in the wider community, through space and time, and beyond the human, to other species, to rocks, plants, ocean, atmosphere, and stars (Irwin and White 2019;' Irwin 2022a;

Smith 2000; Kohn 2007; 2013; Mika 2017; Māhina 2010). But it is quantum physics that helps to unpick the minutiae, the 'spiritual' layer, of how these elements all knit together in mutual formations. Quantum physics reveals that matter entangles over vast distances. It demonstrates how everything is 'co-constructed'. Most cultures have words to hint at these deep-set intra-actions, but the way the ontology of intraconnection has been framed, varies considerably.

'The One' in Plato's *Republic* (427–347BC/1965: 596a) and in Aristotle's *Metaphysics* (384–322 BC/1972) presumes that there is a permanent continuity throughout the natural world, and discrete objects can be distinguished out of the Oneness. Aristotle theorises the ways that beings participate in the One. From that metaphysical basis, Descartes invokes cascades of causality and rationality. Under Newton and Leibniz, this became the mechanical universe. Albeit with important differences between them. Newton thought that space is empty, and atoms have mass and velocity, producing forces that impact other forces. For Leibniz, space is relational, it is always *between* two particles or stars, or objects. Both Leibniz and Newton assume a mechanical universe of cause and effect. Quantum physics complicates and interrupts the seamless story of the mechanical universe. It challenges cause and effect as the fundamental mechanics of the universe, reigniting the relationships between and within entities in ways that disallow the modern dualism between self and other.

It is a mistake to characterise the Enlightenment as a totally consistent field of thought. Substantial differences emerge between Plato (427–347 BC/1965) and Aristotle (384–322 BC/1972). Amongst other influences, these two genealogies of thought continue to play out in Enlightenment philosophy. Peter Evans (2021) refers to the controversy between Newton (1687) and Leibniz to help explain the relationality of quantum physics. Newton's understanding of time and space is that it acts as a 'container for the contents of the universe'. If all the galaxies, stars, planets, and cosmic debris were removed, empty space and time would continue to stretch out to infinity. 'This is the "absolute" view of space and time'. Leibniz, by contrast, argues that 'space and time were nothing more than the sum total of distances and durations between all the objects and events of the world'. Removing objects would remove space and time as well. There 'are only spatial and temporal *relations* between objects and events' (Evans 2021). Evans (2021) suggests that Leibniz's ideas helped Einstein later develop the concept of general relativity (Einstein and Minkowski 1920). The 'between' also assumes concrete objects and events, and quantum field theory is not reliant on such concreteness. It is possibly more apt to imagine an ocean, of which lumpy viscous 'objects' slow or intensify the flow of energy/matter in the field.

QUANTUM ENTANGLEMENT AND WAVE ONTOLOGY

But we leap ahead of ourselves. Major breakthroughs in quantum physics happened in the first half of the twentieth century. It was complicated by the application of atomic theory into warfare and the atomic bomb. The ethics of theoretical particle physics was turbocharged as a contribution to the Second World War, and the subsequent Cold War. But while the technology was immediately put to use, and the ethics hotly debated, the consequences of quantum physics for political philosophy has been almost completely ignored, except perhaps by the early physicist Nils Bohr (Bohr 1987; Faye and Folse 2017).

Not only does the identity of tangible objects break down at subatomic levels but there is a whole lot of 'weirdness' that is associated with quantum physics. Weird is a word used by physicists when they consider many of the modes of connection taking place at this minute level. For example, quantum phenomena can behave either as a wave, or as a particle – despite the ontological difference between these two modes of being. Under Newtonian physics, a particle would fly through one aperture and arrive directly opposite, on the screen. But quantum particles behave quite differently. The particle behaves like a wave, and moves through both apertures at the same time.

A famous and often-repeated quantum experiment shines a photon through a screen with a double slit aperture before 'collecting' it on a second screen. Strangely, the photon behaves like a wave rather than a molecule. That is, instead of showing up on the far screen, directly opposite one of the double slits, the single photon goes through *both slits at the same time* and creates an interference pattern *with itself*, which then 'spreads' like an oscillation of bars of light, across the final screen. How is this possible? The photon has not 'decided' where it 'is'. It is behaving like a wave instead of a particle. The waves from each aperture move through each other, reinforcing or dampening the rhythm of oscillation, creating a 'constructive' or 'deconstructive' interference pattern. Particles are singular, localised entities that can be located precisely in time and space. The two ontologies each show either velocity or location, respectively, but these two aspects are incommensurable. Yet quantum phenomena flow between one ontology and the other, depending on context.

The wave itself is not straightforward, because the photon particle generates *two* waves, through each slit, and these interfere with each other, making a diffraction pattern. Barad explains the wave behaviour of diffraction,

> Diffraction is a phenomenon that is unique to wave behaviour. Water waves exhibit diffraction patterns, as do sound waves and light waves. Diffraction has to do with the way waves combine when they overlap and the apparent bending

> and spreading out of waves when they encounter an obstruction. Diffraction phenomena are familiar from every day experience. A familiar example is the diffraction or interference pattern that water waves make when they rush through an opening in a breakwater or when stones are dropped in a pond and the ripples overlap. (Barad 2007: 28)

The bars of light on the far screen where photons are eventually collected, show up a probability curve, where the highest probability is the brightest area and low probability is in darkness. While enroute, the particle remains in a 'virtual' 'superposition' which could be any part of the bright bars of light, and is as yet unresolved until it is collected on the far screen. Richard Feynman sees quantum mechanics as 'the physics of the superposition of waves' (Barad 2007: 29).

OBSERVATION AND OBJECT

The observer effect changes the situation dramatically. As soon as a mechanism is introduced that identifies *where* the particle is, then the photon decoheres and behaves like a particle and 'lands' or 'locates' itself as a single photon rather than a spectrum of possibility. Paradoxically, the light propagates like a wave and lands like a particle. The observation of the particle forces it to decide and locate. Observation is not neutral; it is intrinsically caught together with the results. 'Phenomena do not merely mark the epistemological inseparability of observer and observed, or the results of measurements; rather, *phenomena* are the *ontological* inseparability of agentially intra-acting components' (Barad 2007: 33, italics in the original).

The process of measuring the location of a particle actually changes the behaviour of quantum entities from the wave, with its virtual probabilities of where the particle *might* end up located, to a defined position. Measuring and observing the quantum event locates the particle. The neutral observer of Cartesian physics is an impossible concept because it is the process of observation that shifts the ontology of the quantum behaviour, from the velocity wave of virtual probability, to a defined and specific particle location. The process of observation inextricably entwines the observer with the observed.

Clearly, Descartes' alienation of the object from the knowing subject, is collapsed in quantum physics. Barad explains the consequences, in philosophical terms, of these changes in fundamentals.

> Being is not simply present, there to be found, already given. There is no fixed essence or substance simply there for the measuring. Particles aren't inherently bounded and propertied entities running in the void. Mattering is about the (contingent and temporary) becoming-determinate (and becoming-indeterminate) of

> matter and meaning, without fixity, without closure. The conditions of possibility of mattering are also conditions of impossibility: intra-actions necessarily entail constitutive exclusions, which constitute an irreducible openness. Intra-actions are a highly non-classical causality, breaking open the binary of stale choices between determinism and free will, past and future. (Barad 2010: 254)

Quantum particles are in no such psychological quandary – no isolation, or alienation, reduces the state of becoming to rational deduction *ad absurdum*. Quantum effects remain in a virtual state of possibility until forced to 'choose' a location under the constraints of observation. Physicists describe the quantum experiments in tightly controlled laboratory conditions, but when multiple quantum events take place together, some argue the quantum effects cancel each other out. The decoherence of the wave takes place as the particle interacts with the environment. (Everth and Gurney 2022). The agency of the environment interacts with the virtual probability of the wave, and it decoheres into a particle which 'lands' (Everth and Gurney 2022). This makes the Newtonian physics of cause and effect continue to appear normal. The Newtonian physical substrata of forces acting upon other forces is clumsy and simplistic, even if generally it appears as the though the world behaves like that. Quantum physics is more sophisticated than Newtonian physics, but it does not entirely mean the abandonment of cause and effect. Instead, it is a more complex picture, where wave-like 'field' theory replaces linear, unidimensional physics. Virtual possibility is mathematically accessible, and new technologies such as quantum computing are on the cusp of emerging, so these deep changes in the assumptions of physics are about to become better understood by the lay population.

THE COPENHAGEN DEBATE

The nature of quantum phenomena were hotly debated at the turn of the twentieth century. In 1927, Bohr and Heisenberg developed the Copenhagen interpretation. It did away with Newtonian classical physics and stopped expecting subatomic particles to behave according to the laws of linear cause and effect. Despite his work on relativity, photons and Brownian motion, Einstein and others had far more difficulty abandoning classical 'laws' and accepting, as Einstein put it, 'spooky action at a distance'.

Bohr noticed, in experiments to create photons from the shell of electrons of an atom, that the sudden jump of the photon was unpredictable and seemed to rely on the 'agential realism' of the electron itself. Bohr (1987) was comfortable attributing agency to external 'objects', even microscopic ones. The practice of observation changes the outcome of quantum experiments, another indicator that agency is not limited to the human. It demonstrates the ways that humanity are inextricably entwined with natural phenomena. For Bohr, the implications collapse the dualism between subjective knower and object of knowledge; at the 'heart of the lesson of quantum physics: *we are a part of that nature that we seek to understand*' (Barad 2007: 26, italics in the original).

Einstein on the other hand, held fast to the notion of causality where forces are acting upon other forces, and there are no other wilful agents in the universe. He spluttered, 'I find the idea quite intolerable that an electron exposed to radiation should choose of its own free will, not only its moment to jump off, but also its direction. In that case, I would rather be a cobbler, or even an employee in a gaming house, than a physicist' (Albert Einstein, quoted in Shapiro and Epstein 2006: 228, in Barad 2010).

The argument between the Copenhagen interpretation and Einstein's commitment to classical physics remained hot for decades. Einstein remained committed to the God-given universe, where everything can become known. He thought the statistical results of the virtual wave in the two slit experiment was 'incomplete' knowledge and he did not accept that measurement might change the behaviour of the particle. Nils Bohr was much more flexible about the statistical probability of a 'smeared out' virtual wave. The Copenhagen argument is based on Bohr's 'complimentary' thesis, where the photon as a wave is complimentary (and different) from the photon as a particle. As Bohr put it, 'complimentary in the sense that only the totality of the phenomena exhausts the possible information about the objects' (Bohr 1987: 40, in Faye and Folse 2017: 183). Both the particle and the wave need to contribute to a full understanding of quantum physics. Einstein's mathematics continue to make important contributions to the generation of new understanding in physics. Importantly though, Nils Bohr thought about the philosophical consequences of entanglement and other weird quantum phenomena. His concept of complimentary theories is at odds with the entire scientific tradition of noncontradiction. Barad explains,

> What [Bohr] is doing is calling into question an entire tradition in the history of Western metaphysics: the belief that the world is populated with individual things with their own independent sets of determinate properties. The lesson that Bohr takes from quantum physics is very deep and profound: there aren't little things wandering aimlessly in the void that possess the complete set of properties that Newtonian physics assumes (e.g. position and momentum); rather, there is something fundamental about the nature of measurement interactions such that, given a particular measuring apparatus, certain properties *become determinate*, while others are specifically excluded. Which properties become determinate is not governed by the desires or will of the experimenter but rather

by the specificity of the experimental apparatus. (Barad 2007: 19, italics in the original)

Ironically, two important papers written by Einstein (1920) on what he thought were entirely separate subjects – relativity and thermodynamics, have become the basis of the most recent ideas about the nature of space-time. Astrophysicists are now completely comfortable with Bohr's complimentary thesis. Statistics and information are becoming the new frontier of physics.

KNOWABILITY AND ENTANGLEMENT

David Hume (1985/1750; 1958/1739) writes that a scientific hypothesis stands until the anomalies, and paradoxes associated with it, force people to think of a better one (Irwin 2008a; 2010c). When I first came across quantum entanglement, I thought it would supersede classical physics, tossing out the causality of 'forces acting upon other forces' discerned by Newton, as the spooky action of quantum entanglement took over. That has not really happened. Quantum physicists tend to regard classical physics as a clumsy but useful guide about why things happen at the level of ordinary life, and there is a lumpy and awkward relationship between the more sophisticated quantum particle physics and the scope of classical physics in normal everyday actions. However, through decoherence theory quantum physics explains the emergence of classical reality as a result of the collective interaction of the constituent quantum particles in the form of the 'environment' (Zurek 2021). And in the classical reality we perceive, it *seems* that normative cause and effect still operates.

What is important for this discussion is that the underlying premise of cause and effect seen in classical physics can now be understood in a new way: while it may *appear* that the continuity of cause and effect rests upon 'forces acting upon other forces' that move and shape the self-contained 'things' in the mechanical universe, the truth is that objects emerge from the movement and momentum of the underlying field or wave patterns. An 'object' has waves of particles running through it, as much as it has 'resident' molecules that make up the matter of the object. Objects are events, that emerge from far more complex layers of entangled particles that make up the fabric of space/time. Things are not independent entities, because at subatomic level, particles 'belong' to multiple 'things'. 'Dark matter' makes up 27 per cent of the mass of the known universe, and nobody understands what it is or how it works, other than the expansion of the universe is slower than it would be without dark matter (Farnes 2018). Additionally, dark energy (another thing entirely) accounts for 68 per cent of forces at work and is a factor in the expansion of the universe, weakening the nature of intra-action between stars, galaxies and all matter. According to NASA, what is known and measurable by humans is less than 5 per cent of the universe.

The incommensurability between quanta behaving as a probability wave, or a particular location, is further complicated by quantum entanglement. Minute subatomic particles can become entangled with another, producing a complimentary pair with equal and opposite 'spin' behaviour in the particle. Like a pair of gloves, if one particle spins 'up', its complementary pair spins 'down' (Thiemann 1998). However, while the entangled particles are not subject to measurement, they remain 'undecided', both with undefined spins simultaneously up and down. This is called a virtual superposition. It is the process of measurement that 'locates' the spin. But its more complicated, again, because the two subatomic particles are entangled; if just one is measured, the other is instantly located in its complimentary opposite spin – no matter how distant it is.

Quantum entanglement stretches our concepts of knowability. Karen Barad explains the philosophical consequences of quantum entanglement succinctly,

> Quantum entanglements are not the intertwining of two (or more) states/entities/events, but a calling into question of the very nature of two-ness, and ultimately of one-ness as well. Duality, unity, multiplicity, being are undone. 'Between' will never be the same. One is too few, two is too many. No wonder quantum entanglements defy commonsense notions of communication 'between' entities 'separated' by arbitrarily large spaces and times. Quantum entanglements require/inspire a new sense of a-count-ability, a new arithmetic, a new calculus of response-ability. (Barad 2010: 251)

'Intra-action' is the crucial conceptual term to understanding the radical shift from autonomy to quantum entanglement. Barad explains it thus,

> The neologism 'intra-action' *signifies the mutual constitution of entangled agencies.* That is, in contrast to the usual 'interaction', which assumes that there are separate individual agencies that precede their interaction, the notion of intra-action recognizes that distinct agencies do not precede, but rather emerge through, their intra-action. It is important to note that the 'distinct' agencies are only distinct in a relational, not an absolute, senses, that is, *agencies are only distinct in relation to their mutual entanglement; they don't exist as individual elements.* (Barad 2007: 33, italics in the original)

The concept of superpositions, is illuminated with Schrödinger's alarming thought experiment about a cat in a box. Some subatomic particles can be entangled across enormous distances. They are like a pair of gloves; if one is spinning up, the other has to spin down. Schrödinger's experiment has one entangled particle in the box, and its pair outside the box. While the quantum particle in the box is in a 'superposition', the spin of the entangled quantum particles are undecided. The spin is both 'up' and 'down', the cat is still in a potential state unaffected by the experiment. Some deadly gas will be released if the entangled particle triggers a switch if it becomes located to an 'up' spin, but the switch remains off if the particle is in a 'down' spin. Virtual quantum particles only leave their superposition of *both* up and down, upon measurement. So, once the measurement is taken, and the particle decides to be either up or down, then the switch releasing the fatal gas is triggered or not, and the cat is dead or alive. The point is that all the virtual potential probability coexist, as a wave, when the particles are in the 'superposition'. All that potential is looming until the moment of interference, when the measurement is taken, and it is only at that moment that the quantum particles behave in a classical way, and 'decide' a precise location on the probability curve. The action of measurement itself, decides the fate of the cat.

In the interference wave, cause and effect are complicated but not annihilated. The particle will never decide to land outside the statistical probability curve dictated by the constructive interference pattern of the wave. But cause and effect is not a simplistic continuity of force acting upon other forces as assumed by Newton and others. It's a virtual probability curve.

Quantum particles can be entangled in tandem, over vast distances. Measuring one particle of the pair determines not only the measured property of the particle but also the property of its entangled pair. While this effect is instantaneous, the particles do not communicate with each other via the limits of space-time and no information can be transmitted between them. Quantum entanglement only seems to corrupt the laws of special relativity and the speed of causality (which is also the speed of light) of three hundred metres per second. Nevertheless, Einstein regarded this behaviour as 'Spooky action at a distance' and as another quantum phenomena almost impossible to comprehend through a classical lens. He believed that there would be a sensible solution that explains these effects based on classical physics.

Multiple experiments have confirmed that when one entangled quantum particle is measured and the 'spin' of the particle ascertained, the entangled pair of particles is instantly located with entangled opposite complimentary spin – even over thousands, or millions, of miles. How entanglement works amongst multiple pairs or nodes is still unclear. None of this is cause and effect in a classical sense. The defining moment of location of spin does not travel at light speed from one particle to the other. It is already entangled in complimentary spins, but the definition of which spins in what direction is only definitely elucidated in the process of measurement of one aspect of an entangled pair. It's probably better to understand the entangled particles as making up 'one' object, or entity, smeared over vast distances, to understand how it 'knows' when its pair has become located. Before that, the particles have not 'decided' and the spins are in a virtual superposition that are both 'up' and 'down'. All that is causal is the introduction of observation, which alters the context and behaviour of the experiment. Bohr (1987) calls this *inseparability* of the 'object' from the 'agencies of observation'.

AGENCY

Another weird example of quantum strangeness is how photons are created. A photon is the emission of a light packet (Bohr) when an electron makes a quantum leap from one energy state (E2) to a lower energy state (E1). It just suddenly arrives. There is no way to predict if, when or even how this will happen, an occurrence that aggravates Einstein in particular.

The 'problem' is that the electron is initially at one energy state, and suddenly it is at another energy state, with no 'in between' that transported it from one to the other. The evidence of the release of the photon is the shift in light colour of each energy state. Newton's mechanical universe relies on continuity. These discontinuities, these eruptions, these unpredictable emergences contest the foundational assumptions of Newton's universe.

> A quantum leap is a dis/continuous movement, and not just any discontinuous movement, but a particularly queer kind that troubles the very dichotomy between discontinuity and continuity. Indeed, quantum dis/continuity troubles the very notion of dicho-tomy – the cutting into two – itself (including the notion of 'itself'!). All this 'quantum weirdness' (the display of an increasing array of uncanny phenomena) is actually 'quantum queerness', and I don't mean simply strange. Q is for queer – the un/doing of identity. Quantum dis/continuity is at the crux of this im/possible, im/passible, trans/formation. (Barad 2010: 246–47)

For our purposes, the inseparability of the observer effect disrupts Cartesian alienation because the object of knowledge is never objective but *always changes in relation* to the subject who knows. The observer effect and the dynamic entanglement of objects over space-time are fundamental challenges to the mechanical universe so wholeheartedly believed in by Cartesian and Newtonian physics. The new ideas disrupt Enlightenment techno-science and philosophy, along with the economic model built upon them. Barad goes on,

> Let's pause to consider the quantum dis/continuity further. This discontinuity that queers our presumptions of continuity is neither the opposite of the continuous, nor continuous with it. Quantum 'leaps' are not mere displacements in space through time, not from here-now to there-then, not when it is the rupture

> itself that helps constitute the here's and now's, and not once and for all. The point is not merely that something is here-now and there-then without ever having been anywhere in between, it's that here-now, there-then have become unmoored – there's no given place or time for them to be. (Barad 2010: 247–48)

There are multiple other examples of quantum weirdness. The moment that a photon 'jumps' from its electron shell (Bohr 1987) is one. The enormous distances that quantumly entangled particles occupy – each the mirror image, is another. There is also the question of what happens in a black hole. The interior matter swallowed and crushed by the giant gravitational pressures of a black hole are quantumly entangled with particles outside the black hole. The theory of thermodynamics is that nothing is ever lost, which raises important questions about what happens to the matter in the black hole. Astrophysicists speculate that the entanglement enables information about the crushed interior matter to leave a two dimensional holographic information residue on the black hole's horizon (Maldacena et al. 1997).

A question remains about the relationship between the scale of nanoparticles and the scale of the ordinary world as we usually move through it, and what these new insights lend to 'normal' classical physics. The relationship is lumpy and does not make immediate sense. It is still controversial. The theory I prefer is that entanglement is relatively common, and is not limited to two 'monogamous' paired subatomic particles. Any molecule that shares electrons tends to have entangled particles. As entangled objects get exposed to other objects, they imprint their entanglement on their environmental context (and vice versa) so that the superposition is taken up by multiple elements. Virtuality proliferates. The wave interference patterns can be complex and multilayered with large and small constructive and deconstructive rhythms. This virtuality orients probability. Probability orients what is generally known as causality.

INTRACONNECTION AS FUNDAMENTAL METAPHYSICS

Quantum physics opens up whole new ways of understanding connectivity which is fundamentally relational. The context of the experimental materials, the context of measurement and observation, has a direct effect on the quantum particles. There is no separate natural universe, operating in a simplistic mechanical model of cause and effect, that can be objectively leveraged to mankind's benefit. The leveraging simultaneously impacts both the human and the nonhuman, even if the scales of operation are outside immediate common sense. We can see this of course, with climate change. Jumping in the car to drive three hundred metres for a pint of milk does not *seem* to have any detrimental impacts. The relationship between agency and ecology is on scales that are difficult for the human brain to comprehend.

The neoliberal market claims to hold a universal position that incorporates the aggregation of all human behaviour (Smith 1904/1776). All aspects of the earth, from the forests, rivers, soils and species to human beings themselves, are understood as a potential resource through a lens, as Heidegger (1977) puts it, of consumerism; the technological horizon of disclosure. Thus, the neoliberal response to climate change has been to incorporate it into the rationality of the market through pollution vouchers. Levels of pollution created by an organisation are based on scopes 1, 2 and 3 emissions accounting. Emissions become part of the costs, and on the emissions trading scheme, part of the benefits of the market. Putting in solar panels can earn emissions vouchers, and burning fossil fuels in the fleet can create emissions costs. It is a simple, double-entry incentive scheme. Increasingly, though, climate scientists (and for much longer, environmental theorists like myself) have been arguing that the 'net-zero' discourse, that allows *some* emissions as long as they are 'offset' through a trading scheme, has almost no overall positive impact on overall reduction of greenhouse gases (Skoerseth and Wettestad 2008; Anderson 2012; Ripple et al. 2019; Source Material 2023). When it comes to reducing climate change, for more than thirty years, the market has failed to deliver.

The notion of rational individuals whose aggregated consumer decisions can alter the course of mass demand, without political intervention, falls down at the level of the political economics of climate change. Quantum physics suggests that it also falls down at the level of molecular 'atomism'. The intentions of agents – which used to be understood purely in anthropocentric terms, as *human* will, is much more complicated than classical philosophy had us believe (Leibniz in Evans 2021; Nietzsche 1909; Whitehead 1969; Prigogene and Stengers 1984; Barad 2007). The isolated individualism of Descartes and Kant, that informs Adam Smith's economics is a misconception. Any individual is a density, a centre of gravity, a concretisation of the general milieu. Agential reality emerges by differentiating the 'self' from its environmental milieu. But the agential çut between subject and object occurs in the context of material phenomenal field.

> It is through specific agential intra-actions that the boundaries and properties of the components of phenomenal become determinate and that particular concepts (that is, particular material articulations of the world) become meaningful. . . . The agential cut enacts a resolution *within* the phenomenon of the inherent ontological (and semantic) indeterminacy. (Barad 2007: 139–40)

Thus, the notion of a separate, wilful, rational individual that makes self-interested market decisions, is a fundamentally outdated and inaccurate rendition of agency, decisions, action, and community.

Since Plato (427–347 BC/ 1965) and Aristotle (384–322 BC/1972), humanity has been privileged as the *animale rationalis*. In contrast, indigenous philosophy, such as Māori, has never limited agency to humanity. Marsden (1992) makes the link between Māori philosophy and the intraconnection of quantum physics. Organic and anorganic features all have *mauri*, or 'life force', which generate their own integrity and orientation or intention. Or as Kohn (2013) puts it, in the Amazon, all the species have *runa*, or 'personhood', which is worlding in their own unique and intra-active ways. To survive in the Amazon, people need to be becoming-jaguar, and to understand its prey, the jaguar needs to be becoming-dog or becoming-human (Kohn 2013). Barad unpacks philosophy of mind from the quantum perspective.

> Attending to the complex material conditions needed to specify 'intentions' in a meaningful way prevents us from assuming that 'intentions' are (1) preexisting states of mind, and (2) properly assigned to individuals. Perhaps intentionality might better be understood as attributable to a complex network of human and nonhuman agents, including historically specific sets of material conditions that exceed the traditional notion of the individual. Or perhaps it is less that there is an assemblage of agents than there is an entangled state of agencies. These issues, however, cannot be resolved by reasoning analogically; they require a different kind of analysis. (Barad 2007: 23)

Entangled agencies across a range of 'entities', some of whom are human, is a completely different orientation towards intentionality. Bohr (1987) was interested in interpreting this complicated intra-action where agents can no longer be understood as self-enclosed entities. He argues that the observer who measures the particle in the double slit experiment creates a particular context that changes the behaviour of the subatomic particles. All behaviours are context dependent. Intentionality is no longer limited to the human, nor even to 'living' species. Nor is it possible to continue to believe that human individuals make isolated 'rationally deductive' intentional decisions in a vacuum. Rather, each agent is irreducibly in 'intra-action' with its context at all times, and all decisions are made in relation to the virtual probabilities in play. Barad again puts this clearly, 'we must understand in an *integral* way the roles of human and nonhuman, material and discursive, and natural and cultural factors in scientific and other practices' (2007: 25).

The implications for policy and politics are profound. Everything is intra-active, with entangled subatomic particles forging identical counter spins across 'items' or 'objects'. These complex milieus can extend across nanometres, or millions of kilometres. They are affected by other milieu with similarly imprinted entanglements, and the complexity of the interference patterns become so subtle that they can no longer be differentiated by human technologies. But unidentifiable, they remain entangled.

As a mundane example, the milieu of economics intra-acts with the milieu of the fish (Deleuze and Guattari 1999), as a coastal fishing boat peruses the ocean. The island waters are bereft at the loss. The entangled fish molecules become part of their predator human's body, as Thursday's dinner. My watery bodily entanglements intra-act with theirs as decisions and locatable spin occurs and then reverts into the probability waves of contextual interaction with elements in the milieu. We never shared breath, nor set eyes upon each other, but agency of the fish and I intra-act with the entangled milieu across strata, taking the eddies and flows of the milieu in new directions and contexts.

Philosophically, the intraconnection of physics and *whakapapa* and their related concepts are important shifts in ontology from the segregated individualism of modernity. Intra-action introduces new responsibilities that have not been at play in modern political-economy.

> This materialist approach has important ethical considerations. In terms of the human/ animal interaction, the familiar ego saturation of the past is replaced by a deep bioegalitarianism, a recognition that we humans and animals are in this together. The bond between us is a vital connection based sharing this territory or environment on terms that are no longer hierarchical or self-evident. They are fast evolving and need to be renegotiated accordingly. (Braidotti 2009: 528)

This means that the policy directive of neoliberalism – that relies so heavily on an aggregation of rational, individual consumer decisions – is clumsy at best and philosophically illogical at worst. Individuals are in constant intra-action with their context, and each thought about this is another form of intra-action with the language community. The body breathes in and breathes out. It sweats, oozes, dribbles and quietly reeks with hormonal pheromones and general dirt and mayhem. The subatomic particles are already entangled with a range of unknowns, in complex rhythms that inform the probability curve of the virtual future. Psychological behaviouralism cannot presume a rational response with an emotional add-on (Skinner 1938; 1969). The context is irreducibly intra-acting, in small and large ways, in minute staccato rhythms and longitudinal superwaves. The aggregation of populations might as well be understood contextually as rationally (Code 2006).

This places ethics in a new realm.

A series of diffractive waves orients likely consequences, observation entangles observer and observed, the context helps to form the results and virtual entanglement defies the law of the speed of light, also known as the maximum speed of causality. Philosophically, this means that complexity, probability and contextualised accuracy can be expected in our political arrangements. Contextual intra-action replaces the clumsy narratives emphasising the rational mind capable of objectivity and alienated from the objects that it observes. Intra-action is a core component of intention, and the artificial alienation of an observing individual from surrounding entities is physically impossible.

The distinctive conceptual apparatus of Māori *whakapapa*, feminist embodiment, the technologically embodied transhuman, all entangle the individual (person/particle) in a network of relationships. These relationships are not linear cause and effect. They include, but are not limited to, rational deduction. They owe little to objectivity, although they can meet the standards of reproduction. They do not obey the laws of individualism; neither are they constrained by individual aggregated accountability. The rules of intra-action suffuse everything, but do not assume the character of universalism. At its roots, all aspects of embodied humanity are intra-acting at all times with the surrounding milieu. Philosophy and policy need to shift from simplistic linear cause and effect to complex inter- and intra-actions that are both the wave and the particle; the between and the individuated concretisation we call personhood.

Chapter 11

Postcarbon Futures

I love my house. It's a simple place, fairly small, converted from a boat shed. I would like to think it has Nordic style, but really it's just a wooden shed with enormous verandas. Its high on a cliff, overlooking the Manakau Harbour. Not really ideal for launching boats. But unbelievably soon, it will be. New Zealand just had the heaviest rainfall on record, five months' worth in two days: eighty millimetres an hour. The water careened out of driveways into not just a waterfall but into a spray that leapt and sprang out above my head. The road drains surged like small, and then not so small, rivers. Large chunky rocks and soil spilt out onto the road. The low areas began to flood, creeks turned into lakes and torrents. Nobody was prepared with sandbags, and houses filled up with streams they had not known existed. Storm water drains overflowed and popped their lids, sending fountains of water tumbling into the already saturated earth. It lasted three days. There are landslides everywhere – all three roads into my house were blocked with slips and closed during the storm. Power lines came down, along with large trees. The weather office gathered up new terminology – we are in the midst of an 'atmospheric river'. Climate warming of 1.3°C above the preindustrial global averages has increased the capacity of the atmosphere to carry approximately 10 per cent more water, and downpours are much heavier than ever before. At the same time, polar ice melt is increasing, and below my house, the sea level rise is overtopping the stable rock base and beginning to excavate into the soft clay that makes up the thirty-metre cliff face.

I grieve the looming end of my home as ocean rise, sea surge and rain downpours increase the erosion that eats the cliffs. I grieve, also, my own, inadvertent complicity. When I took out a mortgage in 2006, I was ignorant about the methods of banking. I had no idea that the bank creates a new debt *ex nihilo – out of nothing!* By simply walking into the bank with a permanent job and plenty of years of employment ahead, I, myself, justified the bank's creation of new money, with no deposits backing the loan. What is more, because of double-entry bookkeeping, the banks effectively make *twice* the new money – once as a debt (in my bank) and once as a new deposit (in the seller's bank). The value associated with that new credit line does not actually come from nowhere. The extra money percolates into the real economy and swamps the value of exchange. The new credit line then dilutes the nations monetary fiat value, contributing to inflation. Consumable price inflation robs the value of wages. To keep inflation down to its 2 to 3 per cent goal, governments are forced to stimulate compensatory economic growth. The correlation between economic growth and growth in emissions is hard to miss (Le Quéré et al. 2020; Caporale et al. 2021; Mitić et al. 2022; Onofrei et al. 2022; Noll 2023; Hansen et al. 2023). In a nutshell, mortgage loans and debt creation are the systemic driver of economic and emissions growth.

Since eighteen years old, I have been mulling the social and systemic causes of climate change in modern global society. For a long time, I was caught up in the conundrum outlined by Heidegger's philosophy of technology. Heidegger addresses the shift in sophisticated modern technologies from earlier craft work, and attributes contemporary technology with modern phenomenological worlding. This is the analysis of the modern world, which has created climate change. Certainly, fossil fuels are the source of CO_2 emissions. However, knowing the direct cause and addressing the issue of continuous growth are two separate, if entwined, problems. On the face of it, the direct cause is a simple issue – just legislate and stop using fossil fuels. That has not happened because the commitment to economic growth pervades modern politics, across the spectrum. Over the last five years and, really, since the 2008 crash, I had the intuition that ex nihilo banking was too large a factor to not significantly impact the problem. But as I delved deeper into the methodology of modern banking, during the research for this book, I realised that ex nihilo money creation is the prime driver of the systemic commitment to economic growth. While ex nihilo banking exists, calls for a steady state economic system are unachievable. This is why governments all around the world are completely ignoring the escalating calls for degrowth.

Having said that, most governments, global organisations and even banks themselves have very little comprehension of how ex nihilo credit swamps the entire economic system. Bank managers merely write up their new loans according to prescribed risk criteria. Bank tellers interact with potential new customers and fill out a hundred boxes on a dozen forms. Government officials and treasury personnel ascribe inflation to rising prices rather than the banking industry. They ignore the disjunction between the nation's debt, and the nation's gross domestic product (GDP), and believe that their central bank or treasury can leverage the price of money through the London Interbank offered rate (LIBOR) or overnight cash rate (OCR). The price of money is, after all, just another market (Von Mises 1953/1912).

FLABBY ECONOMICS

The neoliberal economic paradigm has a conceptual apparatus of market equilibrium – the invisible hand. The profit motive is met through innovative, efficient growth of consumerism. It is the very success of this paradigm that has created climate change. Given it is so successful, and seems to support a thriving economic base for a very large global population, any deep-seated critique of neoliberalism or capitalism has just bounced off and gained no traction. This is a failure to comprehend the consequences of emissions, where large amounts of land will be flooded or eroded into the ocean, and the equator region will be far too hot for human habitation. Drought, fire, and severe storms will escalate throughout the remaining habitable zone. The success of the neoliberal paradigm has also set people up to continue having faith in the universal market, as a mechanism to 'distribute' emissions production, fairly. No distribution mechanism solves the need to reduce emissions to zero. The oxymoron falls on deaf ears.

We need a viable alternative narrative for a thriving postgrowth economic system. This book has made firm links between economic growth and emissions growth (Le Quéré et al. 2020; Caporale et al. 2021; Mitić et al. 2022; Onofrei et al. 2022). Tiny increments in degrees Celsius of global warming increases the hydrological turbulence of the climate (Hansen 2023). Abolishing ex nihilo mortgage credit and futures derivative credit entirely will halt inflationary pressure. Then the driver for economic growth will evaporate. Only then, will new technologies that mitigate emissions begin to successfully drive down the pollution that causes global warming.

Viable alternatives for a material economics are emerging. They are based on a metabolic and thermodynamic view of economics. The new economics is grounded in an analysis of energy throughput, and entropy. Cost–benefit assessment will include emissions accounting. Postgrowth economics will have a conception of money as fiat exchange rather than nominal debt. It will recognise that the laws of thermodynamics are an inescapable aspect of economic production.

Classical and neoliberal economics have a poorly articulated theory of money. It seems plausible that the idea of market 'balance', where price point is crucial, attributes value to money (Keynes 1930). Adam Smith wrote, 'The powers attributed to money are not inherent in money, but derive from its functions as a means of exchange' (1904/1776: 215). The stability of macroeconomics teeters on this notion of a balance between monetary supply and demand. I contend that money never actually behaves in this way. Banks keep creating more and more credit, well beyond the physical capacity of absorption by markets. Supply outweighs demand by 250 per cent. Global GDP is 2.5 times smaller than the total amount of nominal debt now circulating (International Monetary Fund (IMF) 2022). Given the 120-trillion-dollars' worth of nominal derivatives as opposed to only 20.7 trillion of actual real trade, the disparity between nominal value on the stock exchange and the real economy is even vaster (2023). It is more useful, I believe, to conceive of money as flow. Under current conditions, money liquidity 'pools' in the low-lying swamp of housing mortgages (and fine art, the stock market, Bitcoin, and other nonproductive markets). The global market of classical and neoliberal economics has never reached 'equilibrium'. Instead, the market is driven by structural arrangements responding to the need for exponential economic growth, to keep stability in the system.

Fractional banking (sometimes called endogenous money, as the exchange of credit occurs within the banking system) and stock markets both require the macroeconomic system to keep growing. This retains some semblance of connection between the value of nominal money and the broad or real economy. In a spiralling feedback loop, ex nihilo or endogenous credit creates new money, and new currency is required to repay the interest and the loans. But that loop is not a direct Ponzi scheme; it has to go 'through' the fiat economy to legitimise its validity. Interest rates and capital repayments require wealth in excess of daily fiat exchange to be extracted out of the real economy and submitted to service the interest and capital bank loans. This embeds a structural necessity for economic growth.

Keynes (1933) argued that the money multiplier stimulates growth. More currency available in economic circulation increases prices, and profit. Small to medium businesses will then spend that profit on more consumables, which stimulates the economy to grow. More consumerism supports the spiral of growth which builds consumer confidence. Consumerism improves the ability for people to services loans. This allows banks to increase their risk exposure and give more mortgages. More mortgages stimulate the money supply, which increases prices, creating more profit and so on. There are, however, frequently shortfalls, or gluts, in various contributing elements to the macroeconomic system. These either inhibit the cycle or create bubbles. These issues increase volatility and instability in the system. At such times, the precariousness of progressive economic growth is more exposed.

> Minsky argues, an endogenous money system is inherently unstable due to the cyclicality it embodies. Since the vast majority of money in circulation exists as a function of bank profit, attempts to stabilize the economy have largely proven inefficacious (Minsky 95–109). Government response to such cyclicality tends

often to force growth, eliminate regulation, and fund profitable rather than desirable business. (Ament 2019: 10)

Fiat currency is the money that is transacted for labour and the like-for-like exchanges of real goods which make up the real economy. The connection between the real economy and the nominal tokens of fractional banking, shares, and cyber currency, are increasingly abstract and tenuous. But the consequences nominal currency creates, of market failure, are very real. Runaway inflation would result from unmediated proliferation of ex nihilo money without compensatory economic growth. Continuous growth occurs through mechanisms such as increasing population, developing a new or larger middle class, new markets (new technologies, new fashions) and short-term productivity gains such as deforestation, monoculture and over-exploitation of raw materials. These techniques worked when the raw volume of exponential growth was still small. But in 2022, the combined public and private global debt was more than 2.35 trillion (Gaspar et al. 2023) and an exponential increase of 3 per cent growth is an annual increment of $75 billion next year. That is an enormous amount of extra materials, population and consumerism to take place, multiplying year on year. The following year will be 3.1 per cent more than that again, and so on.

The IMF economists, Gaspar et al. (2021; 2023), have been monitoring the rate of private and public global debt since 1950. The continuous rise of debt in relation to global GDP is alarming. 'Before the pandemic, global debt-to-GDP ratios had risen for decades. Global public debt tripled since the mid 1970s to reach 92 per cent of GDP (or just above $91 trillion) by end 2022. Private debt also tripled to 146 per cent of GDP (or close to $144 trillion), but over a longer time span between 1960 and 2022' (Gaspar et al. 2023). The COVID-19 pandemic had a downward pressure on global debt from its high in 2020 at $226 trillion, or 256 per cent of GDP, down to $235 trillion by end of 2022 or 238 per cent of GDP. But, Gaspar et al. (2023) warns, the long-term trends are heading back up again. 'Despite the economic growth rebound from 2020 and much higher-than-expected inflation, public debt remained stubbornly high. Fiscal deficits kept public debt levels elevated, as many governments spent more to boost growth and respond to food and energy price spikes even as they ended pandemic-related fiscal support' (Gaspar et al. 2023).

EFFICIENT CAUSES AND FINAL CAUSES

In 'The Question Concerning Technology', Heidegger (1977) argues that the essence of modern technology 'is nothing technological'. It is the demands of consumerism that enframe every aspect of the planet as a potential resource. The essence of technology is what Aristotle (1972) defined as its *final* cause: consumer demand. But although Heidegger spent some time in that text, considering Aristotle's four causes, he did not put much imagination into the *efficient* cause of modernity. The efficient cause is the push factor of ex nihilo money. It is the abstraction of money from its basis in the value that is set by the fiat market economy. The essence of efficiency has become innovation. It is the *construction* of innovation. Innovation has meandered away from the conduits of truth and is increasingly living in the land of the simulacra.

Neoliberalism has been presented as a rational system with 'no alternatives'. However, this is not a natural or inevitable state of affairs. People had to be taught how to become consumers, through the precarity of landlessness (Marx 1887/1867). Land clearances are a fundamental starting point for industrialisation and modernity, starting with Ireland and the Scottish Highlands. It is occurring right now with the Amazon deforestation, where indigenous tribes are getting forced out of their traditional territories. Colonialism involved genocides and culturecides that have denigrated, and often actively destroyed, the multiple diverse ways of producing culture in safer, more ecologically integrated ways (Ioris 2022).

Interrogating modernity to discover the 'essence' of its production of climate change uncovers some surprises. Capitalism has been the focus of critique since Marx wrote *Das Kapital* in 1867. But not all aspects of capitalism are equally problematic.

Global trade is not new. The Silk Road is many thousands of years old. Ancient artefacts suggest a flourishing trade of goods across vast distances, over thousands of years. Nowadays, subsistence lifestyle, where small stable populations have an intimate connection with their territory, are no longer available to hundreds of millions of urban dwellers. Localised subsistence living, as a general aim, is a horse that has bolted. While two hundred years ago there were less than half a billion people on the planet, there is now about 8 billion people – at least half of whom live in urban settings and need global trade to thrive.

Debt, itself, is not new. In Mesopotamia, debt was regulated in sophisticated ways, with infrequent debt jubilees, which made sure that runaway inflation could not divorce the monetary system from the needs of the ecosystem or equity in the population.

The massive disparity we now have between rich and poor is a result of the abstraction of nominal credit. It is the 'capital' in capitalism that is the problem. Banks have gained informal legitimacy to literally create money out of nothing, whereas the labouring classes have the value of their wages absorbed by the swirling currents of extra money circulating the economic system. This creates an upward siphoning of value from the 99 per cent to the 1 per cent

(Piketty 2022). Meanwhile, because of economic growth, climate disasters are becoming commonplace. In the *Society of Fear* (Beck 1992), the ultra wealthy are building bunkers, but the working classes have the expense of repair and adaptation to much harsher conditions. The simulacra of immense nominal value and its detachment with the real economy is widening. It is seen in the schism between displaced, desperate and conflict-driven climate refuges (which include Middle Eastern and Northern African refugees) and the elite successes of capitalism, like Geoff Bezos, Elon Musk, Sam Alton and Bill Gates. Increasingly abstract ideals of the world as a simulacrum (Bostrom 1998; 2003a; Humphreys et al. 2022; Baudrillard 1994) are informing the actions of many of the super-rich elite (Rushkoff 2018). The divergence between fiat trade and nominal value is posing an existential threat to the environment.

The question now is, How can we transform modern, global economics into something more ecologically and socially viable? In my view, an important step is to tweak the concepts of 'efficiency', 'innovation' and 'economy'. The concept of economy emerges from the Greek term *oikos*, which means 'household economy'. The economics of a household utilise efficiency to make the most out of all the household materials. Economy means reducing the need to buy more, by reusing, mending or re-consigning materials until they are completely worn out. An efficient household slows down the entropy of wear and tear on materials for as long as possible, reducing pollution in the process. This often has no attachment to monetary value but, rather, the flourishing of the ecosystem, family and community. 'Efficient innovation' utterly transformed in concept once it was deployed by market economics to promote a system of profit and growth. The economic 'good' of the *nation-state*, as opposed to the household (Smith 1904/1776), is directed at increasing business profit. Profit derives from introducing innovations to produce products more efficiently or increase consumer demand. Another key difference between home economics and business or national economics is double-entry bookkeeping, which interprets both debts and capital as assets. For the bulk of the modern period, pollution was 'externalised' to the planetary commons, either in emissions, toxic and material waste, water and soil pollution, or plundering raw materials or species via fishing, forestry or mining. Cleaning up the disasters that these externalities have caused contributes to the growth of national GDP (Waring 1990). Until very recently, there has been very little accountability for polluting the commons.

We are transitioning from global to *planetary* economics. Planetary economics recognises the finitude of raw materials and that the pace of renewables are constrained by seasonal growth and rates of solar-, wind- and hydro-energy. A combination of global production lines and national state boundaries have enabled corporations to outsource labour, resources and tax without many repercussions. However, as emissions are accounted for, global manufacturing will no longer shelter corporate pollution. Instead, a metabolic measurement of energy throughput attached to all productivity could inform 'price' as a mechanism of fair distribution and a more accurate understanding of resource use and allocation (Georgescu-Roegen 1986; Mayumi 2001). This is starting to happen with scopes 1, 2 and 3 emissions accounting. Accommodating scope 3, which evaluates embedded emissions in the global production of consumer goods, could shift the cost of items substantially. Scope 3 enables localised production to become more viable. But metabolic accounting of emissions is only a start. Without a better ethos of *slowing metabolic throughput* and recognising the needs of both future generations and ecological systems in the allocation of raw materials, global economics will still be subject to Jevons paradox (Hickel 2020). That is, all efficiency gains will be subsumed within innovative modes of producing increasing consumer demand. While economic growth is the driver of global economics, efficiency is always at the service of expanding consumerism. While economic growth dominates, planetary ecology is increasingly at risk.

Growth is predicated on the need of large institutions, corporations, and governments to increase profit. Government relies on growth to offset inflation and build resiliency in the job market so people can cope with higher prices. Banks rely on national economic growth to meet the interest rates they attach to creating and lending credit to borrowers. The stock market hopes for growth to partially elide the growing disparity between real global GDP and the nominal worth of global stocks and shares. Pension funds, in particular, are totally reliant on economic growth to service their commitments.

The impact of economic growth already far outweighs its benefits. The number of extinctions already distinguishes this period as one of the planetary extinction events (Holzmann et al. 2023; Goulson 2019). The costs of coping with climate change are hard to assess. Newman and Noy (2023) estimate it at 16 million US dollars an hour. Climate refugees from Northern Africa and the Middle East are already multiplying. The conflict in these zones is escalating. When, for example, Pakistan had one-third of its land mass flooded in 2023, the national recovery bill is between 30 and 40 billion USD. It is difficult to quantify globally, but climate-event recovery has been estimated to rise to 38 trillion USD by 2025 (Potsdam Institute 2024).

To properly transition to metabolic economics, growth would be considered an increase in entropy and the opposite of what is desired. Instead, increasing the efficiency of metabolic throughput would mean using technological innovation to reduce, reuse, or replace finite raw materials. Metabolic economics will recognise the needs of future generations, other species and ecosystems. It will include the integrity of planetary ecology as a primary driver for decision-making about manufacturing. Government subsidies for the oil industry will cease. National boundaries will not offer loopholes to hide poor manufacturing practices. Circular design can improve repair, reuse, and recycle techniques, which will slow down entropy.

As a cautionary note, the circular economy is not a stand-in for the universal invisible hand. Circular design remains subject to the laws of entropy. Bioeconomics and circular design can be usurped into the modern 'ordering of order' (Heidegger 1977) unless it is tied to a bioegalitarian posthumanism (Barad 2003; Braidotti 2013; Kohn 2013; Dolphijn and van der Tuin 2012). Posthumanism ensures that economics respects the integrity and needs of other species and ecosystems. The implications of this tweak from global economics to planetary economics are important and far reaching. They are also viable and open up a diverse future. In chapter 8, some key concepts were explored that support planetary ecology and metabolic economics as an apparatus to adequately address the existential crisis of climate change, pollution, plastic waste, deforestation and species extinction. Entrophic cascades are an interlinked approach to productivity, negative entropy, energy throughput and resources. Throughputs are optimised and efficiency is unlocked by linking different scales of manufacturing. Waste from one type of production can be utilised as a resource for a different type. Existing ideas in metabolic bioeconomics are adequate to the task of supporting the 8 billion global population in a dynamic but steady state economic system (Daly 1996).

Metabolic bioeconomics can be achieved if, most importantly, this critique of ex nihilo loans as a key driver in exponential economic growth and the increasing accumulation of greenhouse gas emissions in the atmosphere is taken seriously. Once global legislation makes this practice illegal, then continuous exponential growth is no longer a structural driver. Efficiencies and innovation can return to their true task, of reducing entropy. Economics will become life enhancing instead of progressing further into the zone of nihilism.

BANKING

The current system of banking allows bankers to generate credit when a person with the capacity to pay it back walks in and asks for a loan. As far as I have been able to discover, there is no legislation that allows this process (Dunn and Wheelock 2022). It has become merely historically acceptable, since the Italian bankers began the practice in the 1460s (De Roover 1963; Galbraith 1975; Buch and Dages 2018). There is no back room full of money, no gold standard and no deposits necessary for the bank to guarantee liquidity (Simmel 1978/1907; Soddy 1933; Fisher 1936; Tobin 1963; Minsky 1992; Daly 2004; Werner 2014; Turner 2015). The loan is created ex nihilo, with nothing, as a line in the bank's double-entry account ledger. This new money sets off a cascade of interbank transactions, usually based on an initial house mortgage. The new money then spreads to higher-risk, more productive loans and gets integrated into the existing fabric of real fiat transactions. The downstream cascade of loans is called 'fractional' because of the historic Central Reserve Bank fraction requirement, but they should really be called 'fractal', because they proliferate new money throughout the economic system. The Central Reserve regulations dropped to 0.1 per cent, and now 0 per cent, giving no insurance against faulty loans (presumably on the basis that with continuously increasing property prices the mortgagee sale would cover the liability). If house prices are stable or increasing, there is really no way a mortgage bank (as distinct from a trading bank) can lose money. The new ex nihilo credit lines also get called endogenous, because the banks can create new credit lines that circulate endogenously within the baking system. Banks have no real control over how endogenous credit lines get converted into real transactions or real cash.

Many countries have a similar financial make-up as the United Kingdom. According to Jackson and Dyson (2013), 97 per cent of the money supply in the UK is made by banks. Only 3 per cent is cash, printed by the government. If there is a run on the banks, and people try to withdraw the money in their deposit accounts, it is not physically possible for the banks to hold enough cash to meet demand. This exposes banks to the risk that the general public will become openly aware of the extraordinary mode in which privately owned banks have assumed entitlement to create currency. As inflation has gone up steeply after COVID-19, national central and reserve banks are using higher overnight cash or LIBOR rates to attempt to slow down inflation. In 2023, with the sharp upturn in the reserve bank rate OCR and subsequent mortgage interest rates, more customer insolvency will arise. While the Central Bank fractional reserve has not been resurrected, as an example of increasing risk, New Zealand banks have been required to double their onsite cash reserves.

Private owners generating money out of thin air is very audacious and is almost without limits – this is where the bulk of retail banking profit derives. There is no limit to the amount that can be loaned, as long as someone dependable to repay it arrives, asking for a loan. This requires a healthy economy with high employment, good wages, a dependable currency, and a low level of economic risk. The broad economy needs to be stable enough that a large percentage of hard-working people can expect to find work and inflation will remain low enough that not all their wages are absorbed by consumable needs. Investment markets such as housing, shares, fine art, and Bitcoin act as financial reservoirs, by absorbing and attributing value to a lot of the extra money sloshing through the system. New entries into these investment markets are necessary to maintain the upward trend in values making population growth, and a growing middle class, indispensable.

However, real wages have been gradually falling against inflation indexes since the 1970s (Desilver 2018; Measuring Worth 2022; Statista 2023). In the 1970s, the average house price was three times the annual average wage. Today, it is between seven and ten times in many Western countries. If wages cannot provide sufficient deposit and guarantee of loan repayment, then market saturation or a 'bubble' threatens the entire feedback loop between new ex nihilo credit, economic growth, and the capacity of homeowners to continue to service the interest and capital repayment on their loans (BIS 2023b). This is a real issue for banks, because large numbers of defaults contract their debt 'asset,' rapidly reducing profit and increasing the chance that the public will lose faith and have a run on cash deposits in the bank. Given there is very little cash to cover any of the deposits or mortgages owned by the bank, the house of cards collapses.

If ex nihilo new money keeps saturating those markets that are reservoirs of value, runaway inflation will destabilise the macroeconomic system – first, at national and, ultimately, global levels. This is not like the frequent and controlled use of debt jubilees in Mesopotamia, which returned things to an equitable state when it showed early signs of wealth disparity and ecological strain. The entire edifice of the wealth of nations would collapse.

Governments speed up or slow down the upward dynamic stability of the broad economy with a few levers such the amount of cash in circulation, government bonds and treasury reserve interest rates. Government bonds and the treasury reserve interest rates influence the price of money of the OCR or LIBOR. But these are few tools for such a complex and unregulated task.

To maintain stable macroeconomics, a complex set of relationships ensue, tying together currency value, interest rates, LIBOR or OCR, insurance, types of financial instruments, levels of risk, general indicators of the economy, inflation, unemployment, and so forth. Treasury and the central banks, in particular, are tweaking the levers on these elements of the economy to keep the entire monolith reasonably stable, with an aim to maintain the economy as a whole in a sense of dynamic balance and gradually growing at approximately 2 to 3 per cent (Federal Reserve 2019; BIS 2023a; Reserve Bank of Australia 1996; Stevens 2003; Buch 2022; Lane 2021).

When the entire global economy crashes – as it did in 1719, 1929, 1987 and 2008 – the process of ex nihilo money generation is exposed. Electronic money that is 'made' at the point of borrowing is only viable if its accepted systematically by the public. It is only viable if it makes its way into the real economy and is traded between goods and services, bringing real value to return to the banks in the form of interest and loan 're'-payments. (Even the term '*re*payment' is a sleight of hand, because the original money was never paid for in the first place). Bank deposits and economic growth obscure the reality of ex nihilo money generation. The banks are terrified of being discovered as emperors with no clothes (Kent 2018; Rendahl and Freund 2020).

Banks present themselves as a mediation service between the depositors and those who need loans. This is why deposit accounts bear interest rates 'tied' to mortgage interest rates. But there is no actual link between the two activities. The banks merely sacrifice some profit to maintain the discourse of an intermediary status. In 2023, government treasury officials in Australia have had stern words with banks who rapidly hiked up mortgage interest rates (substantially increasing profit) and lagged the increase in deposit interest rates.

Of course, not all economic collapse can be attributed to banking, although banks are quickly swept up in maelstrom. The Great Depression of the 1920s emerged from a long-term drought in America, where monoculture on the Great Plains put soils at risk. The Dust Bowl was the result of drought, dried out and exposed soil, that was swept up and away by the wind. People walked off their farms in large masses, creating an enormous level of mortgage defaults. The level of defaults on mortgages was unprecedented, and multiple banks closed. Trust in the banking system hit an all-time low between 1929 and 1933, when nobody trusted banks with deposits. Faith in banking was only restored when Roosevelt introduced the Glass-Steagall Act in 1933. This legislation was designed as a mechanism to place limits on private, artificial creation of money via leveraging loans. The Glass-Steagall Act (1933) introduced the idea (itself a myth) that the real economy deposits would limit the leveraging capacity of the banks. Hence, the analysis of Steve Keen (2018), who has a much better understanding of the role of mortgages in the broader economy than most economists. Keen (2018) had an acute analysis of the statistical relationship between the mortgage-to-income ratio seen in house prices and private debt in the 2008 global financial crisis (GFC). But he spends unnecessary time justifying bank deposits as a ratio of loans when there is no real connection.

DOUBLE-ENTRY BOOKKEEPING

The beginning of double-entry bookkeeping in 1395 (De Roover 1963: 37–38) was closely followed by the development of ex nihilo loans in Italy during the 1460s (De Roover 1963; Galbraith 1975; Buch and Dages 2018). Although the Medici Banks thrived due to these new techniques, any exponential growth at the outset was so small it was invisible.

Theoretically, double-entry bookkeeping should have a 'balance' between both sides of the ledger – credit and debit (De Roover 1963: 98). In the twenty-first century, this balance is maintained because banks have to balance the ledger every night by selling or borrowing at the OCR. The daily 'wash' balances books between banks. But crucially, somewhere along the way, the ledger balance is only on cash withdrawals and deposits by customers. It has nothing to do with mortgage loan or deposit creation, which is another ledger altogether. The ledger balance in the daily wash is on cash balances only. Cash being sovereign money printed by the state and not under the auspice of private banking generation.

> Central bank reserves serve as the means of final settlement (payment) between banks themselves and between banks and certain other financial institutions. When customers (households and businesses) make payments between each other, the main payment systems (e.g. BACS and FasterPayments in the UK) queue these payments and calculate the net differences between the total payment flows in each direction. Then, at the end of the clearing cycle, the reserves equivalent to the net difference is transferred from banks with net outflows to banks with net inflows. (Dyson et al. 2016: 8)

Double-entry bookkeeping on credit generation also occurs. The seller's bank has a new debit line, and the buyer's bank has a new credit line – so they 'balance' despite having almost nothing to do with the real economy. This is how nominal debt increased to 258 per cent of total global transactions in the real economy (Gaspar et al. 2021). This may already have been happening with Medici Bank in the fourteenth and fifteenth centuries. It is not a question that De Roover (1963) was asking. He did notice, though, that the Medici's were involved in large volumes of trade. 'The firm of Averardo di Francesco de' Medici was engaged in international as well as in local banking and dealt extensively in bills of exchange. It also traded in wool and other commodities and underwrote marine insurance, but exchange dealings predominated' (1963: 38). By the first half of the fifteenth century,

> The volume of business was already too large for the use of a single ledger to be practicable. It was necessary, therefore, to divide it up into sections, for example, into a libro di creditori e debitort, a libro segreto, and a cash book (quaderno di cassa). How these simultaneous ledgers were interconnected, with or without reciprocal accounts, matters very little provided the basic canons of double entry are not violated and each transaction gives rise to two entries: one on the debit side and one on the credit side. By following this rule throughout, books should balance, if correctly kept, whether only one ledger is used or several. (De Roover 1963: 98)

De Roover shows that from the outset of double-entry bookkeeping, there is a disaggregation of different markets on separate ledgers. This means that account-keeping of nominal debt can be abstracted from accounts in the real economy. This is how nominal debt increased to 258 per cent of total global transactions in the real economy (Gaspar et al. 2021).

The daily wash is another technique for introducing new money into the broad economy. Sellers of property often extract some of their new electronic deposit-asset out of the bank to use in daily affairs. Cash is printed by the government, and banks effectively have to pay to import cash into their system. This is one of the reasons they promote electronic card use and discourage cash withdrawals. The daily 'wash' only balances cash going in or out of particular branches because people and small businesses are depositing or extracting cash from their accounts. At closing time, every bank has to settle its cash movement to zero. If more cash was withdrawn than deposited, banks have to 'buy' the difference via the OCR or LIBOR. The wash has nothing to do with new credit generation.

The fact that branches have to 'pay' for the movement in cash reserves through the interbank OCR or LIBOR is completely unconnected to the process of generating ex nihilo loans. The banks' ability to loan is unrelated to the daily settling of cash reserves and the amount sitting in deposit accounts. Nevertheless, reserve or central banks set the interest or cost of money as a 'reference rate' that is translated into mortgage interest rates. When mortgage rates are high, there is less turnover of property and so less new ex nihilo money is generated – which slows down the rate of inflation.

The banks' ability to create mortgages is always refreshed by new mortgage customers rather than any relation between deposits and loans. The whole cascade of endogenous (fractal) or fractional credit is renewed when a new person walks into the bank with the promise to repay the interest and capital he or she needs to borrow. So, the actual limits to new loan creation are the rates of wage growth and low enough inflation so that people are able to repay without defaulting. That amounts to a growing economy, an increasing population and faith.

The price of housing, in particular, is becoming so high that very few young people have any hope of getting on the property ladder. This is very problematic for the current system. Real estate agents and banks unconsciously collude to keep improving house prices. A large turnover within the existing property market will introduce new mortgages and thus, new ex nihilo profit for banks. Continuous growth needs new inputs and upward pressure from first-property buyers to justify the upward pressure in house values. Without that upward pressure, there is a risk to the role of housing as a reservoir of value, and the guarantee that higher property values offer the bank to offset the risks of defaulting. Should property prices start to fall, as they did in the

Great Depression, people defer buying and the rate of new loans plummets. Inflated commodity prices drop, profits dry up in multiple markets, bankruptcies increase and deflation takes hold.

QUANTITATIVE EASING

During COVID-19, governments turbo charged the economy by reducing the 'cost of money' to zero or less. Despite an enormous drop in productivity, money circulated freely. This was not just government quantitative-easing and the printing of money. By improving unemployment welfare provisions and giving cash directly to the general public, governments were able to keep the circulation of consumables buoyant.

Central Reserve interest rates dropped to zero. Bank interest rates dropped, making it very easy for people to apply for a new mortgage. Housing turnover was very high, and each new mortgage is new money created ex nihilo entering the market. Despite the radical slowdown in fiat money trade and services, economies stayed buoyant because new money in nonproductive markets like housing, Bitcoin and elite art kept growing. Inflation was on hold, as companies struggled to maintain supply and remain afloat during COVID-19 lockdowns.

Given the enormous amounts of cash swilling around in the system after COVID-19, the rhetoric by right-wing governments about 'austerity', after the last GFC, seems increasingly hard to swallow. The Keynesian (1933) government multiplier, as an allocation of money to stimulate the economy, had been spent on bailing out the trading banks (not the mortgage banks). After the GFC, state debt radically increased and welfare provision was subsequently starved. Austerity starved public institutions such as hospitals, universities, schools and welfare. Bailing out trading banks was supposedly necessary for economic stability.

This book has not delved deeply into the role of stock markets, except to that, like fine art, Bitcoin and housing, stocks and shares are reservoirs of wealth. The nominal value of stocks and shares and derivatives are six times more than global GDP according to the World Bank (2023) and the Bank of International Settlement (BIS) (2023). This adds to the abstract, nominal public and private debt ratio of 250 per cent global GDP. The abstraction of nominal currency in debt, and futures trading, are all part of the problem. The role of futures trading, in particular, and other financial mechanisms need further research in relation to their role in economic growth and climate change.

As the disparity between nominal debt and real economy grows, the value of billionaires grows immensely. A few of them – Warren Buffett and Bill Gates – have begun to take the problem of redistribution and public good work seriously. But for most, the upward trending – of stocks and housing – has been an important part of belonging to the successful class as distinct from the disenfranchised. In this economy of exuberant over-supply of money, the real economy has to justify the existence of such extreme riches. The real economy is still based on ideals of adding value to the earth's resources through toil (Locke 2003). So, the disenfranchised are obliged to work extremely hard to keep the entire overblown system afloat. Never mind the unemployment, the homelessness, or the suicide rates (Adult Psychiatric Morbidity Survey 2016) – those are collateral damage, in the same way resource exhaustion, ecological genocide, civil unrest, pollution, a greenhouse atmosphere and an acidified ocean are collateral damage.

Given that exponential economic growth is the driver of climate change, the role of money needs to be better articulated. The material purpose of money is in the real economy, as a trading token that creates equivalent value amongst disparate goods and services. It is also useful, though not the only option, for raising capital. Capital is necessary for large-scale community projects to be built, or so families can raise money and build housing. Finally, it is useful for raising millions for large-scale technological developments; developing mechanical plants to produce solar panels is an example. Even in a postcarbon economy, raising capital will still be necessary. But how finance institutions provide those opportunities needs to be reconfigured (Turner 2015).

Banking is responsive to the calls of climate change. The Taskforce on Nature-related Financial Disclosures (TNFD) introduces higher levels of transparency of information. TNDF make an ethical analysis of the banking industry to ascertain their sustainability policies and initiatives. They acknowledge, 'More than half of the world's economic output – US$44tn of economic value generation – is moderately or highly dependent on nature. Nature loss therefore represents significant risk to corporate and financial stability'. This is undoubtedly an important move forward. The TNDF has a focus on investment, insurance underwriting and the availability of ethical credit. This is very important, as banks are major players in funding mining and oil development. But it does not explore the structural machinery of ex nihilo mortgage banking. While the TNFD engages with ethics and the circular economy, it is completely unaware of the structural issues of exponential growth driven by fractional banking. It misses completely the relationship between ex nihilo mortgages, economic growth, and the accumulation of CO_{2e} emissions in the atmosphere (Le Quéré et al. 2020; Caporale et al. 2021; Mitić et al. 2022; Onofrei et al. 2022; Noll 2023).

The TNDF, and global organisations including the World Bank, are highly aware that during this transition to a postcarbon economy there are a number of major breaks with previous normative ways of doing things. There is a need for green infrastructure, to build fifteen-minute cities and reduce the need for cars. Already, cities are dealing with adaptation demands, as heat escalates and flooding is far more frequent. These are medium- and long-term escalations in financial demand for new capital. Stranded assets are likely, as the fossil fuel industry *should* be winding down. The existing retail banking industry needs to become part of the obsolete infrastructure of modernity. It is possible that banking may transform to accommodate new modes of raising capital from within the real economy (Soddy 1933/1926; Fisher 1936; Galbraith 1975; Buch and Dages 2018; Tavlas 2020; Benes and Kumhof 2012; Turner 2015), but it is also possible that these old institutions will be unable to adapt, and new institutions will arise and fill the void. This really should not be a problem. After all, multiple industries have become obsolete in the last two hundred years, and banking is just another industry.

While endogenous fractional banking, ex nihilo, continues, there is no chance of reducing the rate of exponential economic growth. That means that renewable energy will be unable to replace fossil fuels, as all energy sources will be taken up by the demands of economic growth. There is no possibility for the commitment to net zero by 2050 to be met while these conditions prevail.

POSTGROWTH LOAN STRATEGIES

There are a few courageous people already writing important critiques of the banking system and modern macroeconomics. Most of them have focussed on the very significant issues of instability and diverging poverty and wealth. My view is that the problem is far more urgent than even that. Private banks with the legitimacy to make unfettered *ex nihilo* new credit are driving the systemic commitment to economic growth. Until we halt economic growth, there is no way to even stabilise emissions, let alone reduce them.

A lot can be learned from early technologies of debt and time. The practice of attending to earth-centric astrological cycles helps people attune to the seasonal cycles, both annual and metonic (Freeth and Jones 2012). Planetary cycles affect the subtleties of human intraconnection with planetary ecology. Additionally, a global reset from a debt jubilee, or *andurārum* (Charpin 2010; Lemche 1979), could address the enormous disparity between rich and poor that is warping the current economic system.

Another alternative to exponential growth is Irving Fisher's (1936) Chicago Plan. This is where the bank operates as a genuine intermediary between depositors and creditors. No money is created ex nihilo. Banks are only able to loan the amounts they hold from other depositors. Benes and Kumhof explain the outcomes of Irving Fisher's Chicago Plan.

> It envisaged the separation of the monetary and credit functions of the banking system, by requiring 100% reserve backing for deposits. Irving Fisher (1936) claimed the following advantages for this plan: (1) Much better control of a major source of business cycle fluctuations, sudden increases and contractions of bank credit and of the supply of bank-created money. (2) Complete elimination of bank runs. (3) Dramatic reduction of the (net) public debt. (4) Dramatic reduction of private debt, as money creation no longer requires simultaneous debt creation. (Benes and Kumhof 2012: 1)

Interest rates could still accrue if banks held the role of intermediary. Any interest would pay for the costs of the banks and interest rates contributed to savings deposit accounts. That small amount of interest would continue to drive up economic growth, as it is nonproductive exchange. So, if interest rates were still an option, it is quite possible that economic growth would remain a problem, albeit at much smaller rate and scale – making a blend of bank as intermediary and debt jubilee necessary.

Banking might become a genuine intermediary between depositors and borrowers, so that the loans made are backed by actual funds. Fees could serve to pay for the privilege of secure storage of value and the banks role as an intermediary for borrowers and savers. If interest rates were still to apply to loans – to cover costs and pay for the privilege of borrowing money – then at a much more minor scale, the problem of exponential growth would persist. Frequent debt jubilees might annul the effect of exponential amounts of loan interest, even if new ex nihilo money stops making its way into the macroeconomic system. A jubilee eliminates the need for continuous growth to offset continuous inflation. That irregular, periodic reset would dramatically alter the premise of macroeconomics and would stop driving climate emissions.

Adair Turner (2015) has interesting alternatives to current banking. He suggests retaining much of the current status quo, but adding more emphasis on government currency stimulation. He notices the critique of government credit generation is aimed at political interest groups getting favourable access to new funds. But there are ways to avoid the problem of political leverage, including the already existing independent central reserve or treasury officials. Certainly, government stimulus is more sensitive to the demand for currency in the real economy. Government stimulus is less liable to oversupply and create inflation than the present unregulated private banking system of ex nihilo credit. Positive Money make a similar case for government stimulus, but they recommend, as I do, getting rid of private bank ex nihilo credit altogether. It is beyond the scope of this book to lay out detailed viable alternatives to existing modes of finance except to make the case that the structural imperative of growth urgently needs to stop.

If interest rates and ex nihilo money creation are no longer forcing the devaluation of circulating currency, then the demands for higher work output from fewer hours labour would not be as urgent. More equitable distribution would be easier to leverage, as the balancing act between employment rates, inflation, currency exchange valuation, and the OCR would not be as complex and difficult to manage. The claims of Keynesian theorists that the working classes benefit from the multiplier effect would need to be addressed through innovation, efficiency, and equity rather than pumping more money through the system. Nor would the idea of the economy as a growing pie continue to justify the unequal distribution of wealth across sectors of the community. Disparity between rich and poor would likely return to a narrower spread. The billionaires that 'own' vast proportions of the earth's resources would slowly dissipate their wealth. The structural imperative of accelerating growth would no longer support vastly unequal wealth. Without massive amounts of extra money constantly pumping into the system, the reservoirs of wealth, where excess pools, will dry up. Jevons paradox would disappear. Green technology and infrastructure would have the chance to actually reduce energy throughput and emissions. Efficiency and innovation will be at the service of improving ecology and social well-being, rather than merely the reductive notion of profit.

The Paris Agreement (2016) is compromised by 196 countries that are committed to keep emissions low enough that global warming does not go over 1.5°C above preindustrial levels. Governments are making increasingly better efforts but are still struggling to reduce emissions, let alone meet their net-zero targets.

The Intergovernmental Panel on Climate Change (IPCC) has been very clear that business as usual (BAU) will drive emissions up so far that the planet will warm by 6 or 7°C. Despite a massive shift towards renewable energy, the emission profile of nation-states continues to rise. This is because, according to Jevons paradox, efficiencies are absorbed and overtaken by continuous exponential expansion of productivity and consumerism. Under BAU, the rate of increase in planetary temperature is speeding up (Hansen et al. 2023; Noll 2023).

TECHNOLOGICAL WORLDING

Adam Smith (1998/1776), 250 years ago, made use of the latest philosophy and science to develop his economic theories. He supported shifting norms from home economics to the wealth of nations. It reforged the way people saw themselves. They came from families and communities that had lived on the same land for thousands of years, to become migrant, landless individuals. They gain more consumer choice to compensate for alienation from nature. As individuals, they need to enhance their skillset so they have more valuable human capital. The universal essence of the individual is assumed to be rational, with no influence from gender, sexuality, class, ethnicity or any other consideration of history, culture, or identity (Benson 2020). Classical and neoliberal economics has maintained a regime based on consumerism and the rational behaviour of individual-utility maximisers. The assumption that individuals act with rational self-interest is essential for predictable market psychology to operate. The time has arrived for a new shift in the economic norm.

From a variety of disciplines, a new approach to intraconnected subjectivity is emerging. The new norm is an embodied, affective, interdependent agent who intra-acts with the environment. Feminist philosophy, continental philosophy, indigenous philosophy, and philosophy of science all offer a much richer and more sophisticated understanding of agency. These ideas are very relevant for sustainability. People individuate from the technological enframing of historical knowledge by thinking critically. The agency of new ideas oscillates back through the collective because individua*tion* is not isolated individua*lism*, it is always in connection with everything else. The embodied, affective, intra-active agent presents a compelling narrative that might replace or reshape the neoliberal policy discourse. At present, policymakers remain neoliberal for want of a better cohesive narrative about expectations and norms. Many people, steeped in the modern horizon of thought, tend to be unconsciously anthropocentric in outlook. But when the normative understanding of subjectivity is intra-action with the ecosystem and each other, then consideration of other species and ecosystems will be absorbed into the future ethics of economics and policymaking.

The rational individual is based on outdated Enlightenment concepts of universal essentialism and alienated individuality. In contrast to the individual who only has a sceptical rationality to observe alienated natural objects, Heidegger (1962) emphasises the way technical objects constitute learning and culture. Heidegger (1977) explores how technology enframes modern subjectivity by producing everything as waiting in reserve (*bestand*) for potential consumerism. In each historical age, technology 'worlds us' – meaning, the way technology is put to work shapes our subjectivity and our assumptions about the wider environmental context. At birth, we are 'thrown' into an 'always-already' existing world (Heidegger 1962). The configuration of meaningful world-making through the 'always-already' ordinary, everyday things – such as table and chairs, pots and pans or bedrooms and a separate lounge – all produce a total immersion in the concrete realisation of the normative understanding of the world. It is easy to miss how important this mundane immersion really is. Heidegger's work stands in stark contrast to Descartes's (1980/1641) proposition that human subjectivity is separate from objects, superior and rationally deductive. Heidegger understands that the making of subjectivity is a result of the surrounding technological milieu. Heidegger (1973a/1935) is one of the strongest critics of Cartesian rational individualism. Yet at some level, Heidegger's technological milieu still invokes a cultural and technological emergence of subjectivity that largely retains the separation of technological culture from ecological nature. The alienation generated by Cartesian dualism remains unbroken (Merchant 1980).

Heidegger (1999a) contemplates modernity as a whole. The historical epoch of modernity begins with its earliest influences in modern Greece, to the looming finitude of nuclear war, and now climate change. The earliest sophisticated mechanics go all the way back to ancient Greece. It is from the culture Heidegger and other classicists regard as the origin of modern civilisation. Showing how nonlinear technological innovation really is, the Antikythera mechanism is the earliest clockwork ever found, around 206 BC. The technology was lost for about 1,400 years, before clocks were (re) invented in Europe (Edmunds and Philip 2000; Freeth 2002b). The Antikythera mechanism is embedded in cycles of the earth-centric cosmology of all Axial cultures. Its accurate time keeping helped regulate public events. The Saros and Metonic cycles predicted important eclipses, bringing together the stars and political fates (Oppenheim 1967; Steele 2000a; George 2013). It used the most sophisticated mathematics available to support an already antique approach to the cycles of seasonal time, to govern the tempo of society.

The Antikythera mechanism demonstrates a continuity from Mesopotamia through to modernity, in the use of timekeeping, mathematics, property, writing, accounting and debt (Freeth et al. 2021). But the geocentric orientation of Axial cultures contrasts starkly with our own. At the origins of writing and debt, Mesopotamians had sophisticated techniques to limit the ecological consequences of exponential economic growth (Lemche 1979; Charpin 2010). Those debt jubilees, or *andurārum*, return the distribution of property to its equitable origins (Kaplan 2019). The Antikythera mechanism, along with the technologies of writing and accounting, demonstrate how sophisticated *technē* can produce a geocentric, planetary economics. Planetary ecology is in tune with the rhythms and cycles of the solar system, and these regulate the tempo of economic, political, and festive events. Technological enframing of the horizon of disclosure can be sophisticated *and* geocentric. This cyclic pace contrasts strongly with the modern exponential and infinite tempo of society. On the one hand, modernity conforms to the demands of consumerism. On the other hand, with ex nihilo credit and infinitely compounding interest, the economy is structurally forced to grow or collapse. Too much currency systemically forces exponential growth into the transactional economy.

Lotka (1925; 1945), Leroi- Gourhan (1993/1943; 1945) and Stiegler (2018) were much more eclectic and open-minded than Heidegger about looking outside the modern epoch. The technological enframing of the horizon of disclosure occurs in all cultures. The surrounding normative understanding of the world is embedded in the *technē* into which a child is 'thrown'. Lotka, Leroi-Gourhan, and Stiegler looked at how exosomatic technology forges cultural worlding across multiple historical periods and locations. Exosomatic technologies, such as a painting, a book, or a computer, enable ideas to be communicated across generations, time and distance.

Nevertheless, they remain ignorant of the eco-*technē* that informs the exosomatic context of indigenous communities. It may not be as obvious as surviving artefacts like rock paintings or carvings, but plants, rocks, water, atmosphere, birds, reptiles, insects, and animals are the normative techno-ecological world into which indigenous people are 'always-already' thrown. With startling simplicity, such a posthumanist ethos cuts the Gordian knot of alienation between rational techno-culture and pure nature.

If modern technology is shaping current global culture, then understanding the 'final causes' of technology might help to examine the technological horizon of disclosure. Classically, humanity has been understood as the tool-wielding animal. This makes a hierarchical distinction with yet another dualism between humans and other species. In fact, multiple species have tool use: all birds make nests, and several birds use rocks to open shells, monkeys use reeds as a straw to suck up termites and a species of octopus and hermit crabs make use of other shells and objects to create their homes. The significance of humanity as the only tool-wielding animal has been undermined. Following the originary myth of Prometheus, Stiegler (2018) argues, that technology – namely the use of fire – constitutes culture in *all* human societies. But not even fire can be isolated as an exclusively human technique, as three species of hawk in Australia drop burning sticks to generate forest fires and drive out prey (Bonta et al. 2017).

Indigenous communities have been in tune with the ecological cycles and other species, landscapes, water systems and climate for hundreds of thousands of years. Colonialism built an ethos of disrespect for these ecologically savvy cultures, that emerged from a monotheistic and mechanistic series of assumptions about Enlightenment. But that colonial racism is beginning to unravel. *Whakapapa* is an important concept in Māori cosmology, which links all species, water, geoformations, atmospheric conditions and the stars together in a taxonomy of kin relationships (Ngata 2011/1944; Marsden 2002; T. Smith 2000; C. Smith 2000; Mika 2017; 2019). *Whakapapa* is all about intraconnection. There is no dualism between nature and culture. Indigenous approaches to bio-egalitarianism (Braidotti 2013) are important pathways to a bioeconomic future.

Quantum physics also demonstrates the intra-action between subjects and objects, collapsing the dualism set out by Descartes (1980/1641) and Kant (1992/1781). A variety of quantum 'weirdness' disrupts Newton's (1687) mechanical universe. Instead of linear cause and effect, quantum entanglement operates across vast distances instantaneously. Particles shapeshift from locatable and measurable to smeared out and virtual. Time/space is more like an ocean with diffractive waves interacting. Newtonian physics of a linear causal relation of a force acting upon another force is simplistic (Barad 2007).

Multiple discipline areas, including indigenous philosophy, feminist embodiment, and philosophy of science, are bringing to light how intra-active the universe really is. The notion of a segregated rational individual is rigid, and unlikely. Each of us are *individuated* from merging with the community (Siegal 2022) but not universally *individualistic* (Descartes 1980/1641; Smith 1904/1776; Kant 1992/1781; Hayek 1960; Friedman 1970; Fukuyama 1992). The new onto-epistemology emerging from these disparate fields converges in the important intra-action of humanity and ecology. This informs materialist economics that registers energy throughput, entropy, and pollution as important limiting factors in productivity.

FUTURES

Modern economists are used to dealing with numbers in the range 1 to 100 per cent. Thus, 1 to 2°C appears minimal. This is a failure in category analysis. The impact of tiny increments of planetary warming are immense. As the CO_{2e} load in the atmosphere continues to *accumulate*, the planet will continue to get warmer. For each degree of warming, the atmosphere increases capacity to hold 7 per cent more water (Coumou and Rahmstorf 2012). Currently the planet is at 1.3°C above the preindustrial average. This vastly increases the volatility of the planet's hydrology. We have already left behind the Holocene epoch and have become embedded in the increasingly turbulent Anthropocene (Crutzen and Stoermer 2000; Crutzen 2002). Stopping emissions accumulation is vital to having a chance to adapt to the rapidly changing climatic conditions that are already underway.

Sudden and chronic climate events are wreaking havoc on communities. Cities and regions are now establishing emergency protocols. They are trying to overcome years of individualism and to improve community cohesion for better resilience to frequent and chronic climate events. Close-knit communities are better able to manage the flooding, storms, drought, and fire that is sweeping through landscapes with more severity and frequency (Hansen et al. 2023).

People are keen to implement real change that will halt climate emissions. So far, the focus has been on greener technology and infrastructure, all of which is important to bring down emissions. But Jevons paradox dictates that all new efficiency gains will be co-opted into the expansion of economic growth. To be effective, people need a clear focus on what needs to change, or they dissolve into despair, inertia, or denial. Effective action involves understanding the correlation between economic growth and emissions growth (Le Quéré et al. 2020; Caporale et al. 2021; Mitić et al. 2022; Onofrei et al. 2022; Noll 2023). Fossil fuels are the cause of CO_{2e} emissions. The driver is economic growth. Banking ex nihilo is systemically forcing economic growth to escalate.

Importantly, one country on its own cannot change the global banking system. That country would go bankrupt while the rest of the world continued to invent new money out of nothing and usurp the value of public circulating wealth in the fiat economy. Climate change is a global phenomenon. Banking is also global. System change has to happen in one consistent moment, across all countries. Any isolated individual or little community, can feel that this change is too monumental. The stakes are so high, the powerful lobbies are beyond reach. Democratic rights have few genuine options beyond 'left' or 'right' BAU. Global politics is beyond the reach of ordinary democracy. But change happens in surprising ways. My favourite example is Gregory Bateson's (1979) allegory of change. There is no telling precisely when water will go from fluid to steam. Even when water is heated to 100°C, it will not spontaneously boil. The surface tension on the skin of the water holds, while the water overheats. It is not until a speck of dust, an insect, a particle, lands and breaks the surface tension, that the metamorphosis occurs. When one person changes their interpretation of the system, that new knowledge reverberates through all the interconnected threads of intra-action. Let us be that speck that lands in the overheated climate and changes the system from exponential growth to the cyclic earth-centric pace of sophisticated civilizations. Let each of us be that change.

References

Adam, Barbara. 2006. 'Time'. Special Issue on Problematizing Global Knowledge. Mike Featherstone, Couze Venn, Ryan Bishop & John Phillips (eds). *Theory, Culture & Society*, 23. no. 2–3: 119–126.

Adult Psychiatric Morbidity Survey. 2016. 'Adult Psychiatric Morbidity Survey: Survey of Mental Health and Wellbeing, England, 2014'. NHS. Accessed 10 May 2023. https://digital.nhs.uk/data-and-information/publications/statistical/adult-psychiatric-morbidity-survey/adult-psychiatric-morbidity-survey-survey-of-mental-health-and-wellbeing-england-2014.

Ahenakew, Cash, Vanessa de Oliveira Andreotti, Garrick Cooper and Hemi Hireme. 2014. 'Beyond Epistemic Provincialism: De-provincialising Indigenous Resistance'. *AlterNative: An International Journal of Indigenous Peoples* 10, no. 3: 216–31.

Altvater, Elmer. 2002. 'The Growth Obsession'. *Socialist Register* 38: 73–92.

———. 2007. 'The Social and Natural Environment of Fossil Capitalism'. *Socialist Register* 43: 37–59.

Ament, Joe. 2019. 'Toward an Ecological Monetary Theory'. *Sustainability* 11: 923.

Anderson, Kevin. 2012. 'The Inconvenient Truth of Carbon Offsets'. *Nature* 484, no. 7: https://doi.org/10.1038/484007a.

Anderson, Kevin, Alice Bows and Sarah Mander. 2008. 'From Long-Term Targets to Cumulative Emission Pathways: Reframing UK Climate Policy'. *Energy Policy* 36, no. 10: 3714–22.

Appadurai, Arjun. 1986. 'Introduction: Commodities and the Politics of Value'. *The Social Life of Things*. Cambridge: Cambridge University Press.

Archimedes. 2011. *The Archimedes Palimpsest Project.* Cambridge: Cambridge University Press. Accessed 10 July 2023. http://www.archimedespalimpsest.org/.

Aristotle. 1972. *The Works of Aristotle*, vol. 8, *Metaphysica*. Oxford: Clarendon Press.

Arrhenius, Svante. 1896. 'On the Influence of Carbonic Acid in the Air Upon the Temperature of the Ground'. *The London, Edinburgh, and Dublin Philosophical Magazine and Journal of Science* 41, no. 251: 237–76.

Asiain, Raquel, Roberto Ontanon and Pedro Saura. 2023. 'Animals Hidden in Plain Sight: Stereoscopic Recording of Palaeolithic Rock Art at La Pasiega Cave, Cantabria'. *Antiquity*: 1–16.

Asveld, Lotte, Rinie van Est and Dirk Stemerding. 2011. 'Executive Summary'. In *Getting to the Core of the Bioeconomy: A Perspective on the Sustainable Promise of Biomass*, edited by Lotte Asveld, Rinie van Est and Dirk Stemerding, 11–14. The Hague: Rathenau Instituut.

Atkinson, Scott, and Tom Tietenberg. 1991. 'Market Failure in Incentive-Based Regulation: The Case of Emissions Trading'. *Journal of Environmental Economics and Management* 21: 17–31.

Ayers, Robert, and Benjamin Warr. 2009. *The Economic Growth Engine: How Energy and Work Drive Material Prosperity*. Cheltenham: Edward Elgar Publishing.

Bacon, Francis. 1696/1597. *The Essays: Or Councils, Civil and Moral*. London: Herrington.

Bank of England. 2014. 'Money Creation in the Modern Economy'. *Quarterly Bulletin*, Q1 [doubled].

Bank of International Settlement (BIS). 2016. *86th Annual Report*. https://www.bis.org/publ/arpdf/ar2016e.pdf.

———. 2019. 'Triennial Central Bank Survey of Foreign Exchange and Over the counter (OTC) Derivatives Markets in 2019'. December. https://www.bis.org/statistics/rpfx19.htm.

———. 2022. 'OTC Derivatives Statistics at End-June 2022'. https://www.bis.org/publ/otc_hy2211.pdf.

———. 2023a. 'OTC Derivatives Statistics at End-December 2022'. Accessed 13 October 2023.https://www.bis.org/publ/otc_hy2305.htm.

———. 2023b. 'Total Credit to Households (Core Debt)'. https://stats.bis.org/statx/srs/table/f3.1.

Barad, Karen. 2003. 'Posthumanist Performativity: Toward an Understanding of How Matter Comes to Matter'. *Signs* 28, no. 3, Gender and Science: New Issues: 801–31.

———. 2007. *Meeting the Universe Halfway: Quantum Physics and the Entanglement of Matter*. Durham, NC: Duke University Press.

———. 2010. 'Quantum Entanglements and Hauntological Relations of Inheritance: Dis/continuities, SpaceTime Enfoldings, and Justice-to-Come'. *Derrida Today* 3, no. 2: 240–68.

Barmash, Pamela. 2020. 'The Legal Authority of the Laws of Hammurabi'. In *The Laws of Hammurabi: At the Confluence of Royal and Scribal Traditions*, by Pamela Barmash, 231–50. Oxford: Oxford University Press.

Barthes, Roland. 1977. 'Death of the Author'. In *Image, Music, Text*, translated by S. Heath, 142–48. London: Fontana.

Bashford Alison, and Joyce. E. Chaplin. 2016. *The New Worlds of Thomas Robert Malthus: Rereading the Principle of Population*. Princeton, NJ: Princeton University Press.

Bataille, Georges. 1991. *The Accursed Share*, vol 1. *Consumption*. New York: Zone Books.

Bateson, Gregory. 1979. *Mind and Nature; A Necessary Unity*. London: Wildwood House.

Baudrillard, Jean. 1994/1981. *Simulacra and Simulation*. Translated by S. Glaser. Ann Arbor: University of Michigan Press.

Bayne, Karen, Anita Wreford, Alan Renwick and Peter Edwards. 2018. *Enabling a Transformation in New Zealand's Bioeconomy: A Thinkpiece*. Sydney: Scion.

Beck, Ulrich. 1992/1986. *Risk Society: Towards a New Modernity*. Translated by Mark Ritter. London: Sage.

———. 2009. 'Critical Theory of World Risk Society: A Cosmopolitan Vision'. *Constellations* 16, no. 1: 1–20.

———. 2016. *The Metamorphosis of the World: How Climate: A Change Is Transforming Our Concept of the World*. Boston: Polity.

Bendell, Jem. 2023. *Breaking Together: A Freedom-Loving Response to Collapse*. Cornwall: Earth Works.

Bender, Emily, Timnit Gebru, Angelina McMillan-Major, Shmargaret Scmitchell. 2021. 'On the Dangers of Stochastic Parrots: Can Language Models Be Too Big?' *FAccT'21: Proceedings of the 2021 ACM Conference on Fairness, Accountability, and Transparency:* 610–623.

Benes, Jaromir, and Michael Kumhof. 2012. *The Chicago Plan Revisited*. IMF Working Paper, International Monetary Fund. https://www.imf.org/external/pubs/ft/wp/2012/wp12202.pdf.

Benson, Jeremy. 2020. '"We Don't Care Who You Are": Race, Space, and Dispossession in New York's Charter School Co-Location Reform'. *Policy Futures in Education* 18, no. 5: 662–80.

Bentham, Jeremy. 1995/1787. *Panopticon; Or the Inspection-House Containing the New Principle of Construction Applicable to Any Sort of Establishment, in Which Persons of Any Description Are to Be Kept Under Inspection*. London: Verso.

Blok, Vincent. 2011. 'An Indication of Being – Reflections on Heidegger's Engagement with Ernst Jünger'. *Journal of the British Society for Phenomenology* 42, no. 2: 194–208.

Bobulescu, Roxana. 2017. 'The Original Time Approach of Georgecu-Roegen'. *Oeconomis; History/Methodology/Philosophy*: 87–109.

Bohr, Niels. 1987/1957. *The Philosophical Writings of Niels Bohr*, vol. 3. Woodbridge, CT: Ox Bow Press.

Bonaiuti, Mauro. 2015. 'Bioeconomics'. In *Degrowth: A Vocabulary for a New Era*, edited by Giacomodo D'Alisa, Federico Demaria and Giorgos Kallis, chapter 2. London: Routledge.

Bonta, Mark, Robert Gosford, Dick Eussen, Nathan Ferguson, Erana Loveless and Maxwell Witwer. 2017. 'Intentional Fire-Spreading by "Firehawk" Raptors in Northern Australia'. *The Journal of Ethnobiography* 37, no. 4: 700–18.

Bostrom, Nick. 1998. 'How Long Before Superintelligence?'. *International Journal of Futures Studies* 2: n.p.

———.2003a. 'Are You Living in a Computer Simulation?'. *Philosophical Quarterly* 53, no. 211: 243–55.

———. 2003b. 'Astronomical Waste: The Opportunity Cost of Delayed Technological Development'. *Utilitas* 15, no. 3: 308–14.

Braidotti, Rosi. 2009. 'Animals, Anomalites, and Inorganic Others'. *Theories and Methodologies* 124, no. 2: 526–32.

———. 2013. *Posthuman Ecologies*. Cambridge: Polity Press.

———. 2018. 'A Theoretical Framework for the Critical Posthumanities' *Theory, Culture & Society* 36, no. 6: 31–61.

Bright, Maria, and Chris Eames. 2021. 'From Apathy through Anxiety to Action: Emotions as Motivators for Youth Climate Strike Leaders'. *Australian Journal of Environmental Education* 38, no. 1: 13–25.

Broadman, Kenneth. 1966. 'The Economics of the Coining Spaceship Earth'. In *Environmental Quality in a Growing Economy*, edited by H. Jarrett, 3–14. Baltimore: John Hopkins Press.

Brockway, Paul, Anne Owen, Lina Brand-Correa and Lukas Hardt. 2019. 'Estimation of Global Final Stage Energy-Return-on-Investment for Fossil Fuels with Comparison to Renewable Energy Sources'. *Nature Energy* 4: 612–21.

Brundtland, Gro. 1987. *Our Common Future*. World Commission On Environment and Development. Oxford: Oxford University Press.

Buch, Claudia. 2022. 'Time for Structural Change in Central Bank Statistics? How to Support the Transition to a Climate-Friendly Economy'. Vice-President of the Deutsche Bundesbank, 11 Biennial Irving Fisher committee Conference, Basel. https://www.bis.org/review/r220914a.pdf.

Buch, Claudia, and B. Gerard Dages. 2018. 'Structural Changes in Banking after the Crisis'. Bank of International Settlements: Report prepared by a Working Group established by the Committee on the Global Financial System. https://www.bis.org/publ/cgfs60.pdf.

Budiselic, Chris, Andrew Thoeni, Michael Dubno and Andrew Ramsey. 2020. 'Antikythera Mechanism Shows Evidence of Lunar Calendar'. *SocArXiv*, June 19.

Bugge, Markus M., Tels Hansen and Antje Klitkou. 2016. 'What Is the Bioeconomy? A Review of the Literature'. *Sustainability* 8, no. 7. https://doi.org/10.3390/su8070691.

Buis, Alan. 2019. 'The Atmosphere: Getting a Handle on Carbon Dioxide: Sizing Up Humanity's Impacts on Earth's Changing Atmosphere'. In *Global Climate Change; Vital Signs of the Planet*. Nasa's Jet Propulsion Laboratory: California Institute of Technology. https://climate.nasa.gov/news/2915/the-atmosphere-getting-a-handle-on-carbon-dioxide/.

Bundesbank. 2017. 'The Role of Banks, Non-Banks and the Central Bank in the Money Creation Process'. Monthly Report 2017. Accessed 13 October 2023. https://www.bundesbank.de/resource/blob/667334/de28e82ef38090619f3af34e1442d269/mL/2017-04-monatsbericht-data.pdf.

Cadwalladr, Carol, and Emma Graham-Harrison. 2018. 'Revealed: 50 Million Facebook Profiles Harvested for Cambridge Analytica in Major Data Breach'. *The Guardian*, 18 March. Accessed 4 July 2021. https://www.theguardian.com/news/2018/mar/17/cambridge-analytica-facebook-influence-us-election.

Caporale, Guglielmo Maria, Gloria Claudio-Quiroga and Luis Gil-Alana. 2021. 'Analysing the Relationship between CO Emissions and GDP in China: A Fractional Integration and Cointegration Approach'. *The Journal of Innovation and Entrepreneurship* 10: article 32.

Campbell, Hugh. 2020. *Farming Inside Invisible Worlds: Modernist Agriculture and Its Consequences*. London: Bloomsbury Publishing.

Campbell, Hugh, Rob Burton, Mark Cooper, Matthew Henry, Richard Le Heron, Nicolas Lewis, Eric Pawson, Harvey C. Perkins, Mike Roche, Chris Rosin and Toni White. 2009. 'From Agricultural Science to "Biological Economies"?'. *New Zealand Journal of Agricultural Research* 52, no. 1: 91–97.

Carbon Border Adjustment Mechanism EU. n.d. https://taxation-customs.ec.europa.eu/green-taxation-0/carbon-border-adjustment-mechanism_en.

Carman, Christián, and James Evans. 2014. 'On the Epoch of the Antikythera Mechanism and Its Eclipse Predictor'. *Archive for History of Exact Sciences* 68, no. 6: 693–774.

Carmona, Salvador, and Mahmoud Ezzamel. 2007. 'Accounting and Accountability in Ancient Vivilizations: Mesopotamia and Ancient Egypt'. *Accounting, Auditing, and Accountability* 20, no. 2: 177–209.

Carroll, Rory. 2013. 'Elon Musk's Mission to Mars'. *The Guardian*, 17 July. https://www.theguardian.com/technology/2013/jul/17/elon-musk-mission-mars-spacex.

Carsen, Rachel. 1962. *Silent Spring*. Boston: Houghton Mifflin.

Centre for Sustainable Systems, University of Michigan. 2022. 'Greenhouse Gases Factsheet'. Michigan, UIM.

Charpin, Dominique. 2010. *Writing Law, and Kingship in Old Babylonian Mesopotamia*. Translated by Jane Marie Todd. Chicago: Chicago University Press.

Clausius, R. 1856. 'On a Modified Form of the Second Fundamental Theorem in the Mechanical Theory of Heat'. *Philosophical Magazine* 12, no. 4: 81–98.

Code, Lorraine. 2006. *Ecological Thinking; The Politics of Epistemic Location*. New York: Oxford University Press.

Commoner, Barry. 1971. *The Closing Circle: Nature, Man, and Technology*. Michigan: Knopf.

Copernicus. 1543. *De Revolutionibus Orbium Coelestium* [Six books concerning the revolutions of the heavenly orbs].

Coumou, Dim, and Stefan Rahmstorf. 2012. 'A Decade of Weather Extremes'. *Nature; Climate Change* 2: 491–96.

Crutzen, Paul. 2002. 'Geology of Mankind'. *Nature* 415, no. 6867: 23. https://doi.org/10.1038/415023a.

Crutzen, Paul, and Eugene Stoermer. 2000. 'The Anthropocene'. *Global Change Newsletter*, 41: 17–18.

Cutting, Grace. 2021. 'Simulating Cambrian Climate: The Significance of Atmospheric Co2 Concentration and Continental Position'. *Ursidae: The Undergraduate Research Journal at the University of Northern Colorado* 10, no. 1: article 2.

Dale, Roger. 1989. *The State and Education Policy*. Milton Keynes: Open University Press.

———. 1997. The State of the Governance of Education: An Analysis of the Restructuring of the State-Education Relationship. In *Education, Culture, Society*, edited by A. Halsey, H. Lauder, P. Brown and A. Stuart Wells, 273–82. Oxford: Oxford University Press.

D'Alisa, Giacomo, Federico Demaria and Giorgos Kallis (eds.). 2015. *Degrowth: A Vocabulary for a New Era*. London: Routledge.

Daly, Herman. 1973. *Toward a Steady-State Economy*. Victoria: Abe Books.

———. 1996. *Beyond Growth: The Economics of Sustainable Development*. Boston: Beacon Press.

———. 2004. *Ecological Economics: Principles and Applications*. Washington: Island Press.

Davidson, James Dale, and William Rees-Mogg. 1997. *The Sovereign Individual: How to Survive and Thrive During the Collapse of the Welfare State*. London: Simon & Schuster.

Deeson, Eric. 2007. *Collins Dictionary of Physics*. London: Collins.

Deleuze, Gilles. 1990. *Postscript on the Societies of Control*. New York: Columbia University Press.

———. 1994. *Difference and Repetition*. Translated by Paul Patton. London: Athlone.

Deleuze, Gilles, and Felix Guattari. 1999/1980. *A Thousand Plateaus: Capitalism and Schizophrenia*. Translated by Brian Massumi. London: Athlone Press.

———. 2004. *Anti-Oedipus; Capitalism and Schizophrenia*. Translated by R. Hurley, M. Seem and H. Lane. Preface by Michel Foucault. London: Continuum.

Derrida, Jacques. 1978. *Writing and Difference*. Translated by Alan Bass. Chicago: University of Chicago Press.

———. 1985/1982. *The Ear of the Other: Otobiography, Transference, Translation*. Edited by Christie McDonald. Translated by Peggy Kamuf. Lincoln: University of Nebraska Press.

De Roover, Raymond. 1963. *The Rise and Decline of the Medici Bank, 1397–1494*. Cambridge, MA: Harvard University Press.

De Saussure, Ferdinand. 1966/1916. *Course in General Linguistics*. New York: McGraw Hill.

Descartes, René. 1980/1641. *Discourse on Method and the Mediations*. Translated by F. E. Sutcliffe. Great Britain: Penguin Books.

Desilver, Drew. 2018. 'For Most US Workers, Real Wages Have Barely Budged in Decades'. Pew Research Center. Accessed 19 August 2023. https://www.pewresearch.org/short-reads/2018/08/07/for-most-us-workers-real-wages-havegunbe-barely-budged-for-decades/.

Devine, Nesta. 2004. *Education and Public Choice: A Critical Account of the Invisible Hand in Education*. Westport: Praeger Press.

Diamond, Jared. 2004. *Collapse: How Societies Choose to Fail or Survive*. London: Penguin.

Dicks, H. 2011. 'The Self-Poetizing Earth: Heidegger, Santiago Theory, and Gaia Theory'. *Environmental Philosophy* 8, no. 1: 41–61.

Dolphijn, Rick, and Iris van der Tuin. (2012). *New Materialism: Interviews and Cartographies*. Michigan: Open Humanities Press.

Doorman, Frans. 2015. *Our Money: Towards a New Monetary System*. Lulu Internet Publishers.

Dreyfus, Herbert. 1998. 'Heidegger and Foucault on the Subject, Agency and Practices'. http://ist-socrates.berkeley.edu/~hdreyfus/html/paper_heidandfoucault.html.

Dreyfus, Hubert, and Harrison Hall (eds.). 1992. *Heidegger: A Critical Reader*. Oxford: Blackwell.

Dreyfus, Hubert, and Paul Rabinow (eds.). 1983. *Michel Foucault: Beyond Structuralism and Hermeneutics*, second edition. Chicago: Chicago University Press.

Dunn, Jason, and David Wheelock. 2022. 'National Banking Acts of 1863 and 1864'. https://www.federalreservehistory.org/essays/national-banking-acts.

Dupay, Jean-Pierre. 2011. 'Cybernetics Is an Antihumanism: Advanced Technologies and the Rebellion Against the Human Condition'. *Essay, Transhumanism and its Critics*, 1 September. Accessed 19 June 2023. https://metanexus.net/h-cybernetics-antihumanism-advanced-technologies-and-rebellion-against-human-condition/.

Dyke, James, Robert Watson and Wolfgang Knorr. 2021. 'Climate Scientists Concept of Net Zero Is a Dangerous Trap'. *The Conversation*, 22 April. Accessed 2 October 2021. https://theconversation.com/climate-scientists-concept-of-net-zero-is-a-dangerous-trap-157368.

Dyson, Ben, Graham Hodgson and Andrew Jackson. 2015. *Would a Sovereign Money System be Flexible Enough?* Positive Money. Accessed 13 October 2023. http://positivemoney.org/wp-content/uploads/2015/01/Would_A_Sovereign_Money_System_Be_Flexible_Enough_WEB20140113.pdf.

Dyson, Ben, Graham Hodgson and Frank van Lerven. 2016. *Sovereign Money: An Introduction*. London: Positive Money.

Edmans, Alex. 2020. *Grow the Pie: How Great Companies Deliver Both Purpose and Profit*. Cambridge: Cambridge University Press.

Edmunds, M. G. 2014. 'The Antikythera Mechanism and the Mechanical Universe'. *Contemporary Physics* 55: 263–85. doi:10.1080/00107514.2014.927280.

Edmunds, Mike, and Philip Morgan. 2000. 'The Antikythera Mechanism: Still a Mystery of Greek Astronomy'. *Astronomy and Geophysics* 41, no. 6: 6–10.

Ehrlich, Paul. 1971/1968. *The Population Bomb*. New York: Ballantine Books.

Einstein, Alfred, and H. Minkowski. 1920. *The Principle of Relativity: Original Papers*. Translated by M.N. Saha and S. N. Bose. Calcutta: University of Calcutta.

Escobar, Arturo. 1995. *Encountering Development: The Making and Unmaking the Third World*. Princeton, NJ: Princeton University Press.

Evans, Greg. 2010. 'Kurzarbeit: 'Living Dead Capitalism' and the Future of the New Left'. *Focus on the Czech Republic*. http://czechfocus.cz/art/54238.html.

Evans, Peter. 2021. 'Is Reality a Game of Quantum Mirrors? A New Theory Suggests It Might Be'. *The Conversation*, 28 June. Accessed 31 December 2021. https://theconversation.com/is-reality-a-game-of-quantum-mirrors-a-new-theory-suggests-it-might-be-162936.

Everth, Thomas, and Laura Gurney. 2022. 'Emergent Realities: Diffracting Barad within a Quantum-Realist Ontology of Matter and Politics'. *European Journal for Philosophy of Science 12, no.* 3: 1–20.

Farnes, Jamie S. 2018. 'A Unifying Theory of Dark Energy and Dark Matter: Negative Masses and Matter Creation within a Modified LamdaCDM Framework'. *Astronomy and Astrophysics* 620, no. A92: https://doi.org/10.1051/0004-6361/20832898.

Farquhar, Brodie. 2021. 'Wolf Reintroduction Changes Ecosystem in Yellowstone'. *Outside*.

Faye, Jan, and Henry Folse (eds.). 2017. *Niels Bohr and the Philosophy of Physics: Twenty-First-Century Perspectives*. London: Bloomsbury.

Federal Reserve. 2019. *Reserve Maintenance Manual*. Board of Governors of the Federal Reserve System. https://www.federalreserve.gov/monetarypolicy/files/reserve-maintenance-manual.pdf.

Feenberg, Andrew. 1996. *Alternative Modernity: The Technical Turn in Philosophy and Social Theory*. Berkeley: California Press.

Guillen, Fernandez. 2000. *Fire and Memory: On Architecture and Energy*. Translated by Gina Carino. Athens: MIT Press.

Fisher, Irving. 1936. *100% Money*, revised second edition. New York: Adelphi Company.

Foster, John Bellamy. 2000. *Marx's Ecology: Materialism and Nature*. New York: Monthly Review Press.

Foucault, Michel. 1979. 'On Governmentality'. *Ideology and Consciousness* 6: 5–26.

———. 1980. *Power/Knowledge: Selected Interviews and Other Writings 1972–1977*. Edited by Colin Gordon. East Sussex, UK: BrightonHarvester Press.

———. 1990. *The Care of the Self; History of Sexuality*, vol. 3. Middlesex, NJ: Penguin.

———. 1991/1977. *Discipline and Punish: The Birth of the Prison*. Translated by Alan Sheridan. Middlesex: Penguin.

———. 1998. *Aesthetics, Method, and Epistemology*. Edited by James Faubion. Translated by Robert Hurley. New York: The New Press.

———. 2010. *The Birth of Biopolitics: Lectures at the College de France, 1978–1979*. London: Picador.

Freeth, Tony. 2002a. 'The Antikythera Mechanism: 1. Challenging the Classic Research'. *Mediterranean Archeology and Archeaometry* 2, no. 1: 21–35.

———. 2002b. 'The Antikythera Mechanism: Is it Posidonius' Orrery?'. *Mediterranean Archeology and Archeaometry* 2, no. 2: 45–58.

———. 2014. 'Eclipse Prediction on the Ancient Greek Astronomical Calculating Machine Known as the Antikythera Mechanism'. *Plos One*, 30 July. https://doi.org/10.1371/journal.pone.0103275.

Freeth, Tony, and Alexander Jones. 2012. 'The Cosmos in the Antikythera Mechanism'. ISAW papers, February. New York University Library. http://doi.org/2333.1/xgxd26r7.

Freeth, Tony, David Higgon, Aris Dacanalis, Lindsay MacDonald, Myrto Georgakopoulou and Adam Wojcik. 2021. 'A Model of the Cosmos in the ancient Greek Antikythera Mechanism'. *Scientific Reports* 11: article 5821.

Freeth, Tony, Yanis Bitsakis, Xenophon Moussas, John Seiradakis, et al. 2006. 'Decoding the Ancient Greek Astronomical Calculator Known as the Antikythera Mechanism'. *Nature* 444, no. 7119: 587–91. doi:10.1038/nature05357.

Friedman, Milton. 1970. 'A Friedman Doctrine: The Social Responsibility of Business to Increase its Profits'. *New York Times Magazine*, 13 September. https://www.nytimes.com/1970/09/13/archives/a-friedman-doctrine-the-social-responsibility-of-business-is-to.html.

Friend, Tad. 2016. 'Sam Altman's Manifest Destiny'. *New Yorker Magazine*, 3 October. https://www.newyorker.com/magazine/2016/10/10/sam-altmans-manifest-destiny.

Fukuyama, Francis. 1992. *The End of History and the Last Man*. Los Angeles, CA: Free Press.

Galbraith, John Kenneth. 1975. *Money Whence it Came – Where it Went*. London: Houghton Mifflin Co.

Galileo. 1632. *Dialogue Concerning the Two Chief World Systems*.

Gaspar, Vitor, Marcos Poplawski-Riberiro and Jiae Yoo. 2023. 'Global Debt Is Returning to Its Rising Trend'. International Monetary Fund.

Gaspar, Vitor, Paulo Medas and Roberto Perrelli. 2021. 'Global Debt Reaches a Record 226 Trillion'. International Monetary Fund.

George, Andrew. 2013. 'Babylonian Divinatory Texts: Chiefly in the Schøyen Collection'. *Cornell University Studies in Assyriology and Sumerology* (*CUSAS*) 18: 1–28.

Georgescu-Roegen, Nicholas. 1971. *The Entropy Law and the Economic Process*. Cambridge, MA: Harvard University Press.

———. 1979. 'Energy Analysis and Economic Valuation'. *Southern Economic Journal* 45, no. 4: 1023–58.

———. 1986. 'The Entropy Law and the Economic Process in Retrospect'. *Eastern Economic Journal* 12, no. 1: 3–25.

Gilgameš. 2000. *The Epic of Gilgamesh: A New Translation*. Translated by Andrew George. London: Penguin.

Goldenberg, Shifra, and George Wittemyer. 2020. 'Elephant Behavior toward the Dead: A Review and Insights from Field Observations'. *Primates* 61: 119–28.

Gorz, André. 2009/1971. *Ecologica*. Kolkata, West Begnal: Seagull Books.

Gough, A. 1997. *Education and the Environment: Policy, Trends and the Problems of Marginalisation*. Melbourne: The Australian Council for Educational Research.

Gough, N. 1987. 'Learning with Environments: Towards an Ecological Paradigm of Education'. In *Environmental Education: Practice and Possibility*, edited by Ian Robottom, 49–67. Victoria, Australia: Deakin University Press.

Goulson, Dave. 2019. 'The Insect Apocalypse, and Why It Matters'. *Current Biology* 29, no. 19: R967–71. https://doi.org/10.1016/j.cub.2019.06.069.

Graeber, David. 2011. *Debt, The First 5000 Years*. London: Melville House.

———. 2019. 'Against Economics'. *New York Review of Books*, 5 December.

Greenfield, Patrick. 2023. 'Revealed: More than 90% of Rainforest Carbon Offsets Are Worthless, Analysis Shows'. *The Guardian*, 18 January.

Grosz, Elizabeth. 1994. *Volatile Bodies: Towards a Corporeal Feminism*. Bloomington and Indianopolis: Indiana University Press.

Grünbein, Durs, 2010. *Descartes' Devil: Three Meditations*. New York: Upper West Side Philosophy Inc.

Guignon, Charles. 2001. 'Being as Appearing: Retrieving the Greek Experience of Phusis'. In *A Companion to Heidegger's Introduction to Metaphysics*, edited by Guignon Charles, 34–56. New Haven, CT: Yale University Press.

Guillen, Fernandez. 2000. *Fire and Memory: On Architecture and Energy*. Translated by Gina Carino. Athens: MIT Press.

Haar, Michael. 1993. *Song of the Earth: Heidegger and the Grounds of History of Being*, Translated by Reginald Lilly. Bloomington: Indiana University Press.

Hagans, Nate J. 2020. 'Economics for the Future – Beyond the Superorganism'. *Ecological Economics* 169: 1–16.

Hall, Charles A. S., Cutler J. Cleveland and Robert Kaufmann. 1986. *Energy and Resource Quality: The Ecology of the Economic Process*. Hoboken, NJ: Wiley Interscience.

Hall, Charles Brockway, Jessica Lambert and Stephen Balogh. 2014. 'EROI of Different Fuels and the Implications for Society'. *Energy Policy*, 64, no. C: 141–52.

Hanaček, Ksenija, Brototi Roy, Sofia Avila and Giorgfos Kallis. 2019. 'Ecological Economics and Degrowth: Proposing a Future Research Agenda From the Margins'. *Ecological Economics* 169. DOI: 10.1016/j.ecolecon.2019.106495.

Hansen, James, Makiko Sato, Leon Simons, Larissa Nazarenko, Isabelle Sangha, Karina von Schuckmann, Norman Loeb, Matthew Osman, Qinjian Jin, Pushker Kharecha, George Tselioudis, Eunbi Jeong, Andrew Lacis, Reto Ruedy, Gary Russell, Junjii Cao and Jing Li. 2023. 'Global Warming in the Pipeline'. *Oxford Open Climate Change*. http://www.columbia.edu/~jeh1/Documents/PipelinePaper.2023.07.05.pdf.

Hansen, James, Makiko Sato, Paul Hearty, Reto Ruedy, Maxwell Kelley, Valerie Masson-Delmotte, Gary Russell, George Tselioudis, Junji Cao, Eric Rignot, Isabella Velicogna, Blair Tormey, Bailey Donovan, Evgeniya Kandiano, Karia von Schuckmann, Pushker Kharecha, Allegra Legrande, Michael Bauer and Kwok-Wai Lo. 2016. 'Ice Melt, Sea Level Rise and Superstorms Evidence from Paleoclimate Data, Climate Modelling, and Modern Observations that 2C Global Warming could be Dangerous'. *Atmospheric Chemistry and Physics* 16: 3761–812.

Haraway, Donna. 1991. 'A Cyborg Manifesto: Science, Technology, and Socialist-Feminism in the Late Twentieth Century'. In *Simians, Cyborgs and Women: The Reinvention of Nature*, by Donna Haraway, 149–81. New York: Routledge.

———. 2016. *Staying with the Trouble: Making Kin in the Cthulucene*. Chicago: Duke University Press.

Hardt, Michael. 1998. 'The Global Society of Control'. *Discourse* 20, no. 3: 139–52.

Harman, Graeme. 2010. *Towards Speculative Realism: Essays and Lectures*. Winchester: Zero Books.

Hausnost, Daniel, Ernst Schriefl, Christian Lauk amd Gerald Kalt. 2017. 'A Transition to which Bioeconomy? An Exploration of Diverging Techno-Political Choices'. *Sustainability* 9, no. 4: 669.

Hayek, Friedrich. 1960. *The Constitution of Liberty*. Chicago and London: Chicago Press.

Heidegger, Martin. (unpublished) 'A Letter to Ernst Jünger'. 18 December. 1950 facsimile at University of Freiburg.

———. 1949. *Existence and Being*. London: Vision Press.

———. 1956. *What is Philosophy*. Lanham: Rowman & Littlefield.

———. 1962/1927. *Being and Time*. London and Southampton: The Camelot Press.

———. 1968a/1954. *What is Called Thinking?* Translated by J. Glenn Gray. New York: Harper & Row.
———. 1968b. *What Is a Thing?* Lanham, MD: Book Network.
———. 1969a. *The Essence of Reasons.* Translated by T. Malick Evanston, IL: Northwestern University Press.
———. 1969b/1957. *Identity and Difference.* Translated by Joan Stambaugh. New York: Harper and Row.
———. 1970. *Discourse on Thinking.* New York: Harper & Row.
———. 1973a/1953. *An Introduction to Metaphysics.* Translated by Ralph Manheim in 1959. Written in 1935. New Haven, CT: Yale University Press.
———. 1973b/1954. *The End of Philosophy.* Translated by Joan Stambaugh. New York: Harper & Row.
———. 1975a. *Early Greek Thinking.* New York: Harper & Row.
———. 1975b. *Poetry, Language, Thought.* Translated by Albert Hofstadter. New York: Harper & Row.
———. 1977/1954. *The Question Concerning Technology and Other Essays.* Translated by William Lovitt. New York: Harper & Row.
———. 1979. *Holzwege.* Frankfurt am Main: Vittorio Klostermann.
———. 1982a. *Nietzsche*, vols. 1–4. Edited by Farrell Krell. Translated by Frank A. Capuzzi. San Francisco: Harper & Row.
———. 1982b. *On the Way to Language.* New York: Harper Collins.
———. 1985. *History of the Concept of Time, Prolegomena.* Written in 1924. Bloomington: Indiana University Press.
———. 1989/1976. *Neuzeitliche Naturwissenschaft und Moderne Technik.* Chicago: Chicago Colloquium.
———. 1990/1966. 'Nur noch ein Gott kann uns retten: Der Spiegal Interview with Martin Heidegger'. Interviewed by Rudolf Augstein and Georg Wolff. In *Martin Heidegger and National Socialism: Questions and Answers*, edited by Günter Neske and Emil Kettering, translated by Lisa Harries, 41–66. New York: Paragon House.
———. 1993. *Basic Writings.* Edited by David Farrell Krell. Oxford: Routledge.
———. 1996a. *Hölderlin's Hymn 'The Ister'.* Translated by William McNeill and Julia Davis. Bloomington: Indiana University Press.
———. 1996b. *The Principle of Reason.* Translated by Reginald Lilly. Bloomington: Indiana University Press.
———. 1996c. *Bemerkungen zu Kunst – Plastik – Raum.* Erker-Verlag St.: Gallen.
———. 1996d. *Erläuterung zu Hölderlins' Dichtung* Klostermann.
———. 1999a/1955. 'On the Question of Being'. In *Pathmarks*, edited by William McNeill, translated by Frank Capuzzi, 291–322. Cambridge: Cambridge University Press.
———. 1999b. *Contributions to Philosophy (From Enowning).* Bloomington: Indiana University Press.
———. 1999c. 'Letter on Humanism'. In *Pathmarks*, edited by William McNeill, translated by Frank Capuzzi, 239–76. Cambridge: Cambridge University Press.

———. 1999d/1939. ‘On the Essence and Concept of *Physis* in Aristotle’s *Physics* B’. In *Pathmarks*, edited by William McNeill, translated by Frank Capuzzi, 183–230. Cambridge: Cambridge University Press.

———. 2012. ‘Bremen and Feinberg Lectures’. Translated by Andrew J. Mitchell. Bloomington: Indiana University Press.

Heidegger, Martin, and Ernst Jünger. 2016. *Correspondence 1949–1975*. Translated by Timothy Quinn. London: Rowman & Littlefield.

Heidegger, Martin, and Eugen Fink. 1979. *Heraclitus Seminar 1966–1967*. Translated Charles H. Seibert. Tuscaloosa, AL: University of Alabama Press.

Hermans, Cornelis Andreas Marie. 2017. ‘Ernst Jünger’s World of Fire’. In *Interbellum Literature: Writing in a Season of Nihilism*, edited by Cornelia Hermans, chapter 19. Leiden: Brill.

Hickel, Jason. 2016. ‘Clean Energy Won’t Save Us – Only a New Economic System Can’. *The Guardian*, 15 July. https://www.theguardian.com/global-development-professionals-network/2016/jul/15/clean-energy-wont-save-us-economic-system-can.

———. 2018. ‘The Nobel Prize for Climate Catastrophe’. *Foreign Policy*, 6 December.

———. 2019. ‘The Contradiction of the Sustainable Development Goals: Growth versus Ecology on a Finite Planet’. *Sustainable Development* 27, no. 5: 873–84.

———. 2020. *Less Is More: How Degrowth will Save the World*. London: Penguin.

Higgs, Kerryn. 2014. *Collision Course: Endless Growth on a Finite Planet*. Cambridge: MIT Press.

———. 2023. ‘How Sustainable Are the UN Sustainable Development Goals?’. In *Sustainability and the New Economics*, edited by S. Williams and R. Taylor, 69–86. Cham, Switzerland: Springer.

Hodgskiss, Malcolm, and Erik Sperling. 2022. ‘A Prolonged, Two-Step Oxygenation of Earth’s Early Atmosphere: Support from Confidence Intervals’. *Geology* 50, no. 2: 158–62.

Hoekstra, Arjen, and Thomas Wiedmann. 2014. ‘Humanity’s Unsustainable Environmental Footprint’. *Science* 34, no. 6188: 1114–17.

Holzmann, Kim, Ramona Walls and John Wiens. 2023. ‘Accelerating Local Extinction Associated with Very Recent Climate Change’. *Ecology Letters*, September.

Hudson, Michael. 1993. *The Lost Tradition of Biblical Debt Cancellations*. New York: Henry George School of Social Science.

———. 2002a. ‘The New Economic Archaeology of Debt’. Michaels Hudson website, 23 April. http://michael-hudson.com/2002/04/the-new-economic-archaeology-of-debt/.

———. 2002b. ‘Reconstructing the Origins of Interest Bearing Debt’. In *International Scholars Conference on Ancient Eastern Economics*, vol. 3. Institute for the Study of Long-Term Economic Trends.

Hui, Yuk. 2017a. ‘On the Unhappy Consciousness of Neoreactionaries’. *e-flux Journal*, no. 81: https://www.e-flux.com/journal/81/125815/on-the-unhappy-consciousness-of-neoreactionaries/.

———. 2017b. ‘Cosmotechnics as Cosmopolitics’. *e-flux Journal*, no. 86: https://www.e-flux.com/journal/86/161887/cosmotechnics-as-cosmopolitics/.

Hume, David. 1985/1739. *A Treatise of Human Nature*. Edited by L. A. Selby-Brigge. Oxford: Clarendon Press.

———. 1985/1777. *Enquiries Concerning Human Understanding and Concerning the Principles of Morals*. Oxford: Clarendon Press.

Humphreys, Chloe, Sean Blenkinsop and Bob Jickling. 2022. 'Education, Sustainable or Otherwise, as Simulacra: A Symphony of Baudrillard'. *Educational Philosophy and Theory* 54, no. 3: 310–23.

International Energy Agency (IEA). 2019. 'Global Energy Review: CO2 Emissions in 2021. Global emissions rebound sharply to highest ever level' International Energy Agency.

———. 2021. 'Net Zero Roadmap: A Global Pathway to Keep the 1.5°C Goal in Reach'. Accessed 13 October 2023. https://www.iea.org/reports/net-zero-roadmap-a-global-pathway-to-keep-the-15-0c-goal-in-reach.

———. 2023. *World Energy Outlook 2023*.

International Monetary Fund. 2022. *World Economic Outlook*. Accessed 21 May 2024. https://www.imf.org/en/Publications/WEO

Ioris, Antonio A. R. 2022. 'Say Agribusiness but Mean Genocide: Grabbing the Guarani-Kaiowa World'. In *Agriculture, Environment and Development: International Perspectives on Water, Land and Politics*, second edition, edited Antonio Ioris and Bernardo Fernandes, 181–205. Cham: Palgrave Macmillan.

Intergovernmental Panel on Climate Change (IPCC). 2013. *Climate Change 2013: The Physical Science Basis. Contribution of Working Group I to the Fifth Assessment Report of the Intergovernmental Panel on Climate Change*. Edited by T. F. Stocker, D. Qin, G.-K. Plattner, M. Tignor, S. K. Allen, J. Boschung, A. Nauels, Y. Xia, V. Bex and P. M. Midgley. Cambridge, UK: Cambridge University Press.

———. 2018. *Global Warming of 1.5°C. An IPCC Special Report on the Impacts of Global Warming of 1.5°C above Pre-Industrial Levels and Related Global Greenhouse Gas Emission Pathways, in the Context of Strengthening the Global Response to the Threat of Climate Change, Sustainable Development, and Efforts to Eradicate Poverty*. Edited by V. Masson-Delmotte, P. Zhai, H. -O. Pörtner, D. Roberts, J. Skea, P. R. Shukla, A. Pirani, W. Moufouma-Okia, C. Péan, R. Pidcock, S. Connors, J. B. R. Matthews, Y. Chen, X. Zhou, M. I. Gomis, E. Lonnoy, T. Maycock, M. Tignor and T. Waterfield.

———. 2021. *Climate Change 2021: The Physical Science Basis. Contribution of Working Group I to the Sixth Assessment Report of the Intergovernmental Panel on Climate Change*. Edited by V. Masson-Delmotte, P. Zhai, A. Pirani, S. L. Connors, C. Péan, S. Berger, N. Caud, Y. Chen, L. Goldfarb, M. I. Gomis, M. Huang, K. Leitzell, E. Lonnoy, J. B. R. Matthews, T. K. Maycock, T. Waterfield, O. Yelekçi, R. Yu and B. Zhou. Cambridge, UK: Cambridge University Press.

Intergovernmental Science-Policy Platform on Biodiversity and Ecosystem Services (IPBES). 2019. *Global Assessment Report on Biodiversity and Ecosystem Services*. Edited by E. S. Brondizio, J. Settele, S. Díaz and H. T. Ngo. IPBES Secretariat. Accessed 21 April 2023. https://www.ipbes.net/global-assessment.

Irigaray, Luce. 1985. *This Sex Which is Not One*. Ithaca, NY: Cornell University Press.

Irwin, Geoffrey. 1992. *The Prehistoric Exploration and Colonization of the Pacific*. Cambridge: Cambridge University Press.

Irwin, Geoffrey, and Richard Flay. 2015. 'Pacific Colonisation and Canoe Performance: Experiments in the Science of Sailing'. *The Journal of the Polynesian Society* 124, no. 4: 419–43.

Irwin, Ruth. 2001. 'Nietzsche: Deleuze, Foucault and Genealogy as a Method for Education'. In *Nietzsche's Legacy for Education*; Past and Present Values, edited by M. Peters, J. Marshall and P. Smeyers, 35–56. Westport: Bergin & Garvey.

———. 2003. 'Heidegger and Nietzsche: Nihilism and the Question of Value in relation to Education'. *Studies in Philosophy of Education* 22: 227–44..

———. 2005. 'The Technology Environment: Subjectivity, Language, and Machine'. *International Journal of Technology and Design Education,* Kluwer 15, no. 2: 99–110.

———. 2008a. *Heidegger, Politics and Climate Change*. London: Continuum.

———. 2008b. 'Education for Sustainable Development versus Environmental Education: Emergence of the New Trope for Neoliberal Consumerism'. In *Handbook for Environmental Education*, edited by Michael Peters and Edgar Gonzalez-Gaudiano, 171–94. London: Sense.

———. (ed.). 2010a. *Philosophy and Climate* Change. London: Continuum.

———. 2010b. 'Reflections on Modern Climate Change and Finitude'. In *Climate Change and Philosophy: Transformational Possibilities*, edited by Ruth Irwin, 48–72. London: Sense.

———. 2010c. 'Climate Change and Heidegger's Philosophy of Science'. *Essays in Philosophy* 11, no. 1: 16–30.

———. 2013. 'Does a Failure in Global Leadership Mean It's All Over? Climate, Population, and Progress'. In *Environmental Ethics*, second edition, edited by Michael Boylan, 243–56. Hoboken, NJ: Wiley-Blackwell.

———. 2014. 'Neoliberalism'. In *Beyond the Free Market; Rebuilding a Just Society in New Zealand*, edited by David Cooke, Claire Hill, Pat Baskett, and Ruth Irwin. Auckland, New Zealand: Dunmore Press.

———. 2015a. 'Ecological Ethics in the Context of Climate Change: Feminist and Indigenous Critique of Modernity'. *International Social Science Journal* 64, nos. 211–12: 1–13.

———. 2015b. 'Heidegger, and the Significance of Poetics'. *Policy Futures in Education* 13, no. 1: 57–69.

———. 2017. 'Climate Change and Educational Futures'. *International Studies in Japan: International Yearbook*, no. 11. Japanese Educational Research Association.

———. 2018. 'Rethinking Economics and Education: Exponential Growth and Post-Growth Strategies'. *Educational Theory* 67, no. 4: 379–98. https://doi.org/10.1111/edth.12256.

———. 2019a. 'Heidegger and Stiegler on Failure and Technology'. *Educational Philosophy and Theory* 52, no. 4: 1–15. DOI:10.1080/00131857.2019.1654855.

———. 2019b. 'Climate Change and Education'. *Educational Philosophy and Theory* 52, no. 4: 492–507. https://doi.org/10.1080/00131857.2019.1642196.

———. 2022a. 'Idealist Individualism or Indigenous Cosmology: Finding Entanglement Across Species and Strata'. *Journal of Religions* 13, no. 12: 1193. https://doi.org/10.3390/rel13121193.

———. 2022b. 'Can Education Outgrow the Rhetoric of "Development" Embedded in the UN Sustainability Goals?' *New Zealand Journal of Educational Studies* 58, no. 5: DOI:10.1007/s40841-022-00270-3.

———. 2023b. 'Disaster; Public Health and Climate Change'. In *Climate Change and Global Health*, third edition, edited by Colin Butler and Kerryn Higgs CABI.

———. 2023c. 'Physics, Feminism and Whakapapa: Integrating Eco-Subjectivity after the Enlightenment'. In *Philosophy as Practice in the Ecological Emergency*, Sustainable Development Goal Series, edited by Lucy Weir, n.p. London: Palgrave Macmillan. https://doi.org/10.1007/978-3-030-94391-2_3.

Irwin, Ruth, and Te Haumoana White. 2019. 'Decolonising Technological Futures: A Dialogical Tryptich between Te Haumoana White, Ruth Irwin, and Tegmark's Artificial Intelligence'. *Futures* 112: https://doi.org/10.1016/j.futures.2019.06.003.

———. 2022. 'Negentropy and the Accelerating Anthropocene; Stiegler, Māori, and Exosomatic Memory'. In *Educational Philosophy and Theory*, Joff Bradley (ed.) Special edition in memory of Bernhard Stiegler, pp. 532–44. https://doi.org/10.1080/00131857.2021.1970525

Jackson, Andrew, and Ben Dyson. 2013. *Modernising Money: Why Our Monetary System Is Broken and How It Can Be Fixed.* London: Positive Money.

Jackson, Tim. 2017. *Prosperity without Growth.* London: Routledge.

James, Steven R. 1989. 'Hominid Use of Fire in the Lower and Middle Pleistocene: A Review of the Evidence'. *Current Anthropology* 30, no. 1: 1–26.

Jones, Alexander. 2020. 'The Epoch Dates of the Antikythera Mechanism'. Institute for the Study of the Ancient World (ISAW) Papers 17: http://hdl.handle.net/2333.1/ffbg7m07.

Joy, Mike. 2022. 'Changing Freshwater Management in New Zealand: Looking to the Future We Must Look Back'. *Policy Quarterly* 18, no. 2: 18–22.

Jünger, Ernst. (1947/1939). *On the Marble Cliffs.* Translated by Stuart Hood. Surrey, UK: Guildford and Esher.

———. 1993/1930. 'Total Mobilization'. In *The Heidegger Controversy: A Critical Reader*, edited by Richard Wolin, 119–39. Cambridge, MA: London: MIT Press.

Junod, Tom. 2012. 'Elon Musk: Triumph of His Will'. *Esquire*, 14 November. Accessed 25 August 2023 at https://www.esquire.com/news-politics/a16681/elon-musk-interview-1212/.

Kallis, Giogos. 2018. *Degrowth: The Economy Key Ideas.* London: Agenda Publishing.

Kant, Immanuel. 1992/1781. *Critique of Pure Reason.* Translated by Norman Kemp Smith. London: Macmillan Press.

Kaplan, Jonathan. 2019. 'The Credibility of Liberty: The Plausibility of the Jubilee Legislation of Leviticus 25 in Ancient Israel and Judah'. *The Catholic Biblical Quarterly* 81, no. 2: 183–203.

Keen, Steven. 2018. *Debunking Economics: The Naked Emperor Dethroned?* London: Zed Books.

Kelly, Mark. 2015. 'Discipline Is Control: Foucault Contra Deleuze'. *New Formations* 84: 148–62. https://doi.org/doi:10.3898/NewF:84/85.07.2015.

Kent, Christopher. 2018. 'Money – Born of Credit?' Reserve Bank's Topical Talks Event for Educators. https://www.rba.gov.au/speeches/2018/sp-ag-2018-09-19.html.

Kepler, Johannes. 1995/1621. *Epitome of Copernican Astronomy and Harmonies of the World.* Translated by Charles Glenn Wallis. Amherst, MA: Prometheus Books.

Keyβer, Lorenz, and Manfred Lenzen. 2021. '1.5°C Degrowth Scenarios Suggest the Need for New Mitigation Pathways'. *Nature Communications* 12: article 2676. https://www.nature.com/articles/s41467-021-22884-9.

Keynes, Maynard. 1933. *The Means to Prosperity*. London: MacMillan.

———. (1991/1936). *The General Theory of Employment, Interest, and Money: A Concordance*. London: Rowman & Littlefield.

Klein, Naomi. 2007. *Shock Doctrine*. New York: Penguin

Kohler, Alan. 2023. 'Australia's New Emissions Trading Scheme Is a Carbon Tax'. *The Daily News*, 30 January. https://thenewdaily.com.au/finance/2023/01/30/emissions-labor-carbon-tax-kohler/.

Kohn, Eduardo. 2007. 'How Dogs Dream: Amazonian Natures and the Politics of Transspecies Engagement'. *American Ethnologist* 34, no. 1: 3–25.

———. 2013. *How Forests Think: Toward an Anthropology Beyond the Human.* London: University of California Press.

———. 2017. 'Thinking with Forests'. In *Compatative Metaphysics: Ontology After Anthropology*, edited by Pierre Charbonnier, Gildas Salmon and Peter Skafish, 181–200. Lanham, MD: Rowman & Littlefield.

Koko. 2023. 'Koko Is the Voice of Nature'. YouTube video posted on Kids4Koko website. 01:08. https://www.koko.org/kids4koko/sign-like-koko/videos-voice-of-nature/.

Kondepudi, Dilip, and Ilya Prigogine. 2014. *Modern Thermodynamics; From Heat Engines to Dissipative Structures*, second edition. Hoboken: Wiley.

Koops, Kathelijne, Takeshi Furuichi and Chie Hashimoto. 2015. 'Chimpanzees and Bonobos Differ in Intrinsic Motivation for Tool Use'. *Scientific Reports* 5: article 11356.

Kristeva, Julia. 1982. *Powers of Horror: An Essay on Abjection.* Translated by Leon S. Roudiez. New York: Columbia University Press.

Kyoto Protocol to the United Nations Framework Convention on Climate Change. 1998. Kyoto: UN. Accessed 11 August 2023. https://unfccc.int/resource/docs/convkp/kpeng.pdf.

Lane, Philip R. 2021. 'Structual Change and Central Banking: Some Research Priorities' Fifth Joint Regional Financing Arrangements Research Seminar'. Frankfurt: European Stability Mechanism.

Lazar, Nomi. 2019. *Out of Joint: Power, Crisis, and the Rhetoric of Time.* New Haven, CT: Yale University Press.

Lea, Robert. 2023. 'Doomsday Glacier Melting in Antarctica Means Terrible News for Global Sea Level Rise'. *Space*, 17 February. Accessed 19 August 2023. https://www.space.com/antarctica-doomsday-glacier-melts-in-treacherous-ways.

Leather, Kay, and Richard Hall. 2004. *Work of the Gods: Tatai Arorangi* [Māori astronomy]. Wellington: Seven Seas Ltd.

Leff, Enrique. 2009. 'Degrowth, or Deconstruction of the Economy: Towards a Sustainable World'. *Critical Currents*, no. 6: 101–8.

———. 2010. 'Latin American Environmental Thought: A Heritage of Knowledge for Sustainability'. ISEE Publicación Ocasional, no. 9: 1–16.

———. 2015. 'The Power-Full Distribution of Knowledge in Political Ecology: A View From the South', In *The Routledge Handbook of Political Ecology*, edited by T. Perreault, G. Bridge and J. McCarthy, 64–75. New York: Routledge.

Lemche, Niels. 1979. 'Andurārum and Mīšarum: Comments of the Problem of Social Edicts and Their Application in the Ancient near East'. *Journal of Near Eastern Studies* 38, no. 1: 11–22.

Lendering, Jona. 2020/2000. 'Alexander 2.9'. Livius.org. https://www.livius.org/articles/person/alexander-the-great/alexander-2.9/.

———. 2020/2003. 'Kiddinnu, the Chaldaeans, and Babylonian Astronomy'. Livius .org. https://www.livius.org/articles/person/kidinnu-the-chaldaeans-and-babylonian-astronomy/.

Le Quéré, Corinne, Robert B. Jackson, Matthew W. Jones, Adam J. P. Smith, Sam Abernathy, Robbie M. Andrew, Anthoy De-Gol, David R. Willis, Yuli Shan, Josep G. Canadell, Pierre Friedlingstein, Felix Crutzig ad Glen P. Peters. 2020. 'Temporary Reduction in Daily Global CO2 Emissions During the COVID-19 Forced Confinement'. *Nature Climate Change* 10: 647–53.

Leroi-Gourhan, André. 1945. *Milieu et techniques*. Paris: Albin Michel.

———. 1993/1943. *Gesture and Speech*. Cambridge, MA: MIT Press.

Lewis, David. 1972. *We, the Navigators; The Ancient Art of Landfinding in the Pacific*. Honolulu: Honolulu University of Hawai'i Press.

Leveillee, Nicholas. 2011. 'Copernicus, Galileo, and the Church: Science in a Religious World'. *Inquiries; Social Sciences, Arts, and Humanities* 3, no. 5: 1–2.

Lindberg, Eric. 2017. 'Economic Growth – A Primer'. Resilience.org. Accessed 20 August 2023. https://www.resilience.org/stories/2017-02-22/economic-growth-a-primer/#_ednref16.

Locke, John. 2003. *Two Treatises of Government and a Letter Concerning Toleration*. New Haven, CT: Yale University Press.

Lotka, A. J. 1925. *Elements of Physical Biology,* Baltimore: William & Wilkins Company.

———. 1945. 'The Law of Evolution as a Maximal Principle'. *Human Biology* 88: 188.

Ma, Lin. 2008. *Heidegger on East-West: Anticipating the Event*. New York: Routledge.

Magubane, Nathi. 2023. 'The Hidden Costs of AI: Impending Energy and Resource Strain: Deep Jariwala and Banjamin C. Lee on the Energy and Resource Problems AI Computing Could Bring. Penn State University'. *Penn Today*, 8 March. Accessed 20 August 2023. https://penntoday.upenn.edu/news/hidden-costs-ai-impending-energy-and-resource-strain.

Māhina, Hūfanga 'Okusitino. 2010. 'Tā, Vā, and Moana: Temporality, Spatiality, and indigeneity'. *Pacific Studies* 33, nos. 2/3: 168–202

Mahuika, Nēpia. 2019. 'A Brief History of Whakapapa: Maori Approaches to Genealogy'. *Genealogy* 3, no. 2: 32.

Maldacena, Juan, Andrew Strominger and Edward Witten. 1997. 'Black Hole Entropy in M-Theory'. *Journal of High Energy Physics* 2: https://iopscience.iop.org/article/10.1088/1126-6708/1997/12/002/pdf.

Malthus, Thomas. 1798. 'An Essay on the Principle of Population'. London: Johnston Press.

Manighetti, Barbara. 2001. 'Ocean Circulation: The Planet's Great Heat Engine'. Wellington: National Institute of Water and Atmospheric Research (NIWA), 1 December. Accessed 20 July 2023. https://niwa.co.nz/publications/water-and-atmosphere/vol9-no4-december-2001/ocean-circulation-the-planets-great-heat-engine.

Marchant, Jo. 2006. 'In Search of Lost Time'. *Nature* 444, 30 November. http://www.readcube.com/articles/10.1038/444534a.

Mardirossian, Nora, and Jack Arnold. 2023. 'Nature-Based Insetting: A Harmful Distraction from Corporate Decarbonization' *Advancing LandBased Investment Governance,* Columbia Center on Sustainable Investment. https://ccsi.columbia.edu/sites/default/files/content/docs/publications/align-insetting-harmful-distraction.pdf.

Marks, Gregory. 2019. Metabolic Monstrosities: Vampire Capital in the Anthropocene'. In *The Wasted World*, 14 December. https://thewastedworld.wordpress.com/2019/12/14/metabolic-monstrosities/?fbclid=IwAR3zzi_59Ghj1PrKrLDL8haawlz-b1MMx9Udpb0bEgCzaJ3wonAPqS8e6LE.

Marsden, Maori. 2002. *The Woven Universe: Selected Writings of Rev. Maori Marsden*, edited by Charles Royal, manuscript New Zealand National Library.

Marshall, James (ed.). 1994. *Revisiting the 'Reforms' in Education.* Faculty of Education: University of Auckland.

———. 1996a. *Michel Foucault: Personal Autonomy and Education.* Netherlands: Kluwer Academic Publishers.

———. 1996b. 'The Autonomous Chooser and 'Reforms' in Education' *Studies in Philosophy and Education* 15: 8996.

———. 'Foucault and Neo-liberalism: Biopower and Busno-Power'. In *Philosophy of Education*, edited by Alven Neiman, 320–29. Philosophy of Education Society: University of Urbana-Champagne.

Martinez-Alier, Joan. 2012. 'Reflections' Interviewed by Lorenzo Pellegrini', *Development and Change* 43, no. 1: 341–59. https://onlinelibrary.wiley.com/doi/pdf/10.1111/j.1467-7660.2012.01759.x.

Martinez-Alier, Juan, and Klaus Schlupmann. 1987. *Ecological Economics: Energy, Environment and Society.* Hobokrn, NJ: Blackwell Publishers.

Marx, Karl. 1887/1867. *Capital.* Edited by Frederick Engels, translated by Samuel Moore and Edward Aveling. Moscow: Progress Publishers.

———. 1959/1932. *Economic and Philosophical Manuscripts.* Translated by Martin Mulligan, Moscow: Progress Publishers. Written in 1844. Transcribed by Andy Blunden for Marxist.org. Accessed 17 September 2017. http://www.marxists.org/archive/marx/works/1844/manuscripts/preface.htm.

Masini, Fabio. 2021. 'William Nordhaus: A disputable Nobel [Prize]? Externalities, Climate Change and Governmental Action'. *The European Journal of the History of Economic Thought* 28, no. 6: 985–1004.

Massumi, Brian. 2018. *99 Theses on the Revaluation of Value: A Postcapitalist Manifesto*. London: University of Minnesota Press.

Matikainen, Sini, Emanuele Campiglio and Dimitri Zenghelis. 2017. 'The Climate Impact of Quantitative Easing'. Policy Paper. London: Grantham Institute.

Mauss, Marcel. 1954. *The Gift: The Form and Reason for Exchange in Archaic Societies*. Translated by W. D. Halls. New York: Norton Books.

Mayumi, Kozo. 2001. *The Origins of Ecological Economics: the Bioeconomics of Georgescu-Roegen*. New York: Routledge.

McLeay, Michael, Amar Radia and Ryland Thomas. 2014. 'Money Creation in the Modern Economy'. *Bank of England Quarterly Bulletin* Q1. https://www.bankofengland.co.uk/-/media/boe/files/quarterly-bulletin/2014/money-creation-in-the-modern-economy.

McLuhan, Marshall. 1964. *Understanding Media; The Extensions of Man*. New York: McGraw-Hill.

Meade, J. E. 1973. *The Theory of Economic Externalities: The Control of Environmental Pollution and Similar Social Costs*. Leiden: Sijthoff.

Meadows, Donatella, Dennis L. Meadows, Jorgen Randers and William W. Behrens III. 1972. *Limits to Growth*. New York: Club of Rome, Potomac Associates.

Measuring Worth. 2022. 'The Annual RPI and Average Earnings for Britain, 1209 to Present (New Series)'. Accessed 19 August 2023. https://www.measuringworth.com/datasets/ukearncpi/result2.php.

Mengle, David. 2007. 'Credit Derivatives: An Overview'. International Swaps and Derivatives Association. Federal Reserve Bank of Atlanta.

Merchant, Carolyn. 1980. *The Death of Nature: Women, Ecology, and the Scientific Revolution*. San Francisco: Harper Collins.

Mika, Carl. 2014. 'The Enowning of Thought and Whakapapa: Heidegger's Fourfold'. *Review of Contemporary Philosophy* 13: 48–60.

———. 2017. *Indigenous Education and the Metaphysics of Presence: A Worlded Philosophy*. Oxon: Routledge.

———. 2019. 'When 'Water' Meets Its Limits: A Māori Speculation on the Term *Wai*'. *Dutkansearvi* 3, no. 2. Sámi Language and Culture Association.

Millward-Hopkins, Joel, Julia K. Steinberger, Narasimha D. Rao and Yannick Oswald. 2020. 'Providing Decent Living with Minimum Energy: A Global Scenario'. *Global Environmental Change* 65: https://doi.org/10.1016/j.gloenvcha.2020.102168.

Minsky, Hyman. 1992. 'Financial Instability Hypothesis'. In *Handbook of Radical Political Economy*, edited by Philip Arestis, Malcolm Sawyer and Edward Elgar, 153–57. New York: Aldershot.

Mitić, Peter, Aleksandra Fedajev, Magdalena Radulescu and Abdul Rehman. 2022. 'The Relationship between CO Emissions, Economic Growth, Available Energy, and Employment in SEE Countries'. *Environmental Science and Pollution Research* 30: 16140–55.

Montreal Protocol. 2007. 'About Montreal Protocol'. United Nations Environment Programme. https://www.unep.org/ozonaction/who-we-are/about-montreal-protocol.

Myllyvirta, Lauri. 2022a. 'Analysis: China's CO2 Emissions See Longest Sustained Drop in a Decade'. *Carbon Brief*, 30 May. https://www.carbonbrief.org/analysis-chinas-co2-emissions-see-longest-sustained-drop-in-a-decade/.

———. 2022b. 'China's Carbon Dioxide (CO2) Emissions Fell By a Record 8% in the Second Quarter of 2022, a 230m Tonne (MtCO2) Reduction that is the Largest in at Least a Decade'. *Carbon Brief*, 1 September. https://www.carbonbrief.org/analysis-chinas-co2-emissions-fall-by-record-8-in-second-quarter-of-2022/.

Myrskylä, Mikko, Hans-Peter Kohler and Francesco Billari. 2009. 'Advances in Development Teverse Fertility Declines'. *Nature* 460: 741–43.

National Oceanic and Atmospheric Administration (NOAA). 2007. Global Monitoring Laboratory. https://gml.noaa.gov/ccgg/trends/.

Neaman, Elliot. 2019. 'Ernst Jünger and Storms of Steel'. In *Key Thinkers of the Radical Right: Behind the New Threat to Liberal Democracy*, edited by Mark Sedgwick. Oxford: Oxford University Press.

Nelson, Eric. 2023. *Heidegger and Dao: Things, Nothingness, Freedom*. London: Bloomsbury.

Neurath, Otto. 1973/1919. 'Through War Economy to Economy in Kind'. In *Empiricism and Sociology*, edited by R. S. Cohen and M. Neurath. Dordrecht: Reidel.

Newman, Rebecca, and Ilan Noy. (2023). 'The Global Costs of Extreme Weather that are Attributable to Climate Change'. *Nature Communications* 14: article 6103.

Newton. 1687. *Principia*.

New Zealand Ministry of Business, Innovation and Employment (NZ MBIE). 2022. 'Small Business'. https://www.mbie.govt.nz/business-and-employment/business/support-for-business/small-business.

Ngata, Apirana T. 2011/1944. The Porourangi Māori Cultural School, Rauru-nui-a-Toi Course, Lectures 1–7. Gisborne: Māori Purposes Fund Board/Te Rūnanga o Ngāti Porou.

Nietzsche, Friedrich. 1909. *The Will to Power; An Attempted Transvaluation of all Values*, vol. 2, books 3 and 4. Edited by Dr. Oscar Levy and translated by Anthony Ludovici. London: T. N. Foulis.

———. 1974/1887. *The Gay Science*. Translated by Walter Kaufmann. New York: Random House.

———. 1982/1954. *Thus Spoke Zarathustra*. In *Portable Nietzsche*. Edited and translated by Walter Kaufmann. New York: Penguin.

———. 1985/1889/1895. *Twilight of the Idols* and *The Anti-Christ*. Translated by R. J Hollingdale. Middlesex: Penguin.

———. 1986/1880. *Human, All Too Human*. Translated by R. J. Hollingdale. Cambridge: Cambridge University Press.

———. 1989/1887. *On the Genealogy of Morals and Ecce Homo*. Edited by Walter Kaufmann and translated by R. J. Hollingdale. New York: Vintage Books.

Nissen H. J., Peter Damerow and Robert K. Englund. 1993. *Archaic Bookkeeping – Early Writing Techniques of Economic Administration in the Ancient Near East*. Translated by Paul Larsen. Chicago: University of Chicago Press.

NOAA Global Monitoring Laboratory. nd. Accessed 21 May 2024. https://gml.noaa.gov/publications/browse.php

Noé. 2015. 'Koko the Gorilla Is the Voice of Nature at COP21'. Posted 1 December. YouTube video, 1:14. Accessed 8 March 2019. https://www.youtube.com/watch?v=FVuNTiqHys0.

Noll, Markus. 2023. 'Exponential Life-Threatening Rise of the Global Temperature'. *Earth ArXiv*, 23 May. Accessed 18 June 2023. https://eartharxiv.org/repository/view/5420/.

Nonfinancial Reporting Update (NFRD) European Commission. 2019. 'Non-Binding Guidelines on NonFinancial Reporting Update'. *Official Journal of the European Union.* https://eur-lex.europa.eu/legal-content/EN/TXT/PDF/?uri=CELEX:52019XC0620(01)&from=EN.

Nordhaus, William. 2018. 'Projections and Uncertainties about Climate Change in an Era of Minimal Climate Policies'. *American Economic Journal: Economic Policy* 10, no. 3: 333–60.

Nozick, Robert. 1974. *Anarchy, State, and Utopia.* Oxford: Blackwell.

O'Brien, Karen. 2021. *You Matter More than You Think: Quantum Social Change for a Thriving World.* Oslo: cChange Press.

O'Connell, Mark. 2018. 'Why Silicon Valley Billionaires are Prepping for the Apocalypse in New Zealand'. *The Guardian,* 15 February. https://www.theguardian.com/news/2018/feb/15/why-silicon-valley-billionaires-are-prepping-for-the-apocalypse-in-new-zealand.

O'Malley, Vincent, Bruce Stirling and Wally Penetito. 2010. *The Treaty of Waitangi Companion: Māori and Pākehā from Tasman to Today.* Auckland, NZ: Auckland University Press.

Onofrei, Mihaela, Anca Florentina Vatamanu and Elena Cigu. 2022. 'The Relationship Between Economic Growth and CO2 Emissions in EU Countries: A Cointegration Analysis'. *Frontiers in Environmental Science* 10: article 934885. DOI:10.3389/fenvs.2022.934885.

Oppenheim, A. Leo. 1967. *Letters from Mesopotamia: Official, Business, and Private Letters on Clay Tablets from Two Millennia.* Chicago and London: Chicago University Press.

Oustry, Antoine, Erkan Bunyamin, Romain Svartzman and Pierre-Francois Weber. 2020. 'Climate-Related Risks and Central Banks' Collateral Policy: A Methodological Experiment'. Banque de France Working Paper no. 790. https://papers.ssrn.com/sol3/papers.cfm?abstract_id=3771299.

Oxford University Global Warming Index. (n.d.) Accessed 28 May 2024 https://www.globalwarmingindex.org.

Paine, Chris. (Director). 2006. *Who Killed the Electric Car?* Documentary film. Sony.

Paris (Accord) Agreement (COP21). 2016. The Paris Agreement. United Nations Framework Convention on Climate Change (UNFCCC). 4 November. https://unfccc.int/process-and-meetings/the-paris-agreement.

Pascoe, Bruce. 2014. *Dark Emu: Aboriginal Australia and the Birth of Agriculture.* Broome, AUS: Magabala Books.

Pearce, David W., and R. Kerry Turner. 1990. *Economics of Natural Resources and the Environment*. Hertfordshire: Harvester Press.

Peters, Michael. 2001. 'Environmental Education, Neo-liberalism and Globalisation: "The New Zealand experiment"'. *Educational Philosophy and Theory* 33, no. 2: 203–16.

———. 2007. *Knowledge Economy, Development and the Future of Higher Education*. London: Sense Publishers.

Peters, Michael, and James Marshall. 1996. *Individualism and Community: Education and Social Policy in the Postmodern Condition*. London: The Falmer Press.

Peters, Michael, Marshall James and Bruce Parr. 1993. 'The marketisation of tertiary education in New Zealand', *The Australian Universities' Review* 36, no. 2: 34–39.

Piketty, Thomas. 2022. *A Brief History of Equality*. Translated by Rendell Steven. Cambridge, MA: Harvard University Press.

Plato. 1965. *Republic*. Translated by H. D. P. Lee in 1955. Middlesex: Penguin.

Plumwood, Val. 2002. *Environmental Culture: The Ecological Crisis of Reason*. New York: Routledge.

———. 2010. 'Nature in the Active Voice'. In *Climate Change and Philosophy*, edited by Ruth Irwin, chapter 2. London: Bloomsbury.

Polanyi, Karl. 1968. *Primitive, Archaic, and Modern Economics: Essays of Karl Polanyi*. Edited by George Dalton. London: Doubleday Books.

Popper, Karl. 1974. *Objective Knowledge: An Evolutionary Approach*. Oxford: Clarendon Press.

Potsdam Institute. 2024. '38 Trillion Dollars in Damages Each Year: World Economy Already Committed to Income Reduction of 19% Due to Climate Change'. Downloaded 27 May 2024. https://www.pik-potsdam.de/en/news/latest-news/38-trillion-dollars-in-damages-each-year-world-economy-already-committed-to-income-reduction-of-19-due-to-climate-change

Price, Derek J. de Solla. 1959. 'An Ancient Greek Computer'. *Scientific American* 200, no. 6: 60–67.

Prigogine, Ilya. 1982. 'Only an Illusion'. The Tanner Lectures on Human Values, Jawaharlal Nehru University.

Prigogine, Ilya, and Isabelle Stengers. 1984, *Order Out of Chaos: Man's New Dialogue with Nature*. New York: Bantam Books.

Prigogine, Ilya, and René Lefever. 1973. 'Theory of Dissipative Structures'. In, *Synergetics: Cooperative Phenomena in Multi-Component Systems*, edited by H. Haken, 124–35. Frankfurt: Springer.

Ptolemy. c. AD 150. *Almagest*.

Rahmstorf, Stefan. 2006. 'Thermohaline Ocean Circulation'. In *Encyclopedia of Quaternary Sciences*, edited by S. A. Elias. Amsterdam: Elsevier.

———. 2009. 'Sea Level Rise'. Keynote speech at the Climate Change Congress, Copenhagen, Denmark, 10 March, Potsdam Institute for Climate Impact Research.

Rauber, Ramona, Bart Kranstauber and Marta B. Manser. 2020. 'Call Order within Vocal Sequences of Meercats Contain Temporary Contextual and Individual Information'. *BMC Biology* 18: article 119.

Raworth, Kate. 2018. *Doughnut Economics: Seven Ways to Think Like a 21st Century Economist*. London: Cornerstone Books.

Rendahl, Pontus, and Lukas Freund. 2020. 'Banks Do Not Create Money Out of Thin Air'. Research Institute of Economy, Trade and Industry (RIETI). https://www.rieti.go.jp/en/special/p_a_w/148.html.

Reserve Bank of Australia. 1996. 'Statement of the Conduct of Monetary Policy'. 14 August. https://www.rba.gov.au/monetary-policy/framework/stmt-conduct-mp-1-14081996.html.

Riedl, Rupert. 1992. 'Schrödinger's Negentropy Concept and Biology'. In *Erwin Schrödinger's World View*, vol. 16, *Theory and Decision Library*, edited by J. Götschl, 59–69. Dordrecht, Netherlands: Springer. https://doi.org/10.1007/978-94-011-2428-7_8.

Ripple, William J., Christopher Wolf, Thomas M. Newsome, Phoebe Barnard, William R. Moomaw. 2019. 'World Scientists' Warning of a Climate Emergency'. *BioScience* 70, no. 1: 8–12. https://academic.oup.com/bioscience/article/70/1/8/5610806.

Robins, Nick, Simon Dikau and Ulrich Volz. 2021. 'Net-Zero Central Banking: A New Phase in Greening the Financial System'. The London School of Economics and Political Science and the Grantha Research Institute on Climate Change and the Environment, 15 March. https://www.lse.ac.uk/granthaminstitute/wp-content/uploads/2021/03/Net-zero-central-banking-1.pdf.

Rogelj, Joeri, Michael Schaeffer, Pierre Friedlingstein, Nathan Gillett, Detlef van Vuuren, Keywan Riahi, Myles Allen and Reto Knutt. 2016. 'Differences between Carbon Budget Estimates Unravelled'. *Nature: Climate Change* 6: 245–52.

Rosa, H. 2013. *Social Acceleration: A New Theory of Modernity*. J. Trejo-Mathys (trans.). New York: Columbia University Press.

Rothchild, Emma. 1994. 'Adam Smith and the Invisible Hand'. *The American Economic Review* 84, no. 2: 319–22.

Rousseau, Jean-Jacques. 1892/1762. *Emile: A Treatise on Education*. Translated by William Harold Payne. Ann Arbor, MI: D. Appleton Press.

Rushkoff, Douglas. 2018. 'Survival of the Richest: The Wealthy Are Plotting to Leave Us Behind'. *OneZero Medium*, 5 July. https://onezero.medium.com/survival-of-the-richest-9ef6cddd0cc1.

Sachs, Abraham, and Hermann Hunger (eds.). 1988. *Astronomical Diaries and Related Texts from Babylonia*. Wien, AUS: Austrian Academy of Sciences.

Safian, Syuhaida. 2022. 'Bioeconomy as a Political Project: A New Zealand Caser Study'. MSc Thesis, Auckland University, Auckland, NZ.

Sahlins, Marshall. 2017. 'The Original Political Society'. *Hau: Journal of Ethnographic Theory* 7, no. 2: 91–128.

Schloesser, Fabrian, Tobias Friedrich, Axel Timmermann, Robert M. DeConto and David Pollard. 2019. 'Antarctic Iceberg Impacts on Future Southern Hemisphere Climate'. *Nature Climate Change* 9: 672–77.

Schumacher, Eric. 1973. *Small Is Beautiful: Economics as If People Mattered*. New York: Harper Perennial.

Schumpeter, Joseph A. 1954. *History of Economic Analysis*. London: Allen & Union.

Serov, Pavel, Alexey Portnov, Jurgen Mienert, Peter Senenov and Polina Ilatovskaya. 2015. 'Methane Release from Pingo-Like Features across the South Kara

Sea Shelf, an Area of Thawing Offshore Permafrost'. *Journal of Geophysical Research: Earth Surface* 120, no. 8: 1515–29.

Shiva, Vandana. 1989. *The Violence of the Green Revolution, Third World Agriculture, Ecology, and Politics*. Dehradun: Research Foundation for Science and Ecology.

———. 2015. *Soil Not Oil: Environmental Justice in an Age of Climate Crisis*. New York: North Atlantic Books.

Siegal, Dan. 2022. *IntraConnected: MWe (Me + We) as the Integration of Self, Identity, and Belonging*. New York: Norton.

Simmel, Georg. 1978/1907. *The Philosophy of Money*. Translated by T. B. Bottomore and David Frisby. London: Routledge & Kegan Paul.

Skinner, B. F. 1938. *The Behavior of Organisms*. New York: Appleton-Century-Crofts.

———. 1969. *Contingencies of Reinforcement: A Theoretical Analysis*. New York: Appleton-Century-Crofts.

Skoerseth, Jon Birger, and Jorgen Wettestad. 2008. 'Implementing EU Emissions Trading: Success or Failure?'. *International Environmental Agreements* 8, no. 3: 275–90.

Slotsky, Alice L. 1997. *The Bourse of Babylon: Market Quotations in the Astronomical Diaries of Babylonia*. Bethesda, MD: CDL Press.

Smith, Adam. 1904/1776. *An Inquiry into the Nature and Causes of the Wealth of Nations*. Edited by Edwin Cannan. London: Methuen and Co.

Smith, Cherryl Waerea-I-te-rangi. 2000. 'Straying beyond the Boundaries of Belief: Māori Epistemologies inside the Curriculum'. *Educational Philosophy and Theory* 32, no. 1: 43–51.

Smith, Takirirangi. 2000. 'Nga Tini Ahuatanga o Whakapapa Korero'. *Educational Philosophy and Theory* 32, no.1: 53–60.

Snook, Ivan. 1982. '"Ideology" in Educational Theory'. *Access* 1, no. 1: 12–20.

Soddy, Frederick. 1933/1926. *Wealth, Virtual Wealth and Debt: The Solution of the Economic Paradox*. London: George Allen & Unwin.

Şorman, Alevgül H. 2015. 'Metabolism, Societal'. In *Degrowth: A Vocabulary for a New Era*, edited by Giacomo D'Alisa, Federico Demaria and Giorgos Kallis, chapter 6. London: Routledge.

Source Material. 2023. *The Carbon Con*. Source Material with *Die Zeit* and *The Guardian*. Accessed 20 July 2023. https://www.source-material.org/vercompanies-carbon-offsetting-claims-inflated-methodologies-flawed/.

Spengler, Oswald. 2002/1918. *The Decline of the West*. Edited by Helmut Werner, revised edition in 1922, translated by Charles Francis Atkinson. Oxford University Press: New York. http://www.duke.edu/~aparks/Spengler.html.

———. 1932. *Man and Technics*. Translated by Charles Frances. London: Unwin.

Srnicek, Nick, and Alex Williams. 2015. *Inventing the Future: Postcapitalism and a World without Work*. New York: Verso.

Statista. 2023. 'Difference between the Inflation Rate and Growth of Wages January 2020 to July 2023'. Accessed 19 August 2023. https://www.statista.com/statistics/1351276/wage-growth-vs-inflation-us/.

Steele, John M. 2000a. 'Mesopotamia'. In *Observations and Predictions of Eclipse Times by Early Astronomers*, Archimedes Series, vol. 4, by John M. Steele, 21–83. Dordrecht, Netherlands: Springer.

———. 2000b. ‘Eclipse Prediction in Mesopotamia’. *Archive for History of Exact Sciences* 54, no. 5: 421–54.

Steffen, Will, Wendy Broadgate, Lisa Deutsch, Owen Gaffney and Cornelia Ludwig. 2015. ‘The Trajectory of the Anthropocene: The Great Acceleration’. *The Anthropocene Review* 2, no. 1: 81–98.

Stein, Sharon, Vanessa Andreotti, Rene Suša, Cash Ahenakew and Tereza Čajková. 2022. ‘From ‘Education for Sustainable Development’ to Education for the End of the World as We Know It’. *Educational Philosophy and Theory* 54, no. 3: 274–87.

Stern, Nicholas. 2006. *The Economics of Climate Change: The Stern Review*. London: UK Treasury.

———. 2018. ‘Public Economics as If Time Matters: The Dynamics of Policy’. London: Grantham Research Institute on Climate Change and the Environment. Accessed 11 August 2023. https://www.lse.ac.uk/granthaminstitute/publication/public-economics-as-if-time-matters-climate-change-and-the-dynamics-of-policy/.

Steuart, James. 1966/1767. *An Inquiry into the Principles of Political Economy*. The Strand: A Millar & T. Cadell. https://www.marxists.org/reference/subject/economics/steuart/index.htm.

Stevens, Glenn. 2003. ‘Structural Change and the Conduct of Monetary Policy’. Deputy Governor of Reserve Bank of Australia, Federal Reserve Bank of Kansas City Economic Symposium: Jackson Hole. https://www.bis.org/review/r030905e.pdf.

Stiegler, Bernard. 2015a. ‘Power, Powerlessness, Thinking, and Future’. *Los Angeles Review of Books*, 18 October. https://lareviewofbooks.org/article/power-powerlessness-thinking-and-future/.

———. 2015b. Symptomatology of the Month of January 2015 in France. Accessed 14 December 2016. https://www.academia.edu/12797208/Bernard_Stiegler_Symptomatology_of_the_Month_of_January_2015_in_France_2015.

———. 2016. *The Automatic Society*, vol. 1, *The Future of Work*. Cambridge: Polity.

———. 2018. *The Neganthropocene*. Edited and translated by Daniel Ross. London: Open Humanities Press.

Stockholm Resilience Centre. nd. accessed 27 May 2024. https://www.stockholmresilience.org/publications.html

Stringer, Chris. 2016. ‘The Origin and Evolution of Homo Sapiens’. *Biological Sciences* 371, no. 1698. https://doi.org/10.1098/rstb.2015.0237.

Sveriges Riksbank. 2020a. ‘Sustainability Strategy for the Riksbank’. Stockholm. Accessed 13 October 2023. https://www.riksbank.se/globalassets/media/riksbanken/hallbarhetsstrategi/engelska/sustainability-strategy-for-the-riksbank.pdf.

———. 2020b. ‘What Is Money?’. Stockholm. Accessed 12 October 2023. https://www.riksbank.se/en-gb/payments--cash/what-is-money/.

Swer, Gregory Morgan. 2018. ‘Oswald Spengler and Martin Heidegger on Modern Science, Metaphysics, and Mathematics in Advance’. *Idealistic Studies* 47, nos. 1–2: 1–22.

Taskforce on Nature-related Financial Disclosures (TNFD). 2023. ‘Taskforce on Nature-related Financial Disclosures (TNDF) Recommendations’. Accessed 17 October 2023. https://tnfd.global/publication/recommendations-of-the-taskforce-on-nature-related-financial-disclosures/.

Tavlas, George. 2020. 'On the Controversy Over the Origins of the Chicago Plan for 100 Percent Reserves: Sorry, Frederick Soddy, It Was Knight and (Most Probably) Simons!'. Hoover Institution: Stanford University Press, Economic Working Papers, 1 March.

Tegmark, Max. 2017. *Life 3.0: Being Human in the Age of Artificial Intelligence*. New York: Alfred A. Knopf.

Temin, Peter. 2002. 'Price Behavior in Ancient Babylon'. *Explorations in Economic History* 39, no. 1: 46–60.

Thales, A. P. West, Jan Bőrner, Erin O. Sills and Andreas Kontoleon. 2020. 'Overstated Carbon Emission Reduction from Voluntary REDD+ Projects in the Brazilian Amazon'. *PNAS* (*Proceedings of the National Academy of Sciences*) 117, no. 39: 24188–94. https://www.pnas.org/doi/10.1073/pnas.2004334117.

Thiemann, Thomas. 1998. 'Quantum Spin Dynamics (QSD) V: Quantum Gravity as the Natural Regulator of the Hamiltonian Constraint of Matter Quantum Field Theories'. *Classical and Quantum Gravity* 15: 1281–314.

Tobin, James. 1963. 'Commercial Banks as Creators of "Money"'. In *Banking and Monetary Studies, for the Comptroller of the Currency*, edited by Dean Carson, 408–19. Cowles Foundation for Research in Economics: Yale University. Accessed 21 August 2023. https://elischolar.library.yale.edu/cgi/viewcontent.cgi?article=1387&context=cowles-discussion-paper-series.

Toffler, Alvin. 1970. *Future Shock*. London: Pan.

Tong, Dan, Qiang Zhang, Yixuan Zheng, Ken Caldeira, Christine Shearer, Chaopeng Hong, Yue Qin and Steven J. Davis. 2019. 'Committed Emissions from Existing Energy Infrastructure Jeopardise 1.5°C Climate Target'. *Nature* 572: 373–77.

Toussaint, Eric. 2012. 'The Long Tradition of Debt Cancellation in Mesopotamia and Egypt from 3000 to 1000 BC'. Committee for the Abolition of Illegitimate Debt (CADTM), 2 September. https://www.cadtm.org/The-Long-Tradition-of-Debt.

Turner, Adair. 2013. 'Credit, Money and Leverage: What Wicksell, Hayek and Fisher Knew and Modern Macroeconomics Forgot'. *Positive Money*. Accessed 15 October 2023. https://positivemoney.org/2014/04/former-bank-regulator-adair-turner-says-economics-textbooks-teach-mythological-story-banks-full-transcript/.

———. 2015. *Between Debt and the Devil: Money, Credit, and Fixing Global Finance*. Princeton NJ: Princeton University Press.

Tynan, Aidan. 2017. 'Impotent Signs: Money, Speculative Subjectivity and the Ontological Proof'. In *Credo Credit Crisis: Speculations on Faith and Money*, edited by Aidan Tynan, Laurent Milesi and Christopher Muller, 211–39. London: Rowman & Littlefield.

United Nations. 2015. 'The 17 Goals'. In the *2030 Agenda for Sustainable Development*, Department of Economic and Social Affairs. https://sdgs.un.org/goals.

United Nations Conferences on Environment and Development (UNCED). 1992. *Agenda 21 Earth Summit: The United Nations Programme of Action from Rio*. UN Department of Public Information.

United Nations Department of Economic and Social Affairs, Population Division (UN DESA/POP). 2022. *World Population Prospects 2022: Summary of Results*. UN DESA/POP/2022/TR/NO. 3.

United Nations Educational, Scientific and Cultural Organisation–United Nations Environmental Programme (UNESCO-UNEP). 1977. Tbilisi Final Report, Intergovernmental Conference on Environmental Education, Tbilisi (USSR), 14 to 26 October. https://unesdoc.unesco.org/ark:/48223/pf0000032763.

United Nations Environment Programme (UNEP). 2017. 'Evaluation Synthesis Report 2014–15'. Accessed 17 October 2023. https://www.unep.org/resources/report/unepea2inf13-evaluation-synthesis-report-2014-15.

United Nations International Children's Emergency Fund (UNICEF). 2023. 'Children Displaced in a Changing Climate: Preparing for a Future that's Already Underway'. UNICEF report, October. https://www.unicef.org/reports/children-displaced-changing-climate.

Valdivieso, Joaquin. 2008. 'André Gorz, Michel Bosquet, Precursor of Political Ecology'. *ecologíaPolítica*, 1 January. https://www.ecologiapolitica.info/?p=5686.

Van Eedon, Lily, Dale Nimmo, Michael Mahony, Kerryn Herman, Glenn Ehmke, Joris Driessen, James O'Conner, Gilad Bino, Martin Taylor and Chris Dickman. 2020. *Impacts of the Unprecedented 2019–20 Bushfires on Australian Animals*. WWF Report. Ultimo NSW: WWF. Accessed 23 July 2023. https://assets.wwf.org.au/image/upload/f_pdf/v1/website-media/resources/WWF_Impacts-of-the-unprecedented-2019-2020-bushfires-on-Australian-animals.

Vecellio, Daniel, S. Tony Wolf, Rachel M. Cottle and W. Larry Kenney. 2022. 'Evaluating the 35C Wet-Bulb Temperature Adaptability Threshold for Young, Healthy Subjects (PSU HEAT Project)'. *Journal of Applied Physiology* 132, no. 2: 340–45.

Vergunst, Jo Lee, and Tim Ingold. 2008. *Ways of Walking; Ethnography and Practice on Foot*. Abingdon-on-Thames: Routledge.

Victor, Peter, and Tim Jackson. 2015. 'Toward an Ecological Macroeconomics'. In *Ecological Economics for the Anthropocene: An Emerging Paradigm*, edited by Peter G. Brown and Peter Timmerman, 237–59. New York: Columbia University Press.

Vollset, Stein Emil, Emily Goren, Chun-Wei Yuan, Jackie Cao, Amanda E. Smith, Thomas Hsiao, Catherine Bisignano, Gulrez S. Azhar, Emma Castro, Julian Chalek, Andrew J. Dolgert, Tahvi Frank, Kai Fukutaki, Simon I. Hay, Rafael Lozano, Ali H. Mokdad, Vishnu Nandakumar, Maxwell Pierce, Martin Pletcher, Toshana Robalik, Krista M. Steuben, Han Yong Wunrow, Bianca S. Zlavog and Christopher J. L. Murray. 2020. 'Fertility, Mortality, Migration, and Population Scenarios for 195 Countries and Territories from 2017 to 2100: A Forecasting Analysis for the Global Burden of Disease Study'. *The Lancet* 396, no. 10258: 1285–306.

Von Mises, Ludwig. 1953/1912. *The Theory of Money and Credit*. New Haven, CT: Yale University Press.

Waring, Marilyn. 1990. *If Women Counted: A New Feminist Economics*. San Francisco CA: Harper Collins.

Wachowski, Lana. and Lilly. Wachowski. (Directors). 1999. *The Matrix*. Film. Warner Bros.

Wallach, Arian D., Anthony H. Dekker, Miguel Lurgi, Jose Montoya, Damien A. Fordham and Euan G. Ritchie. 2016. 'Trophic Cascades in 3D: Network Analysis

Reveals How Apex Predators Structure Ecosystems'. *Methods in Ecology and Evolution* 8, no. 91: 135–42.

Wark, McKenzie. 2015. *Molecular Red: Theory for the Anthropocene*. London: Verso.

Waters, Richard. 2014. 'Artificial Intelligence: Machine v Man'. *Financial Times*, 31 October. https://www.ft.com/content/abc942cc-5fb3-11e4-8c27-00144feabdc0.

Watson, Julia. 2020. *Lo-Tek, Design by Radical Indigenism*. Los Angeles, CA: Taschen Books.

Weber, Max. 1948/1919. *From Max Weber: Essays in Sociology*. Translated by H. H. Gerth and C. Wright Mills. London : Routledge and Kegan Paul.

Werner, Richard. 2014. 'Can Banks Individually Create Money Out of Nothing? The Theories and the Empirical Evidence'. *International Review of Financial Analysis* 36: 1–19.

Westman, Robert S. 2022. 'Johannes Kepler: German Astronomer'. In *Encyclopaedia Britannica*. https://www.britannica.com/topic/Epitome-of-Copernican-Astronomy.

Weule, Genelle. 2020. ''Portable' Stone Age Artworks Found for the First Time in South-East Asia'. *ABC Science*, 16 March. https://www.abc.net.au/news/science/2020-03-17/portable-artworks-found-for-first-time-in-southeast-asia/12046794.

Whitehead, Alfred. 1969. *Process and Reality: And Essay in Cosmology*. New York: The Free Press.

Williams, Zoe. 2019. 'How the Actual Magic Money Tree Works'. *The Guardian*, 29 October.

Wilson, Kevin A. 2023. 'Worth the Watt: A Brief History of the Electric Car, 1830 to Present'. *Car and Driver*, 31 March. Accessed 18 June 2023. https://www.caranddriver.com/features/g43480930/history-of-electric-cars.

Wood, David. 1993. 'Reiterating the Temporal: Toward a Rethinking of Heidegger on Time'. In *Reading Heidegger, Commemorations*, edited by John Sallies, 10–16. Bloomington and Indianapolis: Indiana University Press.

World Bank. n.d. 'Birthrate Crude Data'. https://data.worldbank.org/indicator/SP.DYN.CBRT.IN.

World Bank National Accounts Data. 2022. 'Gross Domestic Product 2022'. Data Bank Files. Accessed 17 October 2023. https://databankfiles.worldbank.org/public/ddpext_download/GDP.pdf.

World Wide Fund for Nature (WWF). 2022. *Living Planet Report 2022 – Building a Nature-Positive Society*. Edited by R. E. A. Almond, M. Grooten, D. Juffe Bignoli and T. Petersen. WWF, Gland, Switzerland.

Xue, Ti-Wei, and Guo Zeng-Yuan. 2019. 'What Is the Real Clausius Statement of the Second Law of Thermodynamics?'. *Entropy* 21, no. 10: 926.

Yates, Joshua. 2003. 'An Interview with Ulrich Beck on Fear and Risk Society'. *The Hedgehog Review* 5, no. 3: n.p. https://go.gale.com/ps/i.do?p=AONE&u=anon~843ea22&id=GALE|A133101460&v=2.1&it=r.

Yearly, Steven. 2001. 'Risk, Sociology and Politics of'. *International Encyclopaedia of Social and Behavioural Sciences*, second edition, edited by James D. Wright, 706–10. London: Elsevier.

Yoshikawa, Saeko. 2020. 'The Railway Controversy in Wordsworth's Lake District'. In *William Wordsworth and Modern Travel: Railways, Motorcars and the Lake District, 1830–1940*, by Saeko Yoshikawa, 61–68. Oxford: Oxford University Press.

Young, Julian. 2002. *Heidegger's Later Philosophy*. Cambridge: Cambridge University Press.

Yunkaporta, Tyson. 2019. *Sand Talk: How Indigenous Thinking Can Save the World*. London: The Text Publishing Company.

Zimmerman, Michael E. 1990. *Heidegger's Confrontation with Modernity; Technology, Politics, Art*. Bloomington: Indiana University Press.

———. 1994. *Contesting Earth's Future: Radical Ecology and Postmodernity*. Berkeley: University of California Press.

Zinn, Jens Oliver. 2016. 'Living in the Anthropocene: Towards a Risk-Taking Society'. *Environmental Sociology* 24, no. 4: 385–94.

Zwier, Jochem, Vincent Blok, Pieter Lemmens and Robert-Jan Geerts. 2015. 'The Ideal of a Zero-Waste Humanity: Philosophical Reflections on the Demand for a Bio-Based Economy'. *The Journal of Agricultural Environmental Ethics* 28: 353–74.

Index

About the Author

Ruth Irwin has written on climate change and social transition for more than twenty years. She is the author of *Heidegger, Politics and Climate Change* (2008), *Climate Change and Philosophy* (2010) and more than fifty articles and book chapters.

www.ingramcontent.com/pod-product-compliance
Lightning Source LLC
LaVergne TN
LVHW011704200926
848484LV00004B/1066

* 9 7 8 1 7 8 6 6 1 5 1 1 4 *